George Salmon, Arthur Cayley

A treatise on the higher plane curves

Intended as a sequel to a treatise on conic sections

George Salmon, Arthur Cayley

A treatise on the higher plane curves
Intended as a sequel to a treatise on conic sections

ISBN/EAN: 9783742892300

Manufactured in Europe, USA, Canada, Australia, Japa

Cover: Foto ©berggeist007 / pixelio.de

Manufactured and distributed by brebook publishing software (www.brebook.com)

George Salmon, Arthur Cayley

A treatise on the higher plane curves

A COURSE IN MATHEMATICAL ANALYSIS

DIFFERENTIAL EQUATIONS

BEING PART II OF VOLUME II

BY

ÉDOUARD GOURSAT
PROFESSOR OF MATHEMATICS, THE UNIVERSITY OF PARIS

TRANSLATED BY

EARLE RAYMOND HEDRICK
PROFESSOR OF MATHEMATICS, THE UNIVERSITY OF MISSOURI

AND

OTTO DUNKEL
ASSISTANT PROFESSOR OF MATHEMATICS, WASHINGTON UNIVERSITY

GINN AND COMPANY
BOSTON · NEW YORK · CHICAGO · LONDON
ATLANTA · DALLAS · COLUMBUS · SAN FRANCISCO

PREFACE

The present volume consists of the second half of the second volume of the French edition of Goursat's "Cours d'Analyse Mathématique." As was stated in the preface to the first half of this volume, it seemed best, for purposes of American schools, to issue these two parts separately, and this was done with the approval of Professor Goursat.

It is hoped that the present volume, which is entitled "Differential Equations," will prove serviceable in American universities for courses which bear that name.

E. R. HEDRICK
OTTO DUNKEL

CONTENTS

	PAGE
CHAPTER I. ELEMENTARY METHODS OF INTEGRATION	3

   I. FORMATION OF DIFFERENTIAL EQUATIONS 3

      1. Elimination of constants. 3

   II. EQUATIONS OF THE FIRST ORDER 6

      2. Separation of the variables 6
      3. Homogeneous equations 8
      4. Linear equations 9
      5. Bernoulli's equation 11
      6. Jacobi's equation 11
      7. Riccati's equation 12
      8. Equations not solved for y' 14
      9. Lagrange's equation 16
      10. Clairaut's equation 17
      11. Integration of the equations $F(x, y') = 0, F(y, y') = 0$ . . 18
      12. Integrating factors 19
      13. Application to conformal representation 22
      14. Euler's equation 23
      15. A method deduced from Abel's theorem 28
      16. Darboux's theorems 29
      17. Applications 32

   III. EQUATIONS OF HIGHER ORDER 35

      18. Integration of the equation $d^n y/dx^n = f(x)$ 35
      19. Various cases of depression 36
      20. Applications 39

EXERCISES 42

CHAPTER II. EXISTENCE THEOREMS 45

   I. CALCULUS OF LIMITS 45

      21. Introduction 45
      22. Existence of the integral of a system of differential equations 45
      23. Systems of linear equations 50
      24. Total differential equations 51

		PAGE
25.	Application of the method of the calculus of limits to partial differential equations	53
26.	The general integral of a system of differential equations	57

II. THE METHOD OF SUCCESSIVE APPROXIMATIONS. THE CAUCHY-LIPSCHITZ METHOD ... 61

27.	Successive approximations	61
28.	The case of linear equations	64
29.	Extension to analytic functions	66
30.	The Cauchy-Lipschitz method	68

III. FIRST INTEGRALS. MULTIPLIERS ... 74

31.	First integrals	74
32.	Multipliers	81
33.	Invariant integrals	83

IV. INFINITESIMAL TRANSFORMATIONS ... 86

34.	One-parameter groups	86
35.	Application to differential equations	89
36.	Infinitesimal transformations	91

EXERCISES ... 98

CHAPTER III. LINEAR DIFFERENTIAL EQUATIONS ... 100

I. GENERAL PROPERTIES. FUNDAMENTAL SYSTEMS ... 100

37.	Singular points of a linear differential equation	100
38.	Fundamental systems	102
39.	The general linear equation	106
40.	Depression of the order of a linear equation	109
41.	Analogies with algebraic equations	113
42.	The adjoint equation	115

II. THE STUDY OF SOME PARTICULAR EQUATIONS ... 117

43.	Equations with constant coefficients	117
44.	D'Alembert's method	122
45.	Euler's linear equation	123
46.	Laplace's equation	124

III. REGULAR INTEGRALS. EQUATIONS WITH PERIODIC COEFFICIENTS ... 128

47.	Permutation of the integrals around a critical point	129
48.	Examination of the general case	131
49.	Formal expressions for the integrals	133

CONTENTS

	PAGE
50. Fuchs' theorem	134
51. Gauss's equation	140
52. Bessel's equation	142
53. Picard's equations	143
54. Equations with periodic coefficients	146
55. Characteristic exponents	150

IV. Systems of Linear Equations 152

56. General properties	152
57. Adjoint systems	156
58. Linear systems with constant coefficients	157
59. Reduction to a canonical form	161
60. Jacobi's equation	163
61. Systems with periodic coefficients	164
62. Reducible systems	165

Exercises 167

CHAPTER IV. NON-LINEAR DIFFERENTIAL EQUATIONS . . 172

I. Exceptional Initial Values 172

63. The case where the derivative becomes infinite	172
64. Case where the derivative is indeterminate	173

II. A Study of Some Equations of the First Order ·180

66. Singular points of integrals	180
67. Functions defined by a differential equation $y' = R(x, y)$	182
68. Single-valued functions deduced from the equation $(y')^m = R(y)$	187
69. Existence of elliptic functions deduced from Euler's equation	194
70. Equations of higher order	196

III. Singular Integrals 198

71. Singular integrals of an equation of the first order	198
72. General comments	204
73. Geometric interpretation	207
74. Singular integrals of systems of differential equations	208

Exercises 212

CHAPTER V. PARTIAL DIFFERENTIAL EQUATIONS OF THE FIRST ORDER 214

I. Linear Equations of the First Order 214

75. General method	214
76. Geometric interpretation	218
77. Congruences of characteristic curves	222

II. Total Differential Equations 225

78. The equation $dz = A\,dx + B\,dy$ 225
79. Mayer's method 229
80. The equation $P\,dx + Q\,dy + R\,dz = 0$ 230
81. The parenthesis (u, v) and the bracket $[u, v]$ 234

III. Equations of the First Order in Three Variables 236

82. Complete integrals 236
83. Lagrange and Charpit's method 240
84. Cauchy's problem 246
85. Characteristic curves. Cauchy's method 249
86. The characteristic curves derived from a complete integral 259
87. Extension of Cauchy's method 261

IV. Simultaneous Equations 265

88. Linear homogeneous systems 265
89. Complete systems 267
90. Generalization of the theory of the complete integrals . . 272
91. Involutory systems 274
92. Jacobi's method 277

V. Generalities on the Equations of Higher Order 278

93. Elimination of arbitrary functions 278
94. General existence theorem 283

Exercises 287

Index 291

A COURSE IN
MATHEMATICAL ANALYSIS

VOLUME II. PART II

DIFFERENTIAL EQUATIONS

CHAPTER I

ELEMENTARY METHODS OF INTEGRATION

I. FORMATION OF DIFFERENTIAL EQUATIONS

1. Elimination of constants. Let us consider a family of plane curves represented by the equation

(1) $$F(x, y, c_1, c_2, \cdots, c_n) = 0,$$

which depends upon n arbitrary constants. If we assign to these constants definite but arbitrarily chosen values, the successive derivatives of the function y of the variable x defined by the preceding equation are furnished by the relations

(2) $$\begin{cases} \dfrac{\partial F}{\partial x} + \dfrac{\partial F}{\partial y} y' = 0, \\ \dfrac{\partial^2 F}{\partial x^2} + 2 \dfrac{\partial^2 F}{\partial x \, \partial y} y' + \dfrac{\partial^2 F}{\partial y^2} y'^2 + \dfrac{\partial F}{\partial y} y'' = 0, \\ \cdots \cdots \cdots \cdots \cdots \cdots \cdots \cdots \cdots \cdots \cdots, \\ \dfrac{\partial^n F}{\partial x^n} + \cdots + \dfrac{\partial F}{\partial y} y^{(n)} = 0. \end{cases}$$

If we stop with the equation for calculating the derivative of the nth order, we shall have in all $(n+1)$ relations between $x, y, y', y'', \cdots, y^{(n)}$ and the constants $c_1, c_2, \cdots, c_n$. The elimination of these n constants leads in general to a single relation between $x, y, y', \cdots, y^{(n)}$,

(3) $$\Phi(x, y, y', y'', \cdots, y^{(n)}) = 0.$$

From the very way in which the equation (3) is derived it is clear that every function defined by the relation (1) satisfies this equation (3), whatever may be the values assigned to the constants c_i; hence we say that any such function is a *particular* integral of the differential equation (3). The whole set of these particular integrals is the *general integral* of the same equation. Using geometric language for convenience, we shall also say that every curve represented by the equation (1) is an *integral curve* of the equation (3), or that the equation (3) is the differential equation of the given family of curves (1).

We see that the *order* of the differential equation is equal to the number of arbitrary constants upon which that family of curves depends. It is also clear that the reasoning does not at all prove that the equation (3) has no other integrals than those which are represented by the equation (1). In fact, the equation (3) may have other integrals, as we shall see presently.

The above statements do not apply to the exceptional cases in which the elimination of the n parameters c_i between the $(n+1)$ relations (1) and (2) leads to several distinct relations between $x, y, y', y'', \cdots, y^{(n)}$. We could in those cases find one relation not containing $y^{(n)}$, so that the family of curves considered would be the integral curves of a differential equation of an order less than n. This will occur if these curves depend in reality upon only $n-p$ parameters ($p>0$). For example, the curves represented by the equation $F[x, y, \phi(a, b)] = 0$ apparently depend upon two arbitrary parameters a and b; in reality they depend upon only a single parameter $c = \phi(a, b)$. There is also another way in which the lowering of the order of the differential equation may occur. For example, the curves represented by the equation $y^2 = 2axy + bx^2$ really depend upon the two independent parameters a and b, yet these curves always satisfy the equation $y = xy'$. This is because the preceding equation represents two straight lines through the origin, each of which is an integral curve of the equation $y = xy'$.

Examples. The straight lines passing through a fixed point (a, b) are represented by the equation

(4) $$y - b = C(x - a)$$

and depend upon an arbitrary parameter C. The elimination of this parameter between the preceding relation and the relation $y' = C$ leads immediately to the differential equation of this system of straight lines:

(5) $$y - b = y'(x - a).$$

Conversely, we can write equation (5) in the form

$$\frac{y'}{y-b} = \frac{1}{x-a},$$

and therefore every integral of that equation satisfies the relation

$$\mathrm{Log}\,(y - b) = \mathrm{Log}\,(x - a) + \mathrm{Log}\,C,$$

which is equivalent to the equation (4).

The set of all straight lines in a plane, $y = C_1 x + C_2$, form a two-parameter family whose differential equation is $y'' = 0$. The converse is self-evident.

The circles in a plane

(6) $$x^2 + y^2 + 2Ax + 2By + C = 0$$

form a three-parameter family; the corresponding differential equation must therefore be of the third order. Differentiating the preceding relation three times, we find

(7) $\quad x + yy' + A + By' = 0, \quad 1 + y'^2 + yy'' + By'' = 0,$
$$3y'y'' + yy''' + By''' = 0.$$

The elimination of B between the last two equations leads to the desired equation

(8) $\quad\quad\quad\quad y'''(1 + y'^2) - 3y'y''^2 = 0.$

The only plane curves satisfying this relation are *circles* and *straight lines*. We see first of all that any straight line is an integral curve, for the equation is satisfied if we have $y'' = 0$ and therefore $y''' = 0$. Now let us suppose that $y'' \neq 0$; then we can write the equation (8) in the form

$$\frac{y'''}{y''} = \frac{3 y' y''}{1 + y'^2},$$

from which we derive

$$\text{Log } y'' = \frac{3}{2} \text{Log}(1 + y'^2) + \text{Log } C_1,$$

where C_1 is a constant different from zero. This result may be written in the form

$$\frac{y''}{(1 + y'^2)^{\frac{3}{2}}} = C_1.$$

A second integration gives

$$\frac{y'}{\sqrt{1 + y'^2}} = C_1 x + C_2,$$

or

$$y' = \frac{C_1 x + C_2}{\sqrt{1 - (C_1 x + C_2)^2}};$$

integrating once more, there results finally

$$C_1 y + C_3 = -\sqrt{1 - (C_1 x + C_2)^2},$$

which is the equation of a circle.

The differential equation of all conics may be found easily by the following method, which is due to Halphen. If the conic has no asymptote parallel to the y-axis, its equation solved with respect to y is of the form

$$y = mx + n + \sqrt{Ax^2 + 2Bx + C}.$$

After two differentiations we find

$$y'' = \frac{AC - B^2}{(Ax^2 + 2Bx + C)^{\frac{3}{2}}},$$

or

$$(y'')^{-\frac{2}{3}} = (AC - B^2)^{-\frac{2}{3}}(Ax^2 + 2Bx + C),$$

so that $(y'')^{-2/3}$ is a trinomial of the second degree in x. Hence, to eliminate the three constants A, B, C three differentiations are sufficient, and the desired differential equation can be written in the abridged form

$$\frac{d^3}{dx^3}[(y'')^{-\frac{2}{3}}] = 0.$$

Carrying out the differentiations, we obtain the equation

(9) $\qquad 40\,y'''^3 - 45\,y''y'''y^{\text{iv}} + 9\,y''^2 y^{\text{v}} = 0.$

The differential equation of parabolas may be found by the same method. We have, in fact, for a parabola $A = 0$, and $(y'')^{-2/3}$ is a binomial of the first degree. The differential equation is, therefore, in an abridged form,

$$\frac{d^2}{dx^2}[(y'')^{-\frac{2}{3}}] = 0,$$

or, after carrying out the indicated differentiations,

(10) $\qquad 5\,y'''^2 - 3\,y''y^{\text{iv}} = 0.$

II. EQUATIONS OF THE FIRST ORDER

Every differential equation of the nth order, formed by the elimination of the constants, has an infinite number of integrals that depend upon n arbitrary parameters. But it is by no means evident that a differential equation given a priori has any integrals. This involves a fundamental question to which we shall return in the following chapter. We shall first consider some simple types of differential equations of the first order whose integration can be effected by quadratures. The existence of the integrals will be established by the very method by which we obtain them. If this order of procedure seems subject to criticism from the point of view of pure logic, we may at least observe that it conforms to the historical development of the subject.

2. Separation of the variables. The simplest type of differential equation is the equation already studied,

(11) $\qquad \dfrac{dy}{dx} = f(x),$

where $f(x)$ is a continuous function if the variable x is real, or an analytic function if we regard the independent variable x as complex. We have seen that that equation has an infinite number of integrals which can be represented by the relation

$$y = \int_{x_0}^{x} f(x)\,dx + C,$$

where the lower limit x_0 is considered as fixed, and where C denotes an arbitrary constant. The equation

(12) $\qquad \dfrac{dy}{dx} = \phi(y)$

reduces to the preceding by considering y as the independent variable and x as the unknown function. The equation may then be written in the form $dx/dy = 1/\phi(y)$, and consequently

$$x = \int_{y_0}^{y} \frac{dy}{\phi(y)} + C.$$

In general, when a differential equation is solved with respect to the derivative of the unknown function, it is often convenient to write it in the differential notation,

(13) $\qquad P(x, y)\,dx + Q(x, y)\,dy = 0.$

This form does not commit us in any way as to the choice of the independent variable, which may be either x or y. If we wish to substitute for x and y new variables u and v, we need only replace x, y, dx, dy in the equation (13) by their corresponding expressions in terms of u, v, du, dv. Let us also notice that we may, without changing the integrals of the equation (13), multiply or divide both its terms by the same function of x and y, $\mu(x, y)$, provided that we take account of the solutions of the equation $\mu(x, y) = 0$ which may be made to appear or may be suppressed by the operation. The two cases which we have just treated are particular cases under a more general method, called the *separation of variables*. If a differential equation of the first order is of the form

(14) $\qquad X\,dx + Y\,dy = 0,$

where X and Y depend only upon x and y respectively, we say that *the variables are separated*. The equation is then integrable by quadratures, for if we put

$$U = \int_{x_0}^{x} X\,dx + \int_{y_0}^{y} Y\,dy,$$

the equation can be written in the form $dU = 0$, and the general integral is represented by the relation $U = C$.

The equation

(15) $\qquad XY_1\,dx + X_1Y\,dy = 0,$

where X and X_1 depend only upon x, and where Y and Y_1 depend only upon y, can be reduced to the preceding form by dividing the two terms by X_1Y_1. It should be noticed that in this example the solutions of the two equations, $X_1 = 0$, $Y_1 = 0$, are suppressed. Indeed, it is clear that if $y = b$ is a root of the equation $Y_1 = 0$, $y = b$ is an integral of the proposed equation, while in general it will not be included in the general integral of the new equation.

3. Homogeneous equations.
A differential equation of the first order is said to be *homogeneous* if it can be written in the form

$$\frac{dy}{dx} = f\left(\frac{y}{x}\right), \tag{16}$$

where the right-hand side is a homogeneous function *of degree zero*. It can be reduced to an integrable form by putting $y = ux$, where the new variables are x and u. This substitution gives

$$\frac{dy}{dx} = u + x\frac{du}{dx},$$

and the equation (16) becomes

$$x\frac{du}{dx} + u = f(u).$$

We can now separate the variables by writing the equation in the form

$$\frac{dx}{x} = \frac{du}{f(u) - u},$$

and the general integral is obtained by one quadrature in the form

$$x = Ce^{\int \frac{du}{f(u)-u}}. \tag{17}$$

We have only to replace in it u by y/x in order to obtain the equation of the integral curves.

The general equation of that family of curves is of the form $x = C\phi(y/x)$, where C is an arbitrary constant. These curves are all similar to any one of them, with the origin as center of similitude, the ratio of similitude being alone variable; for we can derive the preceding equation from the equation $x = \phi(y/x)$ by replacing x and y in it by x/C and y/C respectively. Conversely, given a family of curves similar to each other with respect to the origin, the corresponding differential equation of the first order is homogeneous. We can verify this by actual calculation, but the result is evident a priori, for the tangents to the different curves of that family at the points of intersection with a straight line through the origin must be parallel, and therefore the slope of the tangent y' depends only on the ratio y/x.

We can reduce to the homogeneous form any equation of the type

$$\frac{dy}{dx} = f\left(\frac{ax + by + c}{a'x + b'y + c'}\right), \tag{18}$$

where a, b, c, a', b', c' are any constants, except that b and b' are not both zero. In order that this equation be of the desired form, it is sufficient that $c = c' = 0$. Now, if we put

$$x = X + \alpha, \qquad y = Y + \beta,$$

where X and Y are the new variables and where α and β are any two constants, the given equation becomes

$$\frac{dY}{dX} = f\left(\frac{aX + bY + a\alpha + b\beta + c}{a'X + b'Y + a'\alpha + b'\beta + c'}\right),$$

and this new equation will be homogeneous if

$$a\alpha + b\beta + c = 0, \qquad a'\alpha + b'\beta + c' = 0.$$

These two conditions determine α and β if $ab' - a'b$ is not zero. In the particular case in which $ab' - a'b = 0$, suppose $b \neq 0$; we shall have $a'x + b'y = k(ax + by)$, where k is a constant which has a finite value. Putting $ax + by = u$, the equation takes the form

$$\frac{1}{b}\frac{du}{dx} = \frac{a}{b} + f\left(\frac{u + c}{ku + c'}\right),$$

in which the variables are separated.

4. Linear equations. A linear differential equation of the first order is of the form

(19) $$\frac{dy}{dx} + Xy + X_1 = 0,$$

where X and X_1 are functions of x. If $X_1 = 0$, we can write this equation in the form

(20) $$\frac{dy}{y} + X\,dx = 0,$$

and the general integral is obtained by one quadrature in the form

(21) $$y = Ce^{-\int_{x_0}^{x} X\,dx}.$$

In order to integrate the complete equation (19), where X_1 is supposed different from zero, we shall try to satisfy that equation by taking for y an expression of the form (21), considering C no longer as a constant but as an unknown function of x. This amounts to making the change of variable $y = Yz$, where z is the new function to be determined and Y any one of the integrals of the equation (20). After this substitution, the equation (19), by virtue of the relation (20) which Y satisfies, takes the form

$$Y\frac{dz}{dx} + X_1 = 0,$$

which is integrable by one quadrature. We derive from it

$$z = -\int \frac{X_1}{Y} dx + C,$$

where C is an arbitrary constant. The general integral of the equation (19) is therefore obtainable by *two successive quadratures*. Replacing Y by its value, we can again write it in the form

(22) $$y = e^{-\int X dx}\left(C - \int X_1 e^{\int X dx} dx\right),$$

where the lower limits in the two integrals are chosen at pleasure.

The general integral *is an integral linear function of the constant of integration* of the form $y = Cf(x) + \phi(x)$, where $f(x)$ and $\phi(x)$ are definite functions of x. This property characterizes the linear equation, for if we eliminate the constant C between the preceding equation and the equation

$$y' = Cf'(x) + \phi'(x),$$

we are evidently led to a relation that is linear in y and y'.

This result may be stated in another way. Let y_1, y_2, y_3 be three particular integrals of the linear equation, corresponding to the values C_1, C_2, C_3 of the constant C; the elimination of the two functions $f(x)$ and $\phi(x)$ between the three relations,

$$y_1 = C_1 f(x) + \phi(x), \quad y_2 = C_2 f(x) + \phi(x), \quad y_3 = C_3 f(x) + \phi(x),$$

leads to the relation $(y_3 - y_1)/(y_2 - y_1) = (C_3 - C_1)/(C_2 - C_1)$, which shows that the ratio $(y_3 - y_1)/(y_2 - y_1)$ is constant for any three particular integrals of a linear equation. If we know two particular integrals y_1, y_2 of a linear equation, we can then write down immediately the general integral in the form

$$\frac{y - y_1}{y_2 - y_1} = \text{const.}$$

It is also to be noticed that if we know a single particular integral y_1, the general integral can be obtained by a single quadrature; in fact, putting $y = y_1 + u$, we are led to the equation $du/dx + Xu = 0$, which is identical with the equation (20).

5. Bernoulli's equation.
Bernoulli's equation

$$\text{(23)} \qquad \frac{dy}{dx} + Xy + X_1 y^n = 0,$$

where the exponent n may be any number different from zero and from unity, can be reduced to a linear equation by the substitution $z = y^{1-n}$. For then we can write the preceding equation, after dividing all its terms by y^n, in the form

$$\frac{1}{1-n}\frac{dz}{dx} + Xz + X_1 = 0.$$

We can reduce to the preceding type any equation of the form

$$\text{(24)} \qquad \phi\left(\frac{y}{x}\right) dx + \psi\left(\frac{y}{x}\right) dy + kx^m(x\,dy - y\,dx) = 0,$$

where k and m are any two numbers whatever. For if we put $y = ux$, the equation obtained can be written as follows:

$$[\phi(u) + u\psi(u)]\frac{dx}{du} + x\psi(u) + kx^{m+2} = 0,$$

and, putting $z = x^{-(m+1)}$, we are led to a linear equation.

6. Jacobi's equation.
Let us consider the equation

$$\text{(25)} \qquad \begin{cases} (a + a'z + a''y)(x\,dy - y\,dx) \\ \quad - (b + b'x + b''y)\,dy + (c + c'x + c''y)\,dx = 0, \end{cases}$$

where $a, a', a'', b, b', b'', c, c', c''$ are any constant coefficients. If $a = b = c = 0$, the equation comes under type (24), for we have only to divide by $a'z + a''y$ to reduce it to this type. In order to reduce the general case to this particular case, let us put $x = X + \alpha$, $y = Y + \beta$, where X and Y are two new variables and where α and β are two constants. Thus we obtain a new equation of the same form, which can be written as follows:

$$\text{(25')} \qquad \begin{cases} (a'X + a''Y)(X\,dY - Y\,dX) \\ \quad - [B + b'X + b''Y - (A + a'X + a''Y)\alpha - AX]\,dY \\ \quad + [C + c'X + c''Y - (A + a'X + a''Y)\beta - AY]\,dX = 0, \end{cases}$$

where

$$A = a + a'\alpha + a''\beta, \qquad B = b + b'\alpha + b''\beta, \qquad C = c + c'\alpha + c''\beta.$$

This equation (25') will be of the type (24) if we have $A\alpha - B = 0$, $A\beta - C = 0$. We are then led to determine the constants α, β by these two conditions, which may be written in a more symmetric form by introducing an auxiliary unknown λ:

$$A - \lambda = 0, \qquad B - \lambda\alpha = 0, \qquad C - \lambda\beta = 0.$$

The elimination of the unknowns α, β leads to an auxiliary equation of the third degree for the determination of λ:

$$\begin{vmatrix} a-\lambda & a' & a'' \\ b & b'-\lambda & b'' \\ c & c' & c''-\lambda \end{vmatrix} = 0.$$

The integration of Jacobi's equation depends, then, first of all on the solution of this equation of the third degree, as will be seen by other methods a little later.

7. Riccati's equation.
Riccati's equation

(26) $$\frac{dy}{dx} + Xy^2 + X_1 y + X_2 = 0,$$

where X, X_1, X_2 are functions of x, cannot in general be integrated by quadratures. The integrals of this equation, when the coefficients are unrestricted, form new transcendental functions, whose properties we shall study. But this equation is related to the matter which we are discussing on account of the following property: *If we know a particular integral, we can find the general integral by two quadratures.*

Let y_1 be a particular integral. The change of variable $y = y_1 + z$ leads to an equation of the same form which does not contain any term independent of z, since $z = 0$ must be an integral; that equation is, in fact,

(27) $$\frac{dz}{dx} + (X_1 + 2Xy_1)z + Xz^2 = 0,$$

and we have only to put $u = 1/z$ in order to transform it into a linear equation. This proves the proposition just stated.

From this result, several important consequences follow. The general integral of the linear equation in u is of the form (§ 4)

$$u = Cf(x) + \phi(x);$$

hence the general integral of the Riccati equation is of the form

(28) $$y = y_1 + \frac{1}{Cf(x) + \phi(x)} = \frac{Cf_1(x) + \phi_1(x)}{Cf(x) + \phi(x)}.$$

We see that it is a *rational function of the first degree in the constant of integration*. Conversely, every differential equation of the first order which has this property is a Riccati equation. For, let $f(x)$, $\phi(x)$, $f_1(x)$, $\phi_1(x)$ be any four functions of x; all the functions y represented by the expression (28), where C is an arbitrary constant, are integrals of an equation of the first order, which is easily obtained by solving the equation (28) for C and then taking the derivative. This gives

$$C = \frac{\phi_1 - y\phi}{yf - f_1},$$

and the corresponding differential equation is

$$(yf - f_1)(\phi_1' - \phi y' - y\phi') - (\phi_1 - y\phi)(y'f + yf' - f_1') = 0,$$

which is of precisely the form (26).

Let y_1, y_2, y_3, y_4 be four particular integrals corresponding to the values C_1, C_2, C_3, C_4 of the constant C. By the theory of the anharmonic ratio we have the relation

$$\frac{y_4 - y_1}{y_4 - y_2} \div \frac{y_3 - y_1}{y_3 - y_2} = \frac{C_4 - C_1}{C_4 - C_2} \div \frac{C_3 - C_1}{C_3 - C_2},$$

which is easily verified also by direct calculation, and which proves that *the anharmonic ratio of any four particular integrals of Riccati's equation is constant.*

This theorem enables us to find without any quadrature the general integral of a Riccati equation when we know three of its particular integrals y_1, y_2, y_3. Every other integral y must be such that the anharmonic ratio $(y - y_1)/(y - y_2) \div (y_3 - y_1)/(y_3 - y_2)$ is constant. The general integral is then obtained by equating this ratio to an arbitrary constant. It is clear that y will be a rational function of the first degree in this constant, which proves that the preceding property belongs only to the Riccati equations.

Let us observe that if we know only two particular integrals, y_1 and y_2, we can complete the integration by *one* quadrature; for, after the first transformation $y = y_1 + z$, the equation obtained in z has the integral $y_2 - y_1$. The linear equation in u has therefore the known particular integral $1/(y_2 - y_1)$. The general integral of the equation in u will then be found by a single quadrature.*

Application. Let us consider a family of circles in a plane, which depends upon one variable parameter. Let (a, b) be the coördinates of the center of the variable circle and let R be its radius (the axes being rectangular). We shall

---

* The properties of Riccati's equation established in the text can be derived also by observing that the equation is not changed in form by any general linear transformation $y = (fz + \phi)/(f_1 z + \phi_1)$, where f, f_1, ϕ, ϕ_1 are functions of x. If we know one, two, or three integrals of the equation (26), we can always choose the linear transformation in such a way that, in the transformed equation in z, one, two, or three of the coefficients of the polynomial of the second degree in z will be zero. A linear equation may be regarded as a Riccati equation which is satisfied by the particular integral $y = \infty$, that is, such that the equation obtained by putting $y = 1/z$ has the solution $z = 0$.

suppose that a, b, R are known functions of a variable parameter α. Let us try to find the curves which cut each of these circles at a known angle V, which may be constant or a given function of α. The coördinates of any point M of the circle C with the center (a, b) and the radius R can be represented by the equations

$$x = a + R\cos\theta, \qquad y = b + R\sin\theta,$$

where θ is the angle which the radius terminating at the point M makes with the direction Ox. The problem reduces to the determination of the angle θ as a function of the parameter α, so that the curve described by the point M cuts the circle C at the angle V. The differential equation of the problem is therefore

$$\operatorname{ctn} V = \frac{\dfrac{dy}{dx} - \tan\theta}{1 + \tan\theta \dfrac{dy}{dx}},$$

which becomes, after replacing dx and dy by their values and reducing,

$$R\frac{d\theta}{d\alpha} + b'\cos\theta - a'\sin\theta - \operatorname{ctn} V(R' + a'\cos\theta + b'\sin\theta) = 0,$$

where a', b', R' are the derivatives of a, b, R with respect to α. Taking for the new unknown $t = \tan(\theta/2)$, we obtain the Riccati equation

$$(29) \quad 2R\frac{dt}{d\alpha} + b'(1 - t^2) - 2a't - \operatorname{ctn} V[R'(1 + t^2) + a'(1 - t^2) + 2b't] = 0.$$

It will suffice, then, to know a single trajectory in order to obtain all the others by two quadratures.

Let us consider the particular case of *orthogonal trajectories*; the angle V is then a right angle, and the cotangent is zero. If we also suppose that the circles considered have their centers on a straight line, we know a priori two particular integrals of the equation (29), for the line of the centers is an orthogonal trajectory and meets each circle in two points. It is easily shown that the integration requires only one quadrature, for if we take the x-axis for the line of centers, the equation (29) reduces to $R(dt/dx) - a't = 0$.

8. Equations not solved for y'. In the different cases which we have just examined the equation was supposed to be solved with respect to y'. Let us now consider the general equation of the first order $F(x, y, y') = 0$. Let S be the surface represented by the equation $F(x, y, z) = 0$, obtained by replacing y' by z. To every integral $y = f(x)$ of the proposed equation there corresponds a curve Γ, represented by the relations

$$(\Gamma) \qquad y = f(x), \qquad z = f'(x),$$

which lies entirely on the surface S, since we have

$$F[x, f(x), f'(x)] = 0.$$

But this curve Γ is not any curve on the surface S; along this curve, in fact, y and z are functions of x satisfying the relation $dy - z\,dx = 0$, and that relation preserves the same form if we take any independent variable in place of x.

Conversely, let Γ be a curve lying on the surface S; the coördinates x, y, z of a point of that curve are functions of a variable α. If these three functions, $x = \phi_1(\alpha)$, $y = \phi_2(\alpha)$, $z = \phi_3(\alpha)$, satisfy the relation $dy = z\,dx$, we can deduce from them an integral of the given equation; for the first two relations, $x = \phi_1(\alpha)$, $y = \phi_2(\alpha)$, represent a plane curve C. Let $y = f(x)$ be the equation of that curve, supposing it solved for y. Along the entire curve Γ we have $z = f'(x)$, and consequently $F[x, f(x), f'(x)] = 0$; the curve C is therefore an integral curve. There would be an exception only in case the curve C were to reduce to a point, and the curve Γ to a straight line parallel to Oz. The two problems are then equivalent: to integrate the given equation $F(x, y, y') = 0$ or to find the curves of the surface S for which we have

$$dy - z\,dx = 0.$$

This being the case, let us suppose that we can express the coördinates of a point x, y, z of the surface S explicitly as functions of two variable parameters u, v:

$$x = f(u, v), \qquad y = \phi(u, v), \qquad z = \psi(u, v).$$

Every curve Γ of the surface S is obtained by establishing a certain relation between u and v, and, in order that that curve shall define an integral, it is necessary and sufficient that we have $dy = z\,dx$, or

$$\frac{\partial \phi}{\partial u} du + \frac{\partial \phi}{\partial v} dv = \psi(u, v)\left(\frac{\partial f}{\partial u} du + \frac{\partial f}{\partial v} dv\right).$$

We have thus a differential equation $dv/du = \pi(u, v)$, solved with respect to dv/du. It is clear that the preceding discussion applies also to equations which can be solved for y'.

This transformation is immediate for the equations solved for one of the variables x or y. For example, let the equation be

(30) $$y = f(x, y');$$

we can here take for the variable parameters x and $y' = p$. The surface S is then represented by the equations

$$x = x, \qquad z = p, \qquad y = f(x, p),$$

and the relation $dy = z\,dx$ becomes

(31)
$$p = \frac{\partial f}{\partial x} + \frac{\partial f}{\partial p}\frac{dp}{dx}.$$

This result could have been obtained directly by differentiating the relation (30) and replacing y' by p. Let $p = \phi(x, C)$ be the general integral of the equation (31); to deduce from it the general integral of the equation (30), it will only be necessary to replace y' in the equation (30) by $\phi(x, C)$.

9. Lagrange's equation. Let us consider in particular an equation linear in the two variables x and y:

(32)
$$y = x\phi(y') + \psi(y').$$

Differentiating the two sides, and denoting y' by p, we obtain the equation

$$p = \phi(p) + x\phi'(p)\frac{dp}{dx} + \psi'(p)\frac{dp}{dx}.$$

If we consider p as the independent variable, and x as the unknown function, that equation, which can be written in the form

$$[\phi(p) - p]\frac{dx}{dp} + x\phi'(p) + \psi'(p) = 0,$$

is linear and is integrable by two quadratures. Having obtained x as a function of p, by putting that value of x in the expression

$$y = x\phi(p) + \psi(p),$$

we shall have the coördinates x and y expressed as functions of the parameter p and of an arbitrary constant.*

We can readily discuss the general appearance of the family of integral curves by observing that x and y are polynomials of the first degree in the arbitrary constant C:

(33)
$$x = CF(p) + \Phi(p), \qquad y = CF_1(p) + \Phi_1(p).$$

But the functions $F(p)$, $F_1(p)$, $\Phi(p)$, $\Phi_1(p)$ are not arbitrary functions, since the parameter p represents the slope dy/dx of the tangent. On this account we must have $F_1'(p) = pF'(p)$, $\Phi_1'(p) = p\Phi'(p)$. Let Γ_0, Γ_1 be two particular integrals corresponding to the values $C = 0$, $C = 1$ of the constant:

$$\Gamma_0 \begin{cases} x_0 = \Phi(p), \\ y_0 = \Phi_1(p), \end{cases} \qquad \Gamma_1 \begin{cases} x_1 = F(p) + \Phi(p), \\ y_1 = F_1(p) + \Phi_1(p). \end{cases}$$

---

* The equation (32) can also be reduced to a linear equation by means of Legendre's transformation (I, § 62, 2d ed.; § 36, 1st ed.).

A homogeneous equation of the form $y = x\phi(y')$, not solved for y', may be considered as a particular case of Lagrange's equation and integrated in the same way.

The equations (33), which represent any integral Γ, may be written also in the form
$$\Gamma \begin{cases} x = C(x_1 - x_0) + x_0, \\ y = C(y_1 - y_0) + y_0. \end{cases}$$

At the points $M_0(x_0, y_0)$, $M_1(x_1, y_1)$, $M(x, y)$ of the curves Γ_0, Γ_1, Γ, which correspond to the same value of p, the tangents to these curves are parallel. Moreover, we derive from the preceding expressions
$$\frac{y - y_0}{y - y_1} = \frac{x - x_0}{x - x_1} = \frac{C}{C-1},$$
which proves that the three points M, M_0, M_1 are on a straight line and that the ratio MM_0/MM_1 is constant. We have then the following geometric construction: *Given the two curves Γ_0, Γ_1, we join the points M_0, M_1 of these two curves where the tangents are parallel, and we take on the straight line joining these points the point M such that the ratio MM_0/MM_1 will be equal to a given constant K. If the points M_0, M_1 describe the curves Γ_0, Γ_1, the point M describes an integral curve Γ, and we obtain the general integral by varying the constant K.*

10. Clairaut's equation. A remarkable particular case of Lagrange's equation had been treated previously by Clairaut; every equation of the form

(34) $$y = xy' + f(y')$$

is called a *Clairaut equation*. Following the general method, we differentiate the two sides and put $p = y'$; this leads to the equation

(35) $$[x + f'(p)]\frac{dp}{dx} = 0.$$

This equation is satisfied by putting $dp/dx = 0$; whence $p = C$. The general integral of Clairaut's equation is, then,

(36) $$y = Cx + f(C).$$

This equation represents a family of straight lines, and it is readily seen that they are really integral curves. But the equation (35) is also satisfied by causing the first factor $x + f'(p)$ to vanish. From this it follows that there exists a new integral of the equation (34), which is represented by the two equations

$$x + f'(p) = 0, \qquad y = px + f(p).$$

Now the elimination of p between these two equations would lead precisely to the envelope of the straight lines represented by the equation (36). Hence Clairaut's equation has also as an integral curve the *envelope of the straight lines which represent the general integral*. Since we cannot obtain this integral by giving a particular value to the constant C, we say that it is a *singular integral*.

We are led to Clairaut's equation when we undertake to determine a plane curve by a property of its tangents in which the point of contact does not enter. In fact, let $y = f(x)$ be the equation of the desired curve; then the equation of the tangent is $Y = y'X + y - xy'$, and we are led to a relation between y' and $y - xy'$, that is, to Clairaut's equation. It is clear that in this case it is the singular integral which gives the real solution of the problem.

Let us propose, for example, *to find a curve such that the product of the distances from two fixed points F, F' to any one of its tangents is always equal to a constant b^2*. Let $2c$ be the distance FF', let the middle point of the segment FF' be taken for the origin, and let the straight line FF' be the x-axis. This leads to the differential equation

$$(y - xy')^2 - c^2 y'^2 = b^2(1 + y'^2)$$

if we suppose that the two points F, F' lie on the same side of the tangent. This equation reduces to the form $y = xy' \pm \sqrt{b^2 + a^2 y'^2}$; hence the general integral represents the family of straight lines

$$y = Cx \pm \sqrt{b^2 + a^2 C^2}, \qquad a^2 = b^2 + c^2.$$

The singular integral curve, the envelope of these straight lines, is the ellipse

$$\frac{x^2}{a^2} + \frac{y^2}{b^2} = 1,$$

which is the true solution of the problem.

11. Integration of the equations $F(x, y') = 0$, $F(y, y') = 0$. The equations which contain only one of the variables x or y are integrable by a quadrature, provided that we can solve the equation for y' (§ 2). If the equation is algebraic, y is an Abelian integral or the inverse function of an Abelian integral. Whenever the relation is of deficiency *zero* or deficiency *one*, we can express x and y as functions of a variable parameter, either rationally or by means of the classic transcendentals. Let us consider, first, equations of the type $F(y, y') = 0$, of deficiency *zero*; we can express y and y' as rational functions of a parameter u, $y = f(u)$, $y' = f_1(u)$, and the condition $dy = y' dx$ gives us $f'(u) du = f_1(u) dx$. Then the variables x and y are given by the expressions

(37) $$y = f(u), \qquad x = \int \frac{f'(u)}{f_1(u)} du$$

in terms of the variable parameter u. The same procedure is applicable to the equations $F(y, y') = 0$ if the relation is of deficiency *one*; but we must take for $f(u)$ and $f_1(u)$ elliptic functions, and x and y are expressed in terms of the transcendentals p, ζ, σ (Part I, § 75).

We can proceed similarly with the equations $F(x, y') = 0$ if the relation is of deficiency *zero* or *one*; besides, they reduce to the preceding form by interchanging x and y.

Example 1. The equation $y^2(y'-1) = (2-y')^2$ is of deficiency zero. Putting $2 - y' = yu$, we derive from it $y' = 1 + u^2$, $y = 1/u - u$. The relation $dy = y'dx$ becomes here $dx = -du/u^2$. We have, then, $x = 1/u + C$, and the general integral of the given equation is $y = x - C - 1/(x - C)$.

Example 2. The equation $y'^3 - 3y'^2 - 9y^4 - 12y^2 = 0$ represents, if we regard y and y' in it as the coördinates of a point, a unicursal quartic having three double points ($y = 0$, $y' = 0$), ($y = \pm\sqrt{-2/3}$, $y' = 2$). We can, in fact, write the preceding equation

$$(y'-2)^2(y'+1) = (3y^2+2)^2.$$

Putting first $y' = u^2 - 1$, we have $3y^2 = (u+1)^2(u-2)$; if we then put $u - 2 = 3t^2$, we obtain finally the following expressions for y and y' as functions of the parameter t:

$$y = 3(t + t^3), \qquad y' = 3(1 + t^2)(1 + 3t^2).$$

The relation $dy = y'dx$ reduces here to $(1 + t^2)dx = dt$; we derive from it

$$t = \tan(x + C),$$

and the general integral of the given equation is therefore

$$y = 3\tan(x + C) + 3\tan^3(x + C).$$

Example 3. Let $R(y)$ be a polynomial of the third or of the fourth degree, prime to its derivative; let us consider the differential equation

(38) $$y'^2 = R(y).$$

We have seen in § 78, Part I, that we can satisfy this equation of deficiency one by putting $y = f(u)$, $y' = f'(u)$, where $f(u)$ is an elliptic function of the second order. The condition $dy = y'dx$ becomes $du = dx$; the general integral of the equation (38) is therefore an elliptic function $y = f(x + C)$.

If the polynomial $R(y)$ is of lower degree than the third, or if the polynomial, although of the third or of the fourth degree, is not prime to its derivative, the relation (38) is of deficiency zero. We can express y and y' by rational functions of a parameter u, and, by applying the preceding method, we easily show that the general integral is a rational function of x or a rational function of e^{ax}.

12. Integrating factors. The method of integration by the separation of the variables was generalized by Euler. The reasoning of § 2 applies really to every equation of the first order

(39) $$P(x, y)dx + Q(x, y)dy = 0,$$

where the coefficients P and Q contain both x and y, provided that we have $\partial P/\partial y = \partial Q/\partial x$. This condition is necessary and sufficient in order that $Pdx + Qdy$ shall be the total differential of a function $U(x, y)$, and the function $U(x, y)$ is obtained by quadratures, as we have seen (I, § 151). The equation (39) is then identical with the equation $dU = 0$, and the most general solution is given by a relation of the form $U(x, y) = C$ between x and y. The equation (39) is therefore integrable by quadratures whenever the coefficients P and Q satisfy the condition $\partial P/\partial y = \partial Q/\partial x$.

In order that the preceding method may be applied, it is not necessary that we have $\partial P/\partial y = \partial Q/\partial x$; it suffices to know an *integrating factor*, that is, a factor $\mu(x, y)$ such that the product

$$\mu(x, y)[P\,dx + Q\,dy]$$

satisfies the integrability condition $\partial(\mu P)/\partial y = \partial(\mu Q)/\partial x$, or, after developing,

(40) $$P\frac{\partial \mu}{\partial y} - Q\frac{\partial \mu}{\partial x} + \mu\left(\frac{\partial P}{\partial y} - \frac{\partial Q}{\partial x}\right) = 0.$$

The investigation of the integrating factors is thus reduced to the integration of the preceding equation, which is a partial differential equation of the first order. It seems that in proceeding in this way we have made the integration of equation (39) depend on an apparently more difficult problem, but it is to be noticed that it suffices to know *one particular solution* of the equation (40) in order to apply the method, and in many cases we can find a particular integral of the equation (40) by more or less direct processes. Let us see, for example, in what case the equation (39) has an integrating factor depending only on x. If we suppose $\partial \mu/\partial y = 0$, the equation (40) becomes

$$Q\frac{\partial \mu}{\partial x} = \mu\left(\frac{\partial P}{\partial y} - \frac{\partial Q}{\partial x}\right),$$

and the expression $[\partial P/\partial y - \partial Q/\partial x]/Q$ must be independent of y; if it is, we obtain an integrating factor μ by a quadrature. Let us suppose in addition that $Q = 1$; then $\partial P/\partial y$ must be a function X of the variable x, and the equation (39) is a linear equation,

(39') $$dy + (Xy + X_1)\,dx = 0,$$

where X and X_1 denote functions of x alone. In this case, the equation (40) is satisfied by

$$\mu = e^{\int_{x_0}^{x} X\,dx},$$

and it is easy to show that if we multiply the equation (39') by this factor, we have on the left an exact differential

$$e^{\int_{x_0}^{x} X\,dx}(dy + Xy\,dx + X_1\,dx) = d\left(ye^{\int_{x_0}^{x} X\,dx} + \int_{x_0}^{x} X_1 e^{\int_{x_0}^{x} X\,dx}\,dx\right) = 0.$$

The calculations which have to be made for the integration are exactly the same as in the first method (§ 4).

We shall show farther on that the equation (40) has an infinite number of integrals under very general conditions, which are always satisfied in the cases in which we are interested. If we know *one* integrating factor μ_1, we can obtain all others in the following way: Putting $\mu = \mu_1 \nu$, the equation (40) becomes

$$(40') \qquad P\frac{\partial \nu}{\partial y} - Q\frac{\partial \nu}{\partial x} = 0.$$

Now we know one function satisfying this relation: it is the function, $U(x, y)$, whose total differential is $\mu_1(P\,dx + Q\,dy)$, since the partial derivatives $\partial U/\partial x$, $\partial U/\partial y$ are equal to $\mu_1 P$ and to $\mu_1 Q$. We have, then, also $(\partial \nu/\partial y)(\partial U/\partial x) - (\partial \nu/\partial x)(\partial U/\partial y) = 0$, which proves that ν is of the form $\phi(U)$ and that the general expression for the integrating factors is $\mu = \mu_1 \phi(U)$, where ϕ is an arbitrary function of U. It is easy to show that μ is really an integrating factor, for from the identity

$$\mu_1(P\,dx + Q\,dy) = dU$$

we derive, by multiplying by $\phi(U)$,

$$\mu_1 \phi(U)[P(x,y)\,dx + Q(x,y)\,dy] = \phi(U)\,dU,$$

and the right-hand side is the exact differential of the function

$$F(U) = \int \phi(U)\,dU.$$

We deduce from this an interesting consequence: if μ_1 and μ_2 are two integrating factors, the ratio μ_2/μ_1 is a function of U. If this quotient μ_2/μ_1 is not constant, the general integral of the differential equation can then be written in the form $\mu_2/\mu_1 =$ constant.

The preceding theorem is sometimes helpful in finding an integrating factor. Let us consider the differential equation

$$(41) \qquad P\,dx + Q\,dy + P_1\,dx + Q_1\,dy = 0,$$

where P, P_1, Q, Q_1 are functions of x, y, and let us suppose that we know how to find an integrating factor for each of the expressions $P\,dx + Q\,dy$, $P_1\,dx + Q_1\,dy$. The general expression for the integrating factors of $P\,dx + Q\,dy$ is $\mu\phi(U)$, where μ is the known factor, U a function of x and y which we obtain by quadratures, and ϕ an arbitrary function. Similarly, the general expression for the integrating factors of $P_1\,dx + Q_1\,dy$ is $\mu_1\psi(U_1)$, where μ_1 and U_1 are definite functions and ψ an arbitrary function. If we can choose the functions ϕ and ψ in such a way that we have

$$\mu\phi(U) = \mu_1\psi(U_1),$$

we shall have an integrating factor for the given equation (41).

Let us take, for example, the equation

$$ax\,dy + by\,dx + x^m y^n(\alpha x\,dy + \beta y\,dx) = 0,$$

where a, b, α, β are constants. Every integrating factor of $ax\,dy + by\,dx$ is of the form $\phi(x^b y^a)/xy$, and, similarly, every integrating factor of the second part is of the form $\psi(x^\beta y^\alpha)/x^{m+1}y^{n+1}$. In order to have a common integrating factor, it will suffice to find two exponents, p and q, such that we have

$$x^m y^n (x^b y^a)^p = (x^\beta y^\alpha)^q,$$

which leads to the conditions

$$pa - q\alpha + n = 0, \qquad pb - q\beta + m = 0.$$

These conditions are compatible if $a\beta - b\alpha$ is not zero, and determine an integrating factor of the form $x^M y^N$. Multiplying by this integrating factor, the equation takes the form $v^{p-1}dv + v_1^{q-1}dv_1 = 0$, where we have put $v = x^b y^a$, $v_1 = x^\beta y^\alpha$; and this equation is immediately integrable.

In the particular case where $a\beta - b\alpha = 0$, we obtain from it $\alpha/a = \beta/b = k$, and the equation can be written in the form $(ax\,dy + by\,dx)(1 + kx^m y^n) = 0$.

Note. If we know the general integral of a differential equation of the first order, it is quite easy to obtain an integrating factor. For let $f(x, y) = C$ be the general integral of the equation (39). The differential equation of the curves represented by that relation is also $(\partial f/\partial x)\,dx + (\partial f/\partial y)\,dy = 0$; in order that it be identical with the equation (39), we must have

$$\frac{\dfrac{\partial f}{\partial x}}{P} = \frac{\dfrac{\partial f}{\partial y}}{Q},$$

and the common value of the two preceding ratios is evidently an integrating factor for $P\,dx + Q\,dy$. Every other integrating factor is equal to this one multiplied by an arbitrary function of $f(x, y)$.

13. Application to conformal representation. The theory of integrating factors finds an important application in the problem of conformal representation. Let

$$ds^2 = E\,du^2 + 2F\,du\,dv + G\,dv^2$$

be a quadratic form in du, dv whose coefficients E, F, G are *analytic functions* of u and v such that $EG - F^2$ is not zero. We can also write ds^2 in the form

$$ds^2 = (a\,du + b\,dv)(a_1\,du + b_1\,dv),$$

where a, b, a_1, b_1 are also analytic functions of u and v. According to a result which will be rigorously proved later, each of the expressions $a\,du + b\,dv$, $a_1\,du + b_1\,dv$ has an infinite number of integrating factors, which are themselves analytic functions. If μ, μ_1 are two such factors, we have the identities

$$\mu(a\,du + b\,dv) = dU, \qquad \mu_1(a_1\,du + b_1\,dv) = dU_1,$$

and therefore
$$\mu\mu_1 ds^2 = dU dU_1;$$
whence, substituting
$$U = X + Yi, \qquad U_1 = X - Yi, \qquad \mu\mu_1 = \frac{1}{\lambda},$$
we obtain
$$E\,du^2 + 2F\,du\,dv + G\,dv^2 = \lambda(dX^2 + dY^2).$$

Every analytic surface can therefore be represented on a plane conformly; that is, without alteration of the angles between pairs of curves. If the surface is real, we may suppose that the real points of the surface correspond to real values of the variables u, v; the coefficients E, F, G are real, while a and a_1 are conjugate imaginaries, as also b and b_1. We can also take for μ and μ_1, and therefore for U and U_1, conjugate imaginaries, so that to real values of u, v correspond real values of X and of Y. To real points of the surface correspond therefore real points of the plane.

Since it is possible to represent every analytic surface on a plane conformly, we conclude that any analytic surface can be represented conformly on any other analytic surface.

14. Euler's equation. A great many devices have been invented for the integration of differential equations of special forms. A celebrated example, due to Euler and now known by his name, is the equation

(42) $$\frac{dx}{\sqrt{X}} + \frac{dy}{\sqrt{Y}} = 0,$$

where X and Y are two polynomials of the fourth degree in x and y respectively, having the same coefficients:

$$X = a_0 x^4 + a_1 x^3 + a_2 x^2 + a_3 x + a_4,$$
$$Y = a_0 y^4 + a_1 y^3 + a_2 y^2 + a_3 y + a_4.$$

The variables being separated, we obtain the general integral of equation (42) by two quadratures, which introduce two transcendental functions depending respectively upon x and y. Euler's fundamental discovery, which was the starting point of the theory of elliptic functions, consisted in showing that that relation between the variables x and y which in appearance is transcendental is in reality algebraic.

Let us first consider the case where X is a polynomial of the second degree, not a perfect square. A linear substitution enables us to bring it to the form $X = A(x^2 - 1)$, and in this particular case the equation (42) becomes

(43) $$\frac{dx}{\sqrt{1-x^2}} + \frac{dy}{\sqrt{1-x^2}} = 0.$$

Clearing of fractions, we can write this in the form

$$\sqrt{1-y^2}\,dx + \sqrt{1-x^2}\,dy = d\left(x\sqrt{1-y^2} + y\sqrt{1-x^2}\right) + xy\left(\frac{dx}{\sqrt{1-x^2}} + \frac{dy}{\sqrt{1-y^2}}\right) = 0,$$

which shows that we have identically

$$d\left(x\sqrt{1-y^2} + y\sqrt{1-x^2}\right) = \left[\sqrt{(1-x^2)(1-y^2)} - xy\right]\left(\frac{dx}{\sqrt{1-x^2}} + \frac{dy}{\sqrt{1-y^2}}\right).$$

The expression $\sqrt{(1-x^2)(1-y^2)} - xy$ is therefore an integrating factor for the equation (43), and the general integral is given by the relation

(44) $$x\sqrt{1-y^2} + y\sqrt{1-x^2} = C,$$

or by the relation

(45) $$\sqrt{(1-x^2)(1-y^2)} - xy = C',$$

since the equation (43) has the two integrating factors, 1 and the expression on the left-hand side of (45). It is also very easy to verify that the two expressions (44) and (45) are equivalent by means of the identity

$$\left(x\sqrt{1-y^2} + y\sqrt{1-x^2}\right)^2 + \left[\sqrt{(1-x^2)(1-y^2)} - xy\right]^2 = 1.$$

Rationalizing the expression (45), we can write the general integral of the equation (43) in the form

(46) $$x^2 + y^2 + 2C'xy + C'^2 - 1 = 0,$$

where C' denotes an arbitrary constant, and this equation represents the conics tangent to the four straight lines $x = \pm 1$, $y = \pm 1$.

By a bold induction Euler was led to a more general formula of the same kind, which corresponds to the case where X is any polynomial of the third or of the fourth degree (*Institutiones calculi integralis*, Vol. I, chaps. v, vi).

Let $F(x, y)$ be a polynomial of the second degree in each of the variables x and y and symmetrical with respect to these two variables:

(47) $$F(x, y) = A_1 x^2 y^2 + A_2 xy(x+y) + A_3(x^2 + y^2) + A_4 xy + A_5(x+y) + A_6.$$

This polynomial depends upon six arbitrary coefficients A_1, A_2, A_3, A_4, A_5, A_6, and the relation $F(x, y) = 0$ can be written in two equivalent forms:

(48) $\quad \begin{cases} F(x, y) = M y^2 + N y + P = 0, \\ F(x, y) = M_1 x^2 + N_1 x + P_1 = 0, \end{cases}$

where M, N, P are three polynomials of the second degree in x:

$$M = A_1 x^2 + A_2 x + A_3, \quad N = A_2 x^2 + A_4 x + A_5, \quad P = A_3 x^2 + A_5 x + A_6,$$

and where M_1, N_1, P_1 are the polynomials obtained by replacing x by y in M, N, P. From the relation $F(x, y) = 0$ we derive $F'_x dx + F'_y dy = 0$, or, after replacing F'_x and F'_y by their values,

(49) $\quad (2 M_1 x + N_1) dx + (2 My + N) dy = 0.$

We derive, moreover, from the relations (48),

$$2 My + N = \pm \sqrt{N^2 - 4 MP}, \qquad 2 M_1 x + N_1 = \pm \sqrt{N_1^2 - 4 M_1 P_1},$$

and the preceding equation (49) may be written in the form

(50) $\quad \dfrac{dx}{\sqrt{N^2 - 4 MP}} \pm \dfrac{dy}{\sqrt{N_1^2 - 4 M_1 P_1}} = 0.$

This relation will be identical with the given equation (42) if we have $N^2 - 4 MP = X$, which necessarily carries with it the other equality $N_1^2 - 4 M_1 P_1 = Y$. Now, since M, N, P are of the second degree, $N^2 - 4 MP$ is of the fourth degree, and the preceding condition is an identity between two polynomials of the fourth degree, which requires only five conditions. Since we have six coefficients A_i at our disposal, we see that one of these coefficients will remain arbitrary. There are therefore an infinite number of polynomials $F(x, y)$ of the form (47), depending upon an arbitrary constant C and such that the relation

(51) $\quad F(x, y) = 0,$

between the variables x and y, leads to the relation (42). Hence the relation (51) represents the general integral of the proposed equation.

The actual determination of the polynomial $F(x, y)$ requires a calculation by equating coefficients which can be simplified by means of a geometric representation due to Jacobi. Let us consider, in order to take the general case, a polynomial of the fourth degree $R(t)$ prime to its derivative, and let t_1, t_2, t_3, t_4 be the roots of $R(t) = 0$. On the other hand, let Σ be any conic the coördinates of any point of which are rational functions of the second degree of the variable parameter t, so that to a point (x, y) corresponds a single value of t; let us call m_1, m_2, m_3, m_4 the points of Σ which correspond to the values t_1, t_2, t_3, t_4 of the parameter. Finally, let Σ' be a second conic passing through the four points m_1, m_2, m_3, m_4. Every straight line tangent to Σ' meets Σ in two points M and M'; if t and t' are the corresponding values of the parameter, the relation between t and t' is the one desired. It is evident, in fact, that that relation is symmetric in t and t', and that it is of the second degree in each of the variables, for through a point M' we can draw two tangents to Σ', and so to each value of t' correspond only two values of t.

Let

(52) $$F(t, t') = 0$$

be that relation. We can derive from it, as we have just seen, a relation between the differentials dt, dt', of the form

(53) $$\frac{dt}{\sqrt{P(t)}} \pm \frac{dt'}{\sqrt{P(t')}} = 0,$$

where $P(t)$ is a polynomial of the fourth degree. *This polynomial $P(t)$ is identical except for a constant factor with $R(t)$*; for, according to the preceding method for obtaining the polynomial $P(t)$ from $F(t, t') = 0$, the roots of $P(t) = 0$ are the values of t for which the two values of t' coincide. Now the geometric significance of the relation (52) shows immediately that this can only occur if the two tangents from M to Σ' coincide; that is, if the point M is one of the points m_1, m_2, m_3, m_4. We are thus led to the following method, which requires only rational calculations, for obtaining the general integral of the equation

(54) $$\frac{dt}{\sqrt{R(t)}} \pm \frac{dt'}{\sqrt{R(t')}} = 0,$$

where $R(t) = a_0 t^4 + a_1 t^3 + a_2 t^2 + a_3 t + a_4$. This equation differs only in notation from the proposed equation (42). We begin by forming the general equation of the conics Σ' passing through the four points m_1, m_2, m_3, m_4 of Σ; that equation is of the form $f(x, y) + C\phi(x, y) = 0$, where C is an arbitrary constant. We then write the condition that the straight line joining the two points M and M' of Σ, which correspond to the values t, t' of the parameter, shall be tangent to Σ'. The resulting relation, which contains the arbitrary constant C, represents the general integral of Euler's equation.

To carry out the calculations, let us take for Σ the parabola $y^2 = x$, and let us put $x = t^2$, $y = t$. The conic Σ' given by the equation

(55) $$Ax^2 + A'y^2 + 2B''xy + 2B'x + 2By + A'' = 0$$

cuts Σ in four points, given by the equation of the fourth degree in t which is obtained by replacing x by t^2 and y by t. In order that that equation shall be identical with $R(t) = 0$, it is sufficient that

(56) $$A = a_0, \quad A' + 2B' = a_2, \quad 2B'' = a_1, \quad 2B = a_3, \quad A'' = a_4.$$

The coefficient B' remaining arbitrary, we shall put $B' = C$, which gives

$$A' = a_2 - 2C.$$

Let us recall now that the tangential equation of Σ', that is, the condition that the straight line $\alpha x + \beta y + \gamma = 0$ shall be tangent to that conic, is given by the equation

(57) $$\begin{vmatrix} A & B'' & B' & \alpha \\ B'' & A' & B & \beta \\ B' & B & A'' & \gamma \\ \alpha & \beta & \gamma & 0 \end{vmatrix} = 0.$$

The straight line joining the two points (t^2, t) and (t'^2, t') of Σ has for its equation
$$x - (t + t')y + tt' = 0.$$

We can therefore take
$$\alpha = 1, \qquad \beta = -(t + t'), \qquad \gamma = tt'.$$

Substituting the values obtained for A, B, A', B', A'', B'', α, β, γ in the condition (57), and replacing t and t' by x and y respectively, we arrive at the general integral of Euler's equation in the following form, which is due to Stieltjes:

(58) $$\begin{vmatrix} a_0 & \dfrac{a_1}{2} & C & 1 \\ \dfrac{a_1}{2} & a_2 - 2C & \dfrac{a_3}{2} & -(x+y) \\ C & \dfrac{a_3}{2} & a_4 & xy \\ 1 & -(x+y) & xy & 0 \end{vmatrix} = 0.$$

This equation represents a family of curves of the fourth degree, having two double points at infinity on Ox and Oy respectively. The equation being of the second degree with respect to the constant C, through each point of the plane there pass two curves of the family, as we might have foreseen, since the given differential equation gives two equal values, but with opposite signs, for the derivative y' at each point. These two values of y' become equal only if the point (x, y) belongs to the curve $XY = 0$, which is composed of four straight lines D_1, D_2, D_3, D_4 parallel to the axis Oy, and of four straight lines $\Delta_1, \Delta_2, \Delta_3, \Delta_4$ parallel to the axis Ox. Let us write Euler's equation in the rational form $Y dx^2 - X dy^2 = 0$, and let us take a point $M(x, y)$ on one of these straight lines, Δ_1 for example, not belonging to any one of the D lines. For the coördinates of the point M we have $Y = 0$, $X \neq 0$, and Euler's equation gives for y' a double value, $y' = 0$. Hence the straight line Δ_1 itself is an integral curve through M. But it can be verified that the curves represented by the equation (58) have as their envelope the set of eight straight lines given by the equation $XY = 0$. Hence there is a new integral curve tangent to the first one at M. Thus the eight straight lines D_i, Δ_i are singular integral curves, for they are not included among the curves represented by the general integral.

Note. We have supposed, in order to arrive at the equation (58), that the polynomial $R(x)$ was one of the fourth degree and prime to its derivative; but it is clear that the result can be verified directly without the hypothesis that $R(x)$ is prime to its derivative. We could, for example, form the differential equation of the curves represented by the equation (58) by applying the general method of § 1, and the equation obtained would necessarily be identical with Euler's equation, whatever may be the values of the coefficients a_0, a_1, a_2, a_3, a_4, since we reach this result when the coefficients do not satisfy any particular relation. The equation (58) therefore applies to all cases.

15. A method deduced from Abel's theorem. We can also very easily deduce the general integral of Euler's equation from Abel's theorem. Let us now denote by $R(x)$ a polynomial of the third or of the fourth degree, prime to its derivative, and let us consider the curve C which has for its equation $y^2 = R(x)$. If a variable algebraic curve C' meets the curve C in three variable points only, M_1, M_2, M_3, we have shown (Part I, § 103) that the coördinates (x_1, y_1), (x_2, y_2), (x_3, y_3) of these three variable points satisfy the relation

$$(59) \qquad \frac{dx_1}{y_1} + \frac{dx_2}{y_2} + \frac{dx_3}{y_3} = 0.$$

If the variable curve C' depends upon two variable parameters which we can select in such a way that two of the points of intersection, (x_1, y_1), (x_2, y_2), can be brought to coincide with any two points of the curve C given in advance, the coördinates of the third point of intersection, (x_3, y_3), are functions of the coördinates $(x_1, y_1; x_2, y_2)$ of the first two, and satisfy the relation (59). The equation $dx_1/y_1 + dx_2/y_2 = 0$ is therefore equivalent to the equation $dx_3/y_3 = 0$, whose general integral is $x_3 = $ constant. Now, since the points (x_1, y_1), (x_2, y_2) are on the curve C, we have $y_1^2 = R(x_1)$, $y_2^2 = R(x_2)$, and the equation $dx_1/y_1 + dx_2/y_2 = 0$, which may be written in the form

$$(60) \qquad \frac{dx_1}{\sqrt{R(x_1)}} + \frac{dx_2}{\sqrt{R(x_2)}} = 0,$$

is identical with Euler's except in notation. In the expression which gives the general integral

$$(61) \qquad x_3 = F(x_1, y_1; x_2, y_2) = \text{const.}$$

we should replace y_1 and y_2 by $\sqrt{R(x_1)}$ and $\sqrt{R(x_2)}$ respectively, the determinations of the two radicals being the same in the two expressions (60) and (61). We thus obtain for the general integral an expression containing radicals, while the result (58) is rational. But the irrational form is in certain cases the more advantageous.

Let us carry out the calculations, supposing the polynomial $R(x)$ reduced to the normal form of Legendre, $R(x) = (1 - x^2)(1 - k^2 x^2)$, where k^2 is different from zero and from unity. The parabola C',

$$(62) \qquad y = ax^2 + bx + 1,$$

meets the curve C represented by the equation $y^2 = R(x)$ in the point $(x = 0, y = 1)$ and in three variable points whose abscissas x_1, x_2, x_3 are roots of the equation

$$(63) \qquad (a^2 - k^2) x^3 + 2ab x^2 + (b^2 + 2a + k^2 + 1) x + 2b = 0,$$

which is obtained by eliminating y and suppressing the factor x.

We derive from this equation the relations

$$x_1 + x_2 + x_3 = \frac{2ab}{k^2 - a^2}, \qquad x_1 x_2 + x_2 x_3 + x_1 x_3 = \frac{b^2 + 2a + k^2 + 1}{a^2 - k^2},$$

$$x_1 x_2 x_3 = \frac{2b}{k^2 - a^2},$$

whence

$$(64) \qquad x_1 + x_2 + x_3 = a x_1 x_2 x_3.$$

The condition that the parabola C' passes through the two points (x_1, y_1), (x_2, y_2), enables us to determine a and b. We have in particular

$$ax_1 x_2 = 1 + \frac{y_1 x_2 - y_2 x_1}{x_1 - x_2}.$$

Substituting this value of a in the preceding expression, we obtain finally the expression for x_3 in terms of x_1, y_1, x_2, y_2:

$$x_3 = \frac{x_1^2 - x_2^2}{x_2 y_1 - x_1 y_2}.$$

The general integral of Euler's equation,

(65) $$\frac{dx_1}{\sqrt{R(x_1)}} + \frac{dx_2}{\sqrt{R(x_2)}} = 0,$$

is therefore represented by the expression

(66) $$x_3 = \frac{x_1^2 - x_2^2}{x_2 \sqrt{R(x_1)} - x_1 \sqrt{R(x_2)}} = C.$$

16. Darboux's theorems. Let us consider a differential equation of the form

(67) $$- L\,dy + M\,dx + N(x\,dy - y\,dx) = 0,$$

where L, M, N are three polynomials in x, y of at most the mth degree, and where at least one of them is actually of the mth degree. In order that the relation $u(x, y) = $ constant shall represent the general integral, it is necessary and sufficient that the equation (67) be identical with the equation

$$\frac{\partial u}{\partial x} dx + \frac{\partial u}{\partial y} dy = 0,$$

which requires that we have

(68) $$L \frac{\partial u}{\partial x} + M \frac{\partial u}{\partial y} - N\left(x \frac{\partial u}{\partial x} + y \frac{\partial u}{\partial y}\right) = 0.$$

This condition assumes a more symmetric form if we replace x by x/z and y by y/z, where z is a fictitious variable which we shall always suppose equal to unity after the indicated operations have been performed. Then $u(x, y)$ changes into a homogeneous function of degree zero, and we have

$$x \frac{\partial u}{\partial x} + y \frac{\partial u}{\partial y} + z \frac{\partial u}{\partial z} = 0.$$

The condition (68) takes now the form

(69) $$L \frac{\partial u}{\partial x} + M \frac{\partial u}{\partial y} + N \frac{\partial u}{\partial z} = A(u) = 0.$$

Conversely, if we have obtained a homogeneous function of degree zero, $u(x, y, z)$, which satisfies the relation (69), $u(x, y, 1) = $ constant represents the general integral of the equation (67).

Darboux[*] has shown that we could form a function $u(x, y, z)$ satisfying these conditions if we knew a certain number of algebraic integrals of the

---

[*] *Sur les équations différentielles algébriques du premier ordre et du premier degré* (*Bulletin des Sciences mathématiques*, 1878).

equation (67). Suppose that the equation (67) has an algebraic integral defined by the relation $f(x, y) = 0$, where the polynomial $f(x, y)$ is irreducible and of degree h. Repeating the previous work, we find that the relation

(70) $$L\frac{\partial f}{\partial x} + M\frac{\partial f}{\partial y} - N\left(x\frac{\partial f}{\partial x} + y\frac{\partial f}{\partial y}\right) = 0$$

must be a consequence of the equation $f(x, y) = 0$. If we again replace x by x/z, and y by y/z, and then multiply by z^h, $f(x, y)$ becomes a homogeneous function of x, y, z, of degree h, satisfying the relation

$$x\frac{\partial f}{\partial x} + y\frac{\partial f}{\partial y} + z\frac{\partial f}{\partial z} = hf,$$

and the condition (70) becomes

(71) $$\Lambda(f) = L\frac{\partial f}{\partial x} + M\frac{\partial f}{\partial y} + N\frac{\partial f}{\partial z} = Nhf.$$

This condition is not satisfied identically, but by reason of the relation $f(x, y, z) = 0$. Since the last relation is irreducible by hypothesis, it is necessary that we have identically

(72) $$\Lambda(f) = Kf,$$

where K denotes a polynomial in x, y, z which is necessarily of degree $m - 1$, for if f is of degree h, $\Lambda(f)$ is of degree $m + h - 1$.

Let us now suppose that we have found p algebraic solutions of the equation (67), defined by the p following equations:

$$f_1(x, y) = 0, \quad f_2(x, y) = 0, \quad \cdots, \quad f_p(x, y) = 0,$$

where $f_1, f_2, \cdots, f_p$ are irreducible polynomials of the degrees $h_1, h_2, \cdots, h_p$. This requires that we have p identities of the following form:

(73) $\quad \Lambda(f_1) = K_1 f_1, \quad \Lambda(f_2) = K_2 f_2, \quad \cdots, \quad \Lambda(f_p) = K_p f_p,$

where the polynomials $K_1, K_2, \cdots, K_p$ are all of degree $m - 1$.

Let us observe that the symbolic operator $\Lambda(f)$ has properties analogous to those of a derivative. In particular, we can apply to it the rule for the derivative of a function of functions: if $F(u, v, w)$ is any function of u, v, w, we have

$$\Lambda(F) = \frac{\partial F}{\partial u}\Lambda(u) + \frac{\partial F}{\partial v}\Lambda(v) + \frac{\partial F}{\partial w}\Lambda(w).$$

Consequently, if we put $u = f_1^{\alpha_1} f_2^{\alpha_2} \cdots f_p^{\alpha_p}$, where $\alpha_1, \alpha_2, \cdots, \alpha_p$ are any constants, we have

$$\Lambda(u) = \alpha_1 f_1^{\alpha_1-1} f_2^{\alpha_2} \cdots f_p^{\alpha_p} \Lambda(f_1) + \alpha_2 f_1^{\alpha_1} f_2^{\alpha_2-1} \cdots f_p^{\alpha_p} \Lambda(f_2) + \cdots,$$

or, by (73),

$$\Lambda(u) = (\alpha_1 K_1 + \alpha_2 K_2 + \cdots + \alpha_p K_p) u.$$

The function $u(x, y, z)$ is a homogeneous function of degree

$$\alpha_1 h_1 + \alpha_2 h_2 + \cdots + \alpha_p h_p.$$

If we can dispose of the constants $\alpha_1 \cdots \alpha_p$ in such a way that we have

(74) $$\begin{cases} \alpha_1 h_1 + \cdots + \alpha_p h_p = 0, \\ \alpha_1 K_1 + \cdots + \alpha_p K_p = 0, \end{cases}$$

the equation $u(x, y, z) =$ constant will furnish the general integral of the given equation, by what we have established above.

The equations (74) form a system of $m(m+1)/2 + 1$ homogeneous equations in $\alpha_1, \alpha_2, \cdots, \alpha_p$, since the polynomials K_i of degree $m - 1$ contain $m(m+1)/2$ terms. We shall surely be able to satisfy all these equations by values of α_i *not all zero*, and therefore to complete the integration, whenever there are more unknowns than equations; that is, whenever we have

$$(75) \qquad p \geqq \frac{m(m+1)}{2} + 2.$$

This is *Darboux's first theorem*. If the equations (74) are not independent, we can find the solutions without requiring p to reach the preceding limit $m(m+1)/2 + 2$. A large number of examples in which this is the case will be found in Darboux's paper.

If we know only $p = m(m+1)/2 + 1$ particular algebraic integrals, we can, in general, dispose of the p constants α_i in such a way as to satisfy the conditions

$$(76) \qquad \begin{cases} \alpha_1 K_1 + \alpha_2 K_2 + \cdots + \alpha_p K_p = -\dfrac{\partial L}{\partial x} - \dfrac{\partial M}{\partial y} - \dfrac{\partial N}{\partial z}, \\ \alpha_1 h_1 + \alpha_2 h_2 + \cdots + \alpha_p h_p = -m - 2, \end{cases}$$

which are equivalent to a system of $m(m+1)/2 + 1$ linear non-homogeneous equations. The function u thus obtained satisfies the two equations,

$$L \frac{\partial u}{\partial x} + M \frac{\partial u}{\partial y} + N \frac{\partial u}{\partial z} + u \left(\frac{\partial L}{\partial x} + \frac{\partial M}{\partial y} + \frac{\partial N}{\partial z} \right) = 0,$$

$$x \frac{\partial u}{\partial x} + y \frac{\partial u}{\partial y} + z \frac{\partial u}{\partial z} + (m+2) u = 0;$$

whence we derive, by eliminating $\partial u / \partial z$ and replacing z by 1,

$$L \frac{\partial u}{\partial x} + M \frac{\partial u}{\partial y} - N \left[(m+2) u + x \frac{\partial u}{\partial x} + y \frac{\partial u}{\partial y} \right] + u \left(\frac{\partial L}{\partial x} + \frac{\partial M}{\partial y} + \frac{\partial N}{\partial z} \right) = 0.$$

But, since the function N has been made homogeneous by substituting x/z for x and y/z for y, and then multiplying by z^m, we also have, after making $z = 1$,

$$\frac{\partial N}{\partial z} = mN - x \frac{\partial N}{\partial x} - y \frac{\partial N}{\partial y},$$

so that the preceding relation may be written also in the form

$$(77) \qquad \frac{\partial u}{\partial x}(L - Nx) + \frac{\partial u}{\partial y}(M - Ny)$$
$$+ u \left(\frac{\partial L}{\partial x} + \frac{\partial M}{\partial y} - x \frac{\partial N}{\partial x} - y \frac{\partial N}{\partial y} - 2N \right) = 0.$$

It is easily seen that this last condition expresses the fact that u is an integrating factor for the equation (67), and we obtain thus *Darboux's second theorem*:

If $m(m+1)/2 + 1$ particular algebraic integrals of the equation (67) are known, an integrating factor can be determined.

The proof of this last theorem is not complete in the particular case where the determinant of the coefficients of the unknowns α_i in the $m(m+1)/2 + 1$

equations deduced from the relations (76) turns out to be zero. But we can then satisfy the $m(m+1)/2+1$ homogeneous equations, obtained by suppressing the right-hand sides, by values of the α_i not all zero, and therefore obtain the general integral by the first theorem.

Example. Let us consider in particular Jacobi's equation (§ 6); the number m is here equal to 1. Let us look first for the linear integrals of the form $ux + vy + wz = 0$. By the general method we must have identically

$$u(bz + b'x + b''y) + v(cz + c'x + c''y)$$
$$+ w(az + a'x + a''y) = \lambda(ux + vy + wz),$$

where λ is a constant factor. This leads to the three conditions

$$ub + vc + w(a - \lambda) = 0, \quad u(b' - \lambda) + vc' + wa' = 0,$$
$$ub'' + v(c'' - \lambda) + wa'' = 0,$$

and, after eliminating u, v, w, we find again the equation in λ obtained by the first method (p. 12).

Let us limit ourselves to the case in which the equation in λ has three distinct roots $\lambda_1, \lambda_2, \lambda_3$. Each of these roots furnishes a linear integral, and we therefore have three linear functions, X, Y, Z, giving the three identities

$$A(X) = \lambda_1 X, \quad A(Y) = \lambda_2 Y, \quad A(Z) = \lambda_3 Z.$$

By the general theory we can deduce from them the general integral, since in this case $m = 1$. For this purpose it is necessary to determine three numbers α, β, γ satisfying the relations

$$\alpha + \beta + \gamma = 0, \quad \alpha\lambda_1 + \beta\lambda_2 + \gamma\lambda_3 = 0.$$

We may take $\alpha = \lambda_2 - \lambda_3, \beta = \lambda_3 - \lambda_1, \gamma = \lambda_1 - \lambda_2$, and the general integral of Jacobi's equation is therefore

$$X^{\lambda_2 - \lambda_3} Y^{\lambda_3 - \lambda_1} Z^{\lambda_1 - \lambda_2} = \text{const.}$$

17. Applications. When we seek to determine a plane curve by a given relation $F(x, y, m) = 0$ between the coördinates (x, y) of a point on the curve and the slope m of the tangent at this point, the curves desired are evidently obtained by the integration of the differential equation of the first order $F(x, y, y') = 0$, which we obtain from the given relation by replacing in it m by y'. If this equation is of the qth degree in y', there pass in general q such curves through each point of the plane, as will be proved farther on. Let us consider, for example, a family of curves C, represented by the equation $\Phi(x, y, a) = 0$, depending upon an arbitrary parameter, and let us try to find their orthogonal trajectories, that is, the curves C' which cut orthogonally in each of their points a curve C passing through the same point. Let m, m' be the slopes of the tangents to the two orthogonal curves C, C' passing through the same point (x, y). Then m and m' must satisfy the relation $1 + m'm = 0$. On the other hand, let $F(x, y, y') = 0$ be the differential equation of the given curves C. Then we have $F(x, y, m) = 0$, since m is the slope of the tangent to a curve C passing through the point (x, y). It follows that

$$F\left(x, y, -\frac{1}{m}\right) = 0.$$

Moreover, m' is also the slope of the tangent to a curve C' passing through the point (x, y); hence the curve C' satisfies the equation

(78) $$F\left(x, y, -\frac{1}{y'}\right) = 0,$$

and we obtain the differential equation of the orthogonal trajectories of the curves C by replacing y' by $-1/y'$ in the differential equation of the curves C.

In order to obtain the differential equation of the curves C, we must eliminate a between the two equations $\Phi = 0$, $(\partial \Phi/\partial x) + (\partial \Phi/\partial y)y' = 0$. Therefore, *in order to obtain the differential equation of the orthogonal trajectories, it will suffice to eliminate a between the two relations* $\Phi = 0$, $(\partial \Phi/\partial x)y' - (\partial \Phi/\partial y) = 0$.

Let us take, for example, the conics represented by the equation

$$y^2 + 3x^2 - 2ax = 0,$$

where a is a variable parameter. The application of the preceding method leads to the homogeneous differential equation

$$(y^2 - 3x^2)y' + 2xy = 0,$$

which becomes, after putting $y = ux$ and separating the variables,

$$\frac{dx}{x} + \frac{3\,du}{u} - \frac{du}{u+1} - \frac{du}{u-1} = 0.$$

Solving this equation, we find

$$xu^3 = C(u^2 - 1), \quad \text{or} \quad y^3 = C(y^2 - x^2).$$

The orthogonal trajectories are therefore cubics with the origin as a double point.

Let us consider in a more general manner a surface S the coördinates x, y, z of any point of which are expressed as functions of two parameters u, v:

$$x = f(u, v), \qquad y = \phi(u, v), \qquad z = \psi(u, v).$$

We derive from these expressions

$$dx = \frac{\partial f}{\partial u} du + \frac{\partial f}{\partial v} dv, \quad dy = \frac{\partial \phi}{\partial u} du + \frac{\partial \phi}{\partial v} dv, \quad dz = \frac{\partial \psi}{\partial u} du + \frac{\partial \psi}{\partial v} dv.$$

To every value of the ratio dv/du corresponds a tangent to the surface passing through the point (u, v). If we wish to determine the curves of that surface such that the tangent to one of these curves in any point depends only on the position of that point on the surface, we are again led to integrate a differential equation of the first order:

(79) $$F\left(u, v, \frac{dv}{du}\right) = 0.$$

Conversely, every equation of this form establishes a relation between a point of a curve lying on the surface S and the tangent at that point.

Let us, for example, try to find the trajectories at a constant angle V to a family of given curves lying upon the surface. Given two curves, C, C', passing through a point (u, v) and cutting at an angle V, we have the general formula (II, Part I, § 20)

(80) $$\cos V = \frac{E\,du\,\delta u + F(du\,\delta v + dv\,\delta u) + G\,dv\,\delta v}{\sqrt{E\,du^2 + 2F\,du\,dv + G\,dv^2}\sqrt{E\,\delta u^2 + 2F\,\delta u\,\delta v + G\,\delta v^2}},$$

where E, F, G have the usual meanings, where du and dv denote the differentials relative to a displacement on C, and where δu and δv denote the differentials relative to a displacement on C'. The curves C' being given, $\delta v/\delta u$ is a known function of u and v, $\delta v/\delta u = \pi(u, v)$. Replacing $\delta v/\delta u$ by $\pi(u, v)$ in the preceding relation (80), the resulting relation $F(u, v, dv/du) = 0$ is the desired differential equation of the trajectories.

Let us consider in particular the trajectories at a constant angle to the meridians of the surface of revolution,

$$x = \rho \cos \omega, \quad y = \rho \sin \omega, \quad z = f(\rho).$$

We have here

$$u = \rho, \quad v = \omega, \quad E = 1 + f'^2(\rho), \quad F = 0, \quad G = \rho^2, \quad \delta v = 0;$$

hence the equation (80) becomes

$$\cos V = \frac{\sqrt{1 + f'^2(\rho)}\, d\rho}{\sqrt{[1 + f'^2(\rho)]d\rho^2 + \rho^2 d\omega^2}}.$$

Solving for $d\omega$, we find

$$d\omega = \tan V \frac{\sqrt{1 + f'^2(\rho)}\, d\rho}{\rho},$$

whence ω can be obtained by a quadrature.

III. EQUATIONS OF HIGHER ORDER

18. Integration of the equation $d^n y/dx^n = f(x)$. Given a differential equation of the nth order,

(81) $$\frac{d^n y}{dx^n} = F(x, y, y', y'', \cdots, y^{(n-1)}),$$

where $y^{(i)} = d^i y/dx^i$, this equation and those which are obtained from it by repeated differentiation enable us to express all the derivatives, beginning with $y^{(n)}$, in terms of $x, y, y', y'', \cdots, y^{(n-1)}$. If, then, for a particular value x_0 of the independent variable we are given the corresponding values $y_0, y_0', \cdots, y_0^{(n-1)}$ of the unknown function y and of its $n-1$ first derivatives, we can calculate the values of all the derivatives of y for the value x_0 of x, and form a power series,

(82) $$y_0 + (x - x_0) y_0' + \frac{(x - x_0)^2}{2!} y_0'' + \cdots + \frac{(x - x_0)^n}{n!} y_0^{(n)} + \cdots,$$

whose value represents the integral in question, provided that integral can be developed by Taylor's series. Up to the time of Cauchy's work the convergence of this series had been assumed without proof.* We shall see later that the series does converge under certain conditions which will be stated precisely. We shall indicate here only some simple types of differential equations of the nth order whose integration can be reduced to quadratures or to the integration of an equation of lower order than n.

The differential equation

(83) $$\frac{d^n y}{dx^n} = f(x)$$

constitutes the simplest possible type of differential equation of the nth order. It can be integrated by means of n successive quadratures; for, indicating by x_0 any arbitrary constant, we have

$$\frac{d^{n-1} y}{dx^{n-1}} = \int_{x_0}^{x} f(x)\, dx + C_0,$$

$$\frac{d^{n-2} y}{dx^{n-2}} = \int_{x_0}^{x} dx \int_{x_0}^{x} f(x)\, dx + C_0 (x - x_0) + C_1,$$

$$\cdots \cdots \cdots \cdots \cdots \cdots \cdots,$$

$$y = \int_{x_0}^{x} dx \int_{x_0}^{x} dx \cdots \int_{x_0}^{x} f(x)\, dx$$
$$+ \frac{C_0 (x - x_0)^{n-1}}{(n-1)!} + \frac{C_1 (x - x_0)^{n-2}}{(n-2)!} + \cdots + C_{n-1},$$

---

* See, for example, the *Traité* by Lacroix.

where C_{n-1}, C_{n-2}, $\cdots$, C_0 are n arbitrary constants which are equal respectively to the values of the integral and of its first $(n-1)$ derivatives for $x = x_0$.

We can replace the expression

$$Y = \int_{x_0}^{x} dx \int_{x_0}^{x} dx \cdots \int_{x_0}^{x} f(x) dx,$$

which contains n successive signs of integration, by an expression containing only a single quadrature, to be carried out on a function in which the variable x appears only as a parameter. It is easy to verify this fact, which will appear later as a special case of a general theory (§ 39). For if we put

(84) $$Y_1 = \frac{1}{(n-1)!} \int_{x_0}^{x} (x-z)^{n-1} f(z) dz,$$

we obtain successively, by the application of known rules,

$$\frac{dY_1}{dx} = \frac{1}{(n-2)!} \int_{x_0}^{x} (x-z)^{n-2} f(z) dz, \quad \cdots, \quad \frac{d^{n-1}Y_1}{dx^{n-1}} = \int_{x_0}^{x} f(z) dz,$$

and, finally, $d^n Y_1/dx^n = f(x)$. The function Y_1 is therefore an integral of the equation (83). Besides, the two functions Y and Y_1 vanish, as do also their first $(n-1)$ derivatives, for $x = x_0$. Their difference, which is a polynomial of degree equal to $n-1$ at most, cannot be divisible by $(x - x_0)^n$ unless it is identically zero. We have therefore $Y_1 = Y$.

19. Various cases of depression. The most usual cases in which the order of the equation can be depressed are the following:

1) *The equation does not contain the unknown function.* An equation of the form

(85) $$F\left(x, \frac{d^k y}{dx^k}, \frac{d^{k+1} y}{dx^{k+1}}, \cdots, \frac{d^n y}{dx^n}\right) = 0 \qquad (1 \leqq k \leqq n)$$

reduces immediately to one of order $n - k$ by taking $u = d^k y/dx^k$ as a new unknown function. If the auxiliary equation in u can be integrated, we shall then obtain y by quadratures, as has just been explained.

It sometimes happens that we can express x and $u = d^k y/dx^k$ in terms of an auxiliary parameter t,

$$x = f(t), \qquad \frac{d^k y}{dx^k} = \phi(t),$$

where the functions f and ϕ contain also the arbitrary constants introduced by the integration of the equation in u. We can then express y in terms of t also by quadratures. We have first

$$dy^{(k-1)} = \phi(t)\,dx = \phi(t)f'(t)\,dt,$$

whence we derive $y^{(k-1)}$. Continuing in this way, we calculate successively $y^{(k-2)}, \cdots, y'$ up to y.

2) *The equation does not contain the independent variable.* Given an equation of the form

(86) $$F\left(y, \frac{dy}{dx}, \frac{d^2y}{dx^2}, \cdots, \frac{d^ny}{dx^n}\right) = 0,$$

we can reduce it to the preceding form by taking y for the independent variable and x for the unknown function. Then the new equation does not contain x, and, taking dx/dy for the new unknown, we are led to an equation of order $n-1$. But we can carry out these two transformations simultaneously by taking y for the independent variable and $p = dy/dx$ for the dependent variable. This gives

$$\frac{d^2y}{dx^2} = \frac{dp}{dx} = \frac{dp}{dy}\frac{dy}{dx} = p\frac{dp}{dy},$$

$$\frac{d^3y}{dx^3} = \frac{d}{dx}\left(p\frac{dp}{dy}\right) = \frac{d}{dy}\left(p\frac{dp}{dy}\right)p = p\left(\frac{dp}{dy}\right)^2 + p^2\frac{d^2p}{dy^2},$$

and so on. In general, d^ry/dx^r can be expressed in terms of p and of its first $r-1$ derivatives with respect to y. The resulting differential equation is of order $n-1$.

Let us suppose that we have integrated this auxiliary equation of order $n-1$, and for the sake of generality let us suppose that y and p are expressed in terms of a variable parameter t, which may be one of the variables themselves. Then we shall have $y = f(t)$, $p = \phi(t)$, where the functions f and ϕ depend also on arbitrary constants. From the relation $dy = p\,dx$ we derive $f'(t)\,dt = \phi(t)\,dx$, so that x in turn is obtained by a quadrature,

$$x = \int \frac{f'(t)}{\phi(t)}\,dt.$$

This method is especially useful for the equation of the second order,

$$F(y, y', y'') = 0,$$

which is thus reduced to an equation of the first order,

$$F\left(y, p, p\frac{dp}{dy}\right) = 0.$$

Let $p = \phi(y, C)$ be the general integral of this equation of the first order. From the relation $dy/dx = \phi(y, C)$ we obtain x by a quadrature,

$$x = x_0 + \int \frac{dy}{\phi(y, C)}.$$

If the general integral of the equation in p is solved for y and appears in the form $y = f(p, C)$, we have, in the same way,

$$f'(p) dp = p\, dx$$

and therefore

$$x = x_0 + \int \frac{f'(p) dp}{p}.$$

The coördinates of a point of an integral curve are thus expressed in terms of an auxiliary variable p which represents the slope of the tangent to the curve.

3) *The equation is homogeneous in $y, y', y'', \cdots, y^{(n)}$.* If the degree of homogeneity is m, the equation is of the form

(87) $$y^m F\left(x, \frac{y'}{y}, \frac{y''}{y}, \cdots, \frac{y^{(n)}}{y}\right) = 0,$$

and we see that, if y_1 is a particular integral, λy_1 is also an integral for any value of the constant λ. The order of this equation is lowered by unity by putting

$$y = e^{\int u\, dx}.$$

This substitution gives

$$y' = u e^{\int u\, dx}, \qquad y'' = (u' + u^2) e^{\int u\, dx}, \qquad \cdots,$$

and, in general, $y^{(r)}$ is equal to the product of $e^{\int u\, dx}$ and a polynomial in $u, u', u'', \cdots, u^{(r-1)}$. Substituting these values in the given equation, we obtain an equation of order $n - 1$.

4) *The equation is homogeneous in $x, y, dx, dy, d^2y, \cdots, d^n y$.* In this case the equation is not changed by substituting Cx for x, and Cy for y, where C is any constant. Let us now take a new dependent variable $u = y/x$ and a new independent variable $t = \operatorname{Log} x$. The new differential equation does not change if we replace t by $t + \operatorname{Log} C$, leaving u unchanged; hence it does not contain explicitly the variable t. This is readily verified, for it is easy to see that the given equation must be of the form

$$F\left(\frac{y}{x}, y', xy'', x^2 y''', \cdots, x^{n-1} y^{(n)}\right) = 0.$$

If we put $y = ux$, we have, as a general expression,

$$y^{(p)} = xu^{(p)} + pu^{(p-1)},$$

and the quantities y', xy'', x^2y''', $\cdots$ are expressible in terms of u, xu', x^2u'', $\cdots$, $x^nu^{(n)}$, so that the transformed equation takes the form

$$\Phi(u, xu', x^2u'', \cdots, x^nu^{(n)}) = 0.$$

If we now put $x = e^t$, we have successively for the products xu', x^2u'', $\cdots$ certain functions of du/dt, d^2u/dt^2, $\cdots$, and we are led to an equation which does not contain the variable t.*

Note. In the various cases of reduction which precede, it may happen that we can obtain certain integrals of the auxiliary equation without being able to determine the general integral. The preceding methods are still applicable and enable us to obtain by quadratures integrals of the given equation containing less than n arbitrary constants.

20. Applications. 1) Equations of the form $y'' = f(y)$ come under the preceding types. We can integrate them directly without any transformation, for if we multiply the two sides by $2y'$, we deduce from the result, by a first integration,

$$y'^2 = C + \int_{y_0}^{y} 2f(y)\,dy = F(y) + C,$$

and we have next, by a quadrature,

$$x = \int \frac{dy}{\sqrt{F(y) + C}} + C'.$$

Let us consider, for example, the equation

$$y'' = a_0 y^3 + a_1 y^2 + a_2 y + a_3,$$

where one at least of the coefficients a_0, a_1 is not zero. Multiplying the two sides by $2y'$ and integrating, we find

$$y'^2 = \frac{a_0}{2}y^4 + \frac{2}{3}a_1 y^3 + a_2 y^2 + 2a_3 y + C.$$

The general integral of this new equation is an elliptic function (§ 11), which may in special cases reduce to simply periodic functions, or even rational functions, if the constant C has been so chosen that the polynomial on the right has a factor in common with its derivative.

---

* We may proceed in another way by taking u and $v = xu'$ for the variables. This gives $dv/dx = u' + xu''$, and therefore $x^2u'' = (dv/du)\,u'x - xu'$, or

$$x^2u'' = v\frac{dv}{du} - v.$$

Continuing in this way, we are led to a differential equation of order $(n-1)$ in u and v.

2) It may happen that we can apply successively several of the methods of reduction to the same equation. Let us take, for example, the equation of the fourth order $5y'''^2 - 3y''y^{iv} = 0$. If we first put $y'' = u$, we derive from it an equation of the second order, $5u'^2 - 3uu'' = 0$, which is homogeneous in u, u', u''. Let us put
$$u = e^{\int v\,dx};$$
the equation becomes $3v' = 2v^2$, or $v'/v^2 = 2/3$, from which we obtain
$$v = -\frac{3}{2}\frac{1}{x+a},$$
where a is an arbitrary constant. Hence we have
$$u = y'' = b(x+a)^{-\frac{3}{2}},$$
$$y' = -2b(x+a)^{-\frac{1}{2}} + c,$$
$$y = -4b(x+a)^{\frac{1}{2}} + cx + d,$$
where b, c, d are three new constants. We find, therefore, that the general integral represents a system of parabolas (§ 1).

3) Let it be required to determine the plane curves whose radii of curvature are proportional to the portion of the normal included between the foot M and the point of intersection N of that normal with a fixed straight line. Taking the fixed straight line for the x-axis, the differential equation of the problem is

(88) $$1 + y'^2 + \mu y y'' = 0,$$

where the coefficient μ is equal to the ratio of the radius of curvature to the length MN, preceded with the sign $+$ or $-$, according as the direction from M to the center of curvature coincides with the direction MN or with the opposite direction. In order to integrate this differential equation (88), let us put $y' = p$; it becomes
$$1 + p^2 + \mu y p \frac{dp}{dy} = 0,$$
which can be written in the form
$$\frac{dy}{y} + \frac{\mu}{2}\frac{2p\,dp}{1+p^2} = 0,$$
from which we derive, by a first integration,
$$y = C(1+p^2)^{-\frac{\mu}{2}},$$
where C is an arbitrary constant. The relation $dy = p\,dx$ gives us next
$$-\mu C p (1+p^2)^{-\frac{\mu}{2}-1} dp = p\,dx,$$
or
$$x = x_0 - \mu C \int (1+p^2)^{-\frac{\mu}{2}-1} dp.$$

Let us put $p = \tan\alpha$; all the curves obtained by varying C and x_0 result from a translation and an expansion about the origin of the curve Γ represented by the equations

(Γ) $$x = -\mu\int_0^\alpha \cos^\mu\alpha\,d\alpha, \qquad y = \cos^\mu\alpha.$$

It is easy to get an idea of the form of the curve from these equations, whatever may be the value of μ. If μ is an integer, we can carry out the integration.

If μ is a positive integer, the curve has no infinite branches, but it may have two forms that are very different in appearance, according to the character of μ. If μ is an odd integer, x is a periodic function (Part I, § 16), and the curve Γ is an algebraic closed convex curve. If μ is even, x increases by a constant quantity different from zero when α increases by 2π; y is always positive. We have a periodic curve with an infinite number of cusps on the x-axis. The appearance of the curve is that of a cycloid; it is a cycloid for $\mu = 2$.

Note. In the examples which we have just studied we always try to reduce the integration of a differential equation to the integration of an equation of lower order. However singular it may appear at first sight, the reverse process may sometimes succeed. Given, for example, an equation of the first order $f(x, y, y') = 0$, by combining with it a second equation obtained from it by differentiation, we obtain an infinite number of equations of the second order which are satisfied by all the integrals of the original equation. Suppose that we can find thus an equation of the second order which is integrable, and let $y = \phi(x, C, C')$ be the general integral. All the integrals of the original equation of the first order are included in this expression, but since they depend upon only a single arbitrary constant, there must be a relation between the constants C, C'. In order to obtain it, it suffices to write the condition that the function $\phi(x, C, C')$ satisfies the original equation of the first order; we are thus led to a certain number of relations between the constants C, C', and these relations should reduce to a single one.

A most interesting example of this device is due to Monge, who made use of it to find the lines of curvature of an ellipsoid. Let $2a$, $2b$, $2c$ be the three axes; the projections of the lines of curvature on the plane of the major axis and the intermediate axis are determined by the differential equation

$$(89) \quad \begin{cases} Axyy'^2 + (x^2 - Ay^2 - B)y' - xy = 0. \\ A = \dfrac{a^2(b^2 - c^2)}{b^2(a^2 - c^2)}, \quad B = \dfrac{a^2(a^2 - b^2)}{a^2 - c^2}. \end{cases}$$

Differentiating the equation (89), and then eliminating the expression

$$x^2 - Ay^2 - B,$$

we obtain the differential equation of the second order,

$$\frac{y''}{y'} + \frac{y'}{y} - \frac{1}{x} = 0;$$

whence we derive first $yy' = Cx$, then $y^2 = Cx^2 + C'$.

The general integral of the equation (89) will be obtained by establishing between C and C' the relation $ACC' + C' + BC = 0$, as is seen by replacing y^2 by $Cx^2 + C'$ on the right-hand side.*

---

* The equation (89) can also be easily integrated by the classic processes. It suffices, in fact, to put $x^2 = X$, $y^2 = Y$, after having multiplied all the terms by $xy\,dx^2$, in order to transform it into the Clairaut form.

Lagrange and Darboux have employed similar devices to integrate Euler's equation (see J. BERTRAND, *Traité de Calcul intégral*, pp. 569–572). We can also regard a certain theorem of Appell's as an illustration of the same procedure (*Comptes rendus*, Nov. 12, 1888).

EXERCISES

1. Find the differential equation of all conics by starting from the general unsolved equation and eliminating the coefficients between it and the relations obtained by five successive differentiations.

2. Integrate the differential equations

$$(y'^2 - y)^2 = y(y'^2 + y)^2, \qquad y(1 + 2y'^2) + xy' = 0,$$
$$(1 + y'^2)y'y''' = (3y'^2 - 1)y''^2, \qquad (x^2 + y^2)v'' - yv'^2 + xv'^3 + xv' - y = 0,$$
$$x^2 y'^2 + 2xy(y - 2a)y' - 2y^2(y - 2a) = 0, \qquad xyy'' + xy'^2 - yy' = 0,$$
$$y'^3 + 3y'^2 + y^6 - 4 = 0.$$

3. Apply the general methods of depression to the integration of the differential equation of conics.

4. Find the integrals of the equation $y'' = 2y^2(y - 1)$ which are rational functions or simply periodic functions of the variable.

[*Licence*, Paris, 1899.]

5. Given a triangle ABC and a curve Γ in its plane, let a, b, c be the points of intersection of the sides of the triangle with the tangent at m to the curve Γ. Find the curves Γ for which the anharmonic ratio of the four points m, a, b, c is constant when the point m moves on one of them.

The anharmonic ratio of the tangent at m and the straight lines mA, mB, mC is also constant.

6. Given a point O and a straight line D, find a curve such that the portion of the tangent MN included between the point of contact M and the point of intersection N of the tangent and the line D subtends a constant angle at O.

[*Licence*, Besançon, 1885.]

7. Find the projections on the xy-plane of the curves lying on the paraboloid $2az = mx^2 + y^2$, whose tangents make a given constant angle γ with the axis Oz.

[*Licence*, Paris, 1879.]

8. Find the orthogonal trajectories of each of the families of curves represented by one of the following equations:

$$y^2(2a - x) = x^3, \qquad y^2 + mx^2 - 2ax = 0,$$
$$(x^2 + y^2)^2 = a^2 xy, \qquad x^2 + y^2 = a^2 \log\left(\frac{y}{x}\right),$$

where a is the variable parameter.

9. In order that the equation $\theta(x, y) = C$ shall represent a family of parallel curves, it is necessary and sufficient that we have

$$\left(\frac{\partial \theta}{\partial x}\right)^2 + \left(\frac{\partial \theta}{\partial y}\right)^2 = \phi(\theta),$$

where $\phi(\theta)$ is any function of θ.

[Write the condition that the orthogonal trajectories are straight lines.]

10. Find the necessary and sufficient condition that the integral curves of the equation $y' = f(x, y)$ form a family of parallel curves, and show that the integration can be carried out by a quadrature.

[*Licence*, Paris, 1898.]

11*. Form the general equation of the conics which cut a given conic C orthogonally at the four common points. These conics form, in general, several distinct families. Find the orthogonal trajectories of each of these families. Hence derive all the orthogonal systems of which the two families are made up of conics. [If $f = 0$, $\phi = 0$ are the equations of two conics cutting each other orthogonally at each of their four common points, we have an identity of the form

$$\frac{\partial f}{\partial x}\frac{\partial \phi}{\partial x} + \frac{\partial f}{\partial y}\frac{\partial \phi}{\partial y} = \lambda f + \mu \phi,$$

where λ and μ are two constant coefficients.]

12. Find the condition that the integral curves of the differential equation $y' = f(x, y)$ form a family of isothermal curves, and show that an integrating factor can be found.

[Sophus Lie.]

13. Let y_1, y_2 be two particular integrals of Riccati's equation (26) (§ 7). Show that the substitution $(y - y_1)/(y - y_2) = z$ reduces the equation to the linear equation

$$z' + X(y_1 - y_2)z = 0.$$

14. Find a plane curve C such that the triangle formed by any point M of the curve, the corresponding center of curvature, and the foot of the ordinate of the point M, has a constant area. Show that one of the coördinates can be expressed as a function of the other by a quadrature, and that we can obtain a knowledge of the form of the curve without having the definite equation. [The axes of coördinates are supposed to be rectangular.]

[*Licence*, Paris, 1877.]

15. Given a plane curve C, let M be any point of that curve, P the center of curvature of the curve at the point, and MT the tangent. Through the point T where the tangent cuts the axis of x, draw a straight line parallel to the axis of y, meeting the normal MP in a point N. Determine the curve C so that the ratio of MP to MN is constant.

[*Licence*, Toulouse, 1884.]

16. Determine the surfaces of revolution such that in each of their points the radii of curvature of the principal sections are directed in the same sense and have a constant sum a. Sketch a figure of a meridian of the surface.

[*Licence*, Toulouse, 1878.]

17*. Show that the general integral of Euler's equation can be written in the form

$$\left(\frac{\sqrt{X} - \sqrt{Y}}{x - y}\right)^2 - a_0(x + y)^2 - a_1(x + y) - a_2 = C,$$

where $X = a_0 x^4 + a_1 x^3 + a_2 x^2 + a_3 x + a_4$ and where Y has an analogous meaning.

[Lagrange.]

[It suffices to solve the equation (58) (§ 14) with respect to the constant. After a few transformations we obtain Lagrange's form.]

18. The asymptotic lines of the surface represented by the equations

$$x = A(u - a)^m(v - a)^n,$$
$$y = B(u - b)^m(v - b)^n,$$
$$z = C(u - c)^m(v - c)^n$$

are obtained by the integration of Euler's equation when we have $m = n$ or $m + n = 1$. Deduce from this result the asymptotic lines of the tetrahedral surface

$$\left(\frac{x}{a}\right)^m + \left(\frac{y}{b}\right)^m + \left(\frac{z}{c}\right)^m = 1.$$

19. How can we determine whether a differential equation

$$dy - f(x, y)\,dx = 0$$

has an integrating factor of the form XY, where X depends only upon x, and Y depends only upon y, and find this integrating factor when it exists?

[*Licence*, Paris, October, 1902.]

20*. Given a plane curve C, the middle point m is taken of the cord MM' which joins any two points M, M' of that curve. The point M remaining fixed, if the point M' describes the curve C, the point m describes a similar curve c. Prove that the curves c satisfy a differential equation of the first order, which is integrated, like Clairaut's equation, by replacing y' in it by an arbitrary constant. (*Bulletin de la Société mathématique*, Vol. XXIII, p. 88.)

21. Integrate the differential equation

$$F\left(y'', \; y' - xy'', \; y - xy' + \frac{x^2}{2}y''\right) = 0.$$

We observe that y''' appears as a factor in the derivative of the left-hand side. There exist equations of an analogous form and of any order (see DIXON, *Philosophical Transactions*, Vol. CLXXXVI, Part I; RAFFY, *Bulletin de la Société mathématique*, Vol. XXV, p. 71; BOUNITZKY, *Bulletin des Sciences mathématiques*, Vol. XXXI, 2d series, p. 250).

CHAPTER II

EXISTENCE THEOREMS

The first rigorous investigations to establish the existence of the integrals of a system of ordinary differential equations or of partial differential equations are due to Cauchy. That illustrious mathematician gave for analytic equations a type of demonstration based on a method of comparison to which he gave the name of "calculus of limits" (*calcul des limites*). We owe to him also another method which does not assume the functions to be analytic, and which we shall discuss later.

I. CALCULUS OF LIMITS

21. Introduction. The fundamental idea of the calculus of limits consists in the use of dominant functions. The reasoning is quite analogous to that which has already been used to establish the existence of implicit functions (I, § 193, 2d ed.; § 187, 1st ed.). Since every analytic function has an infinite number of dominant functions, we see that the method can be varied in a great many ways. The simplicity of the demonstrations depends largely on the choice of the dominant functions. Since the work of Cauchy, his proofs have been perfected and extended to more general cases by Briot and Bouquet, Weierstrass, Darboux, Méray, Riquier, Madame Kovalevsky, and many others. Even to-day we make use of this same method constantly to treat analogous questions relative to partial differential equations with various initial conditions.

22. Existence of the integrals of a system of differential equations. Let us consider first a single equation,

$$(1) \qquad \frac{dy}{dx} = f(x, y),$$

the right-hand side of which, $f(x, y)$, is an analytic function in the neighborhood of a system of values x_0, y_0. We propose to prove that *this equation has an integral $y(x)$ analytic in the neighborhood of the point x_0 and reducing to y_0 for $x = x_0$.*

Let us suppose for the sake of brevity that $x_0 = y_0 = 0$, which amounts simply to writing x and y in place of $x - x_0$ and $y - y_0$. If the given equation has an integral which is analytic in the neighborhood of the point $x = 0$, and which vanishes with x, and if we can calculate the values of all the successive derivatives of that integral for $x = 0$, we can write the development of that integral in a power series.

The equation (1) gives us first of all $(dy/dx)_0 = f(0, 0)$. On the other hand, the equations which we derive from it by repeated differentiations enable us to calculate the value of a derivative of any order in terms of x, y and of derivatives of lower order,

(2) $$\begin{cases} \dfrac{d^2 y}{dx^2} = \dfrac{\partial f}{\partial x} + \dfrac{\partial f}{\partial y} \dfrac{dy}{dx}, \\ \dfrac{d^3 y}{dx^3} = \dfrac{\partial^2 f}{\partial x^2} + 2 \dfrac{\partial^2 f}{\partial x \partial y} \dfrac{dy}{dx} + \dfrac{\partial^2 f}{\partial y^2} \left(\dfrac{dy}{dx}\right)^2 + \dfrac{\partial f}{\partial y} \dfrac{d^2 y}{dx^2}, \\ \cdots \cdots \cdots \cdots \cdots \cdots \cdots \cdots \cdots \cdots \cdots \cdots \end{cases}$$

Setting in these relations $x = y = 0$, we calculate step by step the initial values $(d^2y/dx^2)_0$, $(d^3y/dx^3)_0$, $\cdots$, $(d^ny/dx^n)_0$, $\cdots$ of the successive derivatives of the desired integral in terms of the coefficients of the development of $f(x, y)$ in a power series in x and y. Until Cauchy's work appeared, mathematicians had assumed without proof that the power series thus obtained,

(3) $$y = \left(\frac{dy}{dx}\right)_0 \frac{x}{1} + \left(\frac{d^2 y}{dx^2}\right)_0 \frac{x^2}{2!} + \cdots + \left(\frac{d^n y}{dx^n}\right)_0 \frac{x^n}{n!} + \cdots,$$

was convergent for values of x near zero.

To establish rigorously this essential point, let us observe that the operations by which we calculate the coefficients of the series (3) reduce precisely to additions and multiplications alone, so that the value obtained for $(d^n y/dx^n)_0$ can be written in the form

(4) $$\left(\frac{d^n y}{dx^n}\right)_0 = P_n(a_{00}, a_{01}, a_{10}, \cdots, a_{0n}, \cdots, a_{n0}),$$

where P_n is a polynomial with positive integral coefficients, and where a_{ik} is the coefficient of $x^i y^k$ in the development of $f(x, y)$. If, then, we replace the function $f(x, y)$ by a dominant function $\phi(x, Y)$, and if we seek to determine an analytic integral of the auxiliary equation

(5) $$\frac{dY}{dx} = \phi(x, Y)$$

vanishing with x, the coefficients of the series obtained for the development of Y will be positive numbers greater than the absolute value of the corresponding coefficients of the same rank in the series (3). If the series obtained for Y is convergent in a certain neighborhood, the same must be true a fortiori of the series (3). Now the series obtained for Y will certainly be convergent if the auxiliary equation has an analytic integral vanishing for $x = 0$.

Let us suppose that the function $f(x, y)$ is analytic when the variables x and y remain in the circles C, C' of radii a and b described in the planes of the two variables about the two origins as centers, and that it is continuous on the circumferences, and let M be the upper limit of $|f(x, y)|$ in this neighborhood. We can take for the dominant function

$$\phi(x, Y) = \frac{M}{\left(1 - \frac{x}{a}\right)\left(1 - \frac{Y}{b}\right)},$$

and, multiplying the two sides by $(1 - Y/b)$, we may write the auxiliary equation (5) in the form

$$(6) \qquad \left(1 - \frac{Y}{b}\right)\frac{dY}{dx} = \frac{M}{1 - \frac{x}{a}}.$$

We can show directly that this equation has an analytic integral which vanishes for $x = 0$. In fact, separating the variables, we obtain the integral of that equation in the form

$$(7) \qquad Y - \frac{Y^2}{2b} = -aM \operatorname{Log}\left(1 - \frac{x}{a}\right).$$

The constant which must be added to the right-hand side to express the general integral of the equation (6) is here zero if we adopt for the determination of the logarithm the one which is zero for $x = 0$. Solving equation (7) for Y, we get

$$(8) \qquad Y = b - b\sqrt{1 + 2a\frac{M}{b}\operatorname{Log}\left(1 - \frac{x}{a}\right)}.$$

If we take for the radical the determination which reduces to 1 for $x = 0$, the result (8) represents precisely an integral of the equation (6) which is zero for $x = 0$. This function Y is also analytic in the neighborhood of the origin, for the function under the radical is analytic in the interior of the circle C of radius a, and is zero for

$$(9) \qquad x = \rho = a\left(1 - e^{-\frac{b}{2aM}}\right).$$

When the variable x remains in the interior of the circle C_ρ of radius ρ described about the origin as center, the absolute value of $(2\,aM/b)\,\mathrm{Log}\,(1-x/a)$ remains less than unity,* and the radical is an analytic function of x in this circle. The series obtained for the development of Y is therefore convergent in the circle of radius ρ, and the same is true a fortiori of the series (3) first obtained.

It is easily seen from the formula (8) that all the coefficients of the development of Y are real and positive, a fact which is evident also a priori. If we give to x any value whose absolute value is less than ρ, the absolute value of Y will be less than the value obtained by replacing x by ρ. We have, then, for every point in the circle C_ρ, $|Y| < b$, and therefore $|y| < b$. If we replace y in $f(x, y)$ by the sum of the series (3), the result of the substitution is therefore an analytic function $\Phi(x)$ in the circle of radius ρ. From the manner in which we have obtained the coefficients of the series (3), the two functions $\Phi(x)$ and dy/dx are equal, as well as all their successive derivatives for $x = 0$. Hence they are identical, and the analytic function y satisfies all the given conditions.

In order to calculate the coefficients of the series (3), we can substitute directly for y in the equation (1) a power series $y = C_1 x + C_2 x^2 + \cdots$ and write the conditions that the two sides are identical. The coefficient of x^{n-1} in dy/dx is nC_n, while the coefficient of x^{n-1} on the right depends evidently only on $C_1, C_2, \cdots, C_{n-1}$ and the coefficients a_{ik}. It is easily seen that the coefficients C_n are calculated in this way by the use of the operations of addition and multiplication alone.

The method can be extended without difficulty to a system of any number of differential equations of the first order. Let

(10) $$\frac{dy_i}{dx} = f_i(x, y_1, y_2, \cdots, y_n) \qquad (i = 1, 2, \cdots, n)$$

be a system of differential equations in which the functions f_i are analytic in the neighborhood of the values $x_0, (y_1)_0, \cdots, (y_n)_0$. *These equations have a system of integrals analytic in the neighborhood of the point x_0 and taking on the values $(y_1)_0, (y_2)_0, \cdots, (y_n)_0$ respectively for $x = x_0$.*

The proof of this theorem can be made to depend on the fact that the system of auxiliary equations

(11) $$\frac{dY_1}{dx} = \frac{dY_2}{dx} = \cdots = \frac{dY_n}{dx} = \frac{M}{\left(1 - \frac{x}{a}\right)\left(1 - \frac{Y_1}{b}\right) \cdots \left(1 - \frac{Y_n}{b}\right)}$$

---

*In fact, all the coefficients of the development of that function in powers of z are real and negative. The absolute value of the preceding expression for $|z| < \rho$ is therefore less than its absolute value when $z = \rho$, that is, less than unity.

has a system of integrals which are analytic in the neighborhood of the origin, and which vanish for $x = 0$. The functions f_i are supposed to be analytic as long as we have $|x - x_0| \leq a$, $|y_i - (y_i)_0| \leq b$, and M denotes again the maximum absolute value of the functions f_i in this neighborhood. These integrals, having their derivatives equal and all vanishing for $x = 0$, must be identical, and it suffices to consider the single equation

$$\frac{dY}{dx} = \frac{M}{\left(1 - \frac{x}{a}\right)\left(1 - \frac{Y}{b}\right)^n},$$

in which we can again separate the variables. This equation has the integral

$$Y = b - b \sqrt[n+1]{1 + \frac{(n+1)Ma}{b} \mathrm{Log}\left(1 - \frac{x}{a}\right)},$$

which is analytic in the circle with the radius

$$\rho = a\left(1 - e^{\frac{-b}{(n+1)Ma}}\right),$$

and which is zero for $x = 0$. Hence the system (10) has a system of integrals that are analytic in the same circle.

A single equation of the nth order,

(12) $$\frac{d^n y}{dx^n} = F\left(x, y, \frac{dy}{dx}, \cdots, \frac{d^{n-1}y}{dx^{n-1}}\right),$$

can be replaced by an equivalent system formed of n equations of the first order,

(13) $$\begin{cases} \dfrac{dy}{dx} = y_1, \quad \dfrac{dy_1}{dx} = y_2, \\ \dfrac{dy_{n-2}}{dx} = y_{n-1}, \quad \dfrac{dy_{n-1}}{dx} = F(x, y, y_1, \cdots, y_{n-1}), \end{cases}$$

by introducing as auxiliary dependent functions the successive derivatives of y up to the $(n-1)$th order. We deduce from the general theorem, then, the proposition that *the equation* (12) *has an analytic integral in the neighborhood of the point* x_0 *and such that that function and its first* $n-1$ *derivatives take on for* $x = x_0$ *the values* $y_0, y_0', \cdots, y_0^{(n-1)}$ *given in advance, provided that the function* F *is analytic in the neighborhood of the system of values* $x_0, y_0, y_0', \cdots, y_0^{(n-1)}$.

From the demonstration it results that there cannot be more than one analytic integral of the equation (1) taking on for $x = x_0$ the value y_0. But nothing enables us to say up to this point that there do not exist non-analytic functions satisfying the same conditions.* This is a point which will be rigorously established farther on (§ 26).

23. Systems of linear equations. We shall find farther on, by another method, a larger value for a lower bound of the radius of convergence of the series which represents the integrals (§ 29). If the functions f_i have special forms, we can sometimes employ more advantageous dominant functions, still making use of the method of the calculus of limits.

In particular, this is what happens in the very important case of linear equations. Let

$$(14) \quad \frac{dy_i}{dx} = a_{i1}y_1 + a_{i2}y_2 + \cdots + a_{in}y_n + b_i \qquad (i = 1, 2, \cdots, n)$$

be a system of linear equations in which the functions a_{ik} and b_i are functions of the single variable x, analytic in the circle C of radius R about the point x_0 as center. *These equations have a system of integrals analytic in the circle C and reducing respectively to $(y_1)_0$, $(y_2)_0$, $\cdots$, $(y_n)_0$ for $x = x_0$.*

We may suppose in the proof that

$$(y_1)_0 = (y_2)_0 = \cdots = (y_n)_0 = 0,$$

for if we change y_i into $(y_i)_0 + y_i$, the system (14) does not change in form, and the new coefficients are again analytic in the circle C. Let M be the maximum value of the absolute values of all the

---

* The following is the reasoning used by Briot and Bouquet to treat this matter. Let y_1 be an analytic integral of the equation (1) taking on the value y_0 for $x = x_0$. Putting $y = y_1 + z$, the equation (1) takes the form

$$(1') \qquad \frac{dz}{dx} = z \psi(x, z),$$

where $\psi(x, z)$ is analytic for $x = x_0$, $z = 0$. Let us suppose that this equation has an integral, other than $z = 0$, approaching zero when the variable x describes a curve C ending in the point x_0. Let x_1, x_2 be two points of this curve to which correspond the two values z_1 and z_2 of z. We obtain from the equation (1')

$$\int_{z_1}^{z_2} \frac{dz}{z} = \int_{x_1}^{x_2} \psi(x, z) \, dx.$$

If x_1 approaches x_0, z_1 approaches zero, and the absolute value of the left-hand side of this equality becomes infinite, while the absolute value of the right-hand side remains finite; there cannot be, then, an integral approaching zero different from $z = 0$. But the reasoning supposes that the point x approaches x_0 along a curve C of finite length.

functions a_{ik}, b_i in a circle C' with the center x_0 and the radius $r < R$. The function

$$\frac{M}{1-\dfrac{x-x_0}{r}}(1+Y_1+Y_2+\cdots+Y_n)$$

is a dominant function for all the functions $a_{i1}y_1+\cdots+a_{in}y_n+b_i$, and we are led to consider the auxiliary system

(15) $\quad \dfrac{dY_1}{dx}=\dfrac{dY_2}{dx}=\cdots=\dfrac{dY_n}{dx}=\dfrac{M}{1-\dfrac{x-x_0}{r}}(1+Y_1+Y_2+\cdots+Y_n).$

Since the functions $Y_1, Y_2, \cdots, Y_n$ are required to be zero for $x=x_0$, and since their derivatives are equal, they are identical, and the system (15) can be replaced by the single equation

(16) $\quad \dfrac{dY}{dx}=\dfrac{M}{1-\dfrac{x-x_0}{r}}(1+nY),$

which can be integrated by separating the variables. The integral which is zero for $x = x_0$ has the form

$$Y=\frac{1}{n}\left[\left(1-\frac{x-x_0}{r}\right)^{-nMr}-1\right],$$

and it is analytic in the circle C'. The same thing is therefore true of the integrals of the system (14), and, since the number r may be taken as near R as we wish, it follows that these integrals are analytic in the circle C.

24. Total differential equations. Let $x_1, x_2, \cdots, x_n$ be a system of n independent variables, let z be an unknown function of these variables, and let $f_1, f_2, \cdots, f_n$ be n given functions of $x_1, x_2, \cdots, x_n, z$. A total differential equation is a relation of the form

(17) $\quad dz = f_1 dx_1 + f_2 dx_2 + \cdots + f_n dx_n;$

it is really equivalent to n distinct equations:

(18) $\quad \dfrac{\partial z}{\partial x_1}=f_1, \quad \dfrac{\partial z}{\partial x_2}=f_2, \quad \cdots, \quad \dfrac{\partial z}{\partial x_n}=f_n.$

Let us suppose that there exists a function z of $x_1, x_2, \cdots, x_n$ satisfying these n relations. We can calculate the second derivative $\partial^2 z/\partial x_i \partial x_k$ ($i \neq k$) in two different ways. Writing the results obtained as identical, we obtain thus $n(n-1)/2$! relations of the form

(19) $\quad \dfrac{\partial f_i}{\partial x_k}+\dfrac{\partial f_i}{\partial z}f_k=\dfrac{\partial f_k}{\partial x_i}+\dfrac{\partial f_k}{\partial z}f_i, \qquad (i, k = 1, 2, \cdots, n)$

and the function z can only be taken from among those functions which satisfy these relations. We are going to consider only the very important case, in which these relations are satisfied *identically*. The equation (17) or the equivalent system (18) is then said to be *completely integrable*.

Given a completely integrable total differential equation in which the functions f_i are analytic in the neighborhood of the system of values $(x_1)_0$, $(x_2)_0$, $\cdots$, $(x_n)_0$, z_0, this equation has an analytic integral in the neighborhood of the system of values $(x_1)_0$, $\cdots$, $(x_n)_0$, which reduces to z_0 when $x_1 = (x_1)_0, \cdots, x_n = (x_n)_0$.

The equations (18) and those which are derived from them by successive differentiations enable us to express all the partial derivatives of the unknown function z in terms of $z, x_1, x_2, \cdots, x_n$; hence we can obtain the values of the coefficients of the development of the analytic integral, if it exists. But, while it is evident that we can calculate such derivatives as $\partial^p z/\partial x_1^p$ in only one way, it requires a little more care to assure ourselves that we shall always obtain the same expression for a derivative of any order, such as $\partial^{p+q} z/\partial x_i^p \partial x_k^q$, which can be calculated in several different ways. This will be the case for the derivatives of the second order, if the conditions (19) are identically satisfied. In order to show that the same property is true in general, it suffices to show that, if it is true up to the partial derivatives of order p, it will also be true for the partial derivatives of order $p+1$. We shall base the proof on the following fact: Let $U(x_1, x_2, \cdots, x_n, z)$ be any function of $x_1, x_2, \cdots, x_n, z$, and let us put

$$\frac{dU}{dx_i} = \frac{\partial U}{\partial x_i} + \frac{\partial U}{\partial z} f_i, \qquad \frac{d^2 U}{dx_i dx_k} = \frac{d}{dx_k}\left(\frac{dU}{dx_i}\right). \qquad (i, k = 1, 2, \cdots, n)$$

From the conditions (19) it follows immediately that we have for any function U the relation

$$\frac{d^2 U}{dx_i dx_k} = \frac{d^2 U}{dx_k dx_i}.$$

Let now u and v be two partial derivatives of the pth order differing only in the fact that a differentiation with respect to x_i in one has been replaced by a differentiation with respect to x_k in the other. The proof depends on showing that we have

$$\frac{\partial u}{\partial x_k} + \frac{\partial u}{\partial z} f_k = \frac{\partial v}{\partial x_i} + \frac{\partial v}{\partial z} f_i,$$

or that $du/dx_k = dv/dx_i$. But u and v have been obtained by taking the partial derivatives of a partial derivative w of order $p-1$ with respect to the variables x_i and x_k respectively. We have therefore $u = dw/dx_i$, $v = dw/dx_k$, and the equality to be established reduces to $d^2 w/dx_i dx_k = d^2 w/dx_k dx_i$, an equality which has already been proved.

To prove the convergence of the series thus obtained, we can therefore replace the functions f_i by dominant functions ϕ_i, provided that we choose these functions ϕ_i so that the resulting auxiliary total differential equation shall itself be completely integrable. For simplicity let us put $(x_1)_0 = (x_2)_0 = \cdots = (x_n)_0 = z_0 = 0$; we can take for the dominant function of all the functions f_i an expression of the form

$$\frac{M}{\left(1 - \dfrac{x_1 + \cdots + x_n}{r}\right)\left(1 - \dfrac{Z}{\rho}\right)},$$

and the auxiliary equation

(20) $$dZ = \frac{M(dx_1 + dx_2 + \cdots + dx_n)}{\left(1 - \frac{x_1 + x_2 + \cdots + x_n}{r}\right)\left(1 - \frac{Z}{\rho}\right)}$$

is completely integrable from the symmetry of the right-hand side relative to the n variables x_i. In order to obtain an analytic integral that vanishes with these variables, we need only seek an integral which is a function of the single variable $X = x_1 + x_2 + \cdots + x_n$. This leads to an ordinary differential equation of the form (6)

$$\left(1 - \frac{Z}{\rho}\right)dZ = \frac{MdX}{1 - \frac{X}{r}}.$$

Since the integral of this equation is represented by a development in a convergent series the coefficient of any term $x_1^{a_1} \cdots x_n^{a_n}$ of which is real and positive, the development obtained for z is a fortiori convergent in the same neighborhood.

The theorem can be extended without difficulty to systems of total differential equations in n independent variables $x_1, x_2, \cdots, x_n$ and m dependent variables $z_1, z_2, \cdots, z_m$:

(21) $$dz_h = f_{1h}dx_1 + \cdots + f_{ih}dx_i + \cdots + f_{nh}dx_n. \qquad \binom{h=1, 2, \cdots, m}{i=1, 2, \cdots, n}$$

By calculating in two different ways the derivatives of the form $\partial^2 z_h/\partial x_i \partial x_k$ we are led to the conditions

(22) $$\frac{\partial f_{ih}}{\partial x_k} + \frac{\partial f_{ih}}{\partial z_1}f_{k1} + \cdots + \frac{\partial f_{ih}}{\partial z_m}f_{km} = \frac{\partial f_{kh}}{\partial x_i} + \frac{\partial f_{kh}}{\partial z_1}f_{i1} + \cdots + \frac{\partial f_{kh}}{\partial z_m}f_{im}.$$

The system (21) is said to be completely integrable if these conditions (22) are satisfied identically, and we have the following theorem which is demonstrated like the preceding:

Every completely integrable system in which the functions f_i are analytic in the neighborhood of a system of values $(x_1)_0, (x_2)_0, \cdots, (x_n)_0, (z_1)_0, \cdots, (z_m)_0$ has a system of integrals analytic in the neighborhood of the point $(x_1)_0, \cdots, (x_n)_0$ and taking on respectively the values $(z_1)_0, (z_2)_0, \cdots, (z_m)_0$ when $x_1 = (x_1)_0, \cdots, x_n = (x_n)_0$.

25. Application of the method of the calculus of limits to partial differential equations. The calculus of limits enables us also to prove the existence of integrals of a system of partial differential equations. Let us consider first an equation of the first order,

(23) $$\frac{\partial z}{\partial x_1} = f\left(x_1, x_2, \cdots, x_n, z, \frac{\partial z}{\partial x_2}, \frac{\partial z}{\partial x_3}, \cdots, \frac{\partial z}{\partial x_n}\right),$$

in which the right-hand side does not contain the derivative $\partial z/\partial x_1$. This equation and those obtained from it by successive differentiation enable us to express all the partial derivatives of z in terms of $x_1, x_2, \cdots, x_n, z$, and of the partial derivatives of z taken with respect to the variables $x_2, x_3, \cdots, x_n$ alone. This property is evident for the derivatives of the form $\partial^{a_2+\cdots+a_n+1}z/\partial x_1 \partial x_2^{a_2} \cdots \partial x_n^{a_n}$, as is seen by differentiating the two sides of the equation (23) a_2 times with respect to $x_2, \cdots$, and then a_n times with respect to x_n. If we differentiate the two sides of the equation (23) once with respect to x_1, and any number of times with respect to the other variables $x_2, x_3, \cdots, x_n$, and if we then replace in the right-hand side of the result the partial derivatives which involve just one differentiation with respect to the variable x_1 by the expressions already obtained, we shall obtain also the derivatives $\partial^{a_2+\cdots+a_n+2}z/\partial x_1^2 \partial x_2^{a_2} \cdots \partial x_n^{a_n}$ expressed in the manner stated above, and it is clear that we can continue to apply the same process indefinitely.

Let us now suppose that the function f is analytic in the neighborhood of a system of values $(x_1)_0, \cdots, (x_n)_0, z_0, (p_2)_0, \cdots, (p_n)_0$, and let $\phi(x_2, x_3, \cdots, x_n)$ be a function of the $(n-1)$ variables $x_2, x_3, \cdots, x_n$ analytic in the neighborhood of the point* $(x_2)_0, (x_3)_0, \cdots, (x_n)_0$ and such that we have for these particular values

$$(\phi)_0 = z_0, \quad \left(\frac{\partial \phi}{\partial x_2}\right)_0 = (p_2)_0, \quad \left(\frac{\partial \phi}{\partial x_3}\right)_0 = (p_3)_0, \quad \cdots, \quad \left(\frac{\partial \phi}{\partial x_n}\right)_0 = (p_n)_0.$$

If these conditions are satisfied, *the equation* (23) *has an integral which is regular in the neighborhood of the point* $(x_1)_0, \cdots, (x_n)_0$ *and which reduces to* $\phi(x_2, x_3, \cdots, x_n)$ *for* $x_1 = (x_1)_0$.

By hypothesis, the function $\phi(x_2, x_3, \cdots, x_n)$ can be developed in a series of positive powers of the variables $x_i - (x_i)_0$, and the coefficients are, except for certain numerical factors, the values of the partial derivatives of that function at the point $(x_2)_0, (x_3)_0, \cdots, (x_n)_0$. Since the function z, the existence of which we wish to prove, must reduce to $\phi(x_2, x_3, \cdots, x_n)$ for $x_1 = (x_1)_0$, we know from that fact alone the values at the point $(x_1)_0, (x_2)_0, \cdots, (x_n)_0$ of all the partial derivatives of the function z which involve no differentiation with respect to the variable x_1. We have just seen how all the other partial derivatives of z can be expressed in terms of these. We can therefore calculate, step by step, all the coefficients of the development of z according to powers of $x_i - (x_i)_0$ in terms of the coefficients of the two developments of the function f and of the function ϕ, and the calculation involves the operations of addition and multiplication alone. We can therefore employ again dominant functions to prove convergence: if the series obtained by replacing, in the preceding

---

* For the sake of brevity we shall designate as a point every system of particular values, real or imaginary, assigned to the variables appearing in the discussion.

calculation, f by a dominant function F, and ϕ by another dominant function Φ, is convergent, the same thing must necessarily be true of the series obtained for z.

We can, first of all, replace the given initial conditions by other simpler conditions by means of a succession of easy transformations. We may suppose $(x_1)_0 = (x_2)_0 = \cdots = (x_n)_0 = 0$, for that amounts to writing x_i in place of $x_i - (x_i)_0$. If we also put

$$z = \phi(x_2, x_3, \cdots, x_n) + u,$$

the new unknown function u must reduce to zero for $x_1 = 0$. We may suppose also that after these transformations the right-hand side of the equation, when developed, does not contain a constant term, for if the development commenced with a constant term a different from zero, it would suffice to put $u = ax_1 + v$ in order to make it disappear. Having made these transformations, if we now replace the right-hand side by a suitable dominant function, the demonstration of the theorem reduces to showing that the equation

$$(24) \quad \frac{\partial Z}{\partial x_1} = \frac{M}{\left(1 - \dfrac{x_1 + x_2 + \cdots + x_n + Z}{r}\right)\left(1 - \dfrac{\dfrac{\partial Z}{\partial x_2} + \cdots + \dfrac{\partial Z}{\partial x_n}}{\rho}\right)} - M,$$

where M, r, ρ are determined positive numbers, has an integral which is analytic in the neighborhood of the origin and which reduces to zero for $x_1 = 0$. If we replace x_1 on the right-hand side by x_1/α, where α is a positive number less than unity, we increase the coefficients, and the theorem will be established a fortiori if we prove the proposition for the new equation

$$(25) \quad \frac{\partial Z}{\partial x_1} = \frac{M}{\left(1 - \dfrac{\dfrac{x_1}{\alpha} + x_2 + \cdots + x_n + Z}{r}\right)\left(1 - \dfrac{\dfrac{\partial Z}{\partial x_2} + \cdots + \dfrac{\partial Z}{\partial x_n}}{\rho}\right)} - M.$$

Indeed, it is sufficient to show that this equation has a regular integral, represented by a power series whose coefficients are all real and positive; for the coefficients of this third development are at least equal to those of the series obtained by supposing that Z vanishes when $x_1 = 0$, since the coefficients are all obtained by means of additions and multiplications of the coefficients of the terms independent of x_1. In order to establish this last point, let us try to satisfy the equation (25) by taking for Z a function of the single variable $X = x_1/a + x_2 + \cdots + x_n$. We are thus led to the differential equation of the first order,

(26) $\quad \left(\dfrac{1}{a} - \dfrac{n-1}{\rho} M\right) \dfrac{dZ}{dX} = \dfrac{n-1}{a\rho} \left(\dfrac{dZ}{dX}\right)^2 + \dfrac{M}{1 - \dfrac{X+Z}{r}} - M.$

Let us suppose that a has been chosen so small that the coefficient of dZ/dX on the left is positive. For $X = Z = 0$ the equation (26) has two distinct roots, one of which is equal to zero. That equation has therefore an analytic integral in the neighborhood of the origin, which, together with its first derivative, is zero for $X = 0$. It is easy to show directly that all the coefficients of the development of this integral are positive; for the equation (26) may be written in the form

$$\dfrac{dZ}{dX} = A \left(\dfrac{dZ}{dX}\right)^2 + \Phi(X, Z),$$

where A is positive and where $\Phi(X, Z)$ denotes a series whose coefficients are all positive. After a first differentiation we find

$$\dfrac{d^2Z}{dX^2} = 2A \dfrac{dZ}{dX} \dfrac{d^2Z}{dX^2} + \dfrac{\partial \Phi}{\partial X} + \dfrac{\partial \Phi}{\partial Z} \dfrac{dZ}{dX}.$$

For $X = 0$, Z and dZ/dX are zero; hence d^2Z/dX^2 is positive. The verification for the following derivatives is similar.

The series obtained for the development of the desired integral z is therefore convergent as long as the absolute values of the differences $x_i - (x_i)_0$ remain less than a positive number r. The value of that series is an analytic function in the neighborhood of the point $(x_1)_0, (x_2)_0, \cdots, (x_n)_0$ and reduces to $\phi(x_2, x_3, \cdots, x_n)$ for $x_1 = (x_1)_0$. That function satisfies the given equation; for if we replace in f the variables $z, \partial z/\partial x_2, \cdots, \partial z/\partial x_n$ by the preceding function and by its partial derivatives, the result is a function $\psi(x_1, x_2, \cdots, x_n)$ which is regular in the neighborhood of the point $(x_1)_0, (x_2)_0, \cdots, (x_n)_0$, and, from the manner in which we have obtained the coefficients of the series z, the two functions ψ and $\partial z/\partial x_1$ are equal, as well as all their partial derivatives, at the point $(x_1)_0, (x_2)_0, \cdots, (x_n)_0$. They are therefore identical.

The proof is the same for a simultaneous system of equations of the first order,

(27) $\quad \dfrac{\partial z_1}{\partial x_1} = f_1, \quad \dfrac{\partial z_2}{\partial x_1} = f_2, \quad \cdots, \quad \dfrac{\partial z_p}{\partial x_1} = f_p,$

whose right-hand sides contain only the variables $x_1, x_2, \cdots, x_n$, the functions $z_1, z_2, \cdots, z_p$, and the partial derivatives of the first order except those with respect to x_1. Supposing the right-hand sides analytic in the neighborhood of a system of particular values $(x_i)_0$, $(z_k)_0$, $(p_i^k)_0$, assigned to all the variables which appear in the function f, *these equations have a system of integrals which are analytic in the neighborhood of the point* $(x_1)_0, \cdots, (x_n)_0$ *and which reduce for* $x_1 = (x_1)_0$ *to p given functions* $\phi_1, \phi_2, \cdots, \phi_p$ *of the $(n-1)$ variables* $x_2, x_3, \cdots, x_n,$ *which are analytic in the neighborhood of the point* $(x_2)_0, (x_3)_0, \cdots, (x_n)_0$ *and are such that the values of ϕ_k and of $\partial \phi_k/\partial x_i$ at that point are precisely* $(z_k)_0$ *and* $(p_i^k)_0$ $(k = 1, 2, \cdots, p \,;\, i = 2, 3, \cdots, n)$.

26. The general integral of a system of differential equations. The preceding theorem enables us to complete the theory of differential equations on several important points. Thus, the existence of an infinite number of integrating factors for an expression of the form $P(x, y)\,dx + Q(x, y)\,dy$ is an immediate consequence of it if P and Q are analytic functions of the variables x and y (§ 12).

Let us consider again the equation of the first order $y' = f(x, y)$, and let (x_0, y_0) be a pair of values for which the function $f(x, y)$ is regular. The analytic integral the existence of which has been established, which takes on the value y_0 for $x = x_0$, may be considered as a function of three independent variables x, x_0, y_0; it is from this point of view that we are going to study it. For definiteness let us suppose that the function $f(x, y)$ is regular in the neighborhood of a point $(x = \alpha, y = \beta)$. We can evidently consider the given equation as a partial differential equation,

$$(28) \qquad \frac{\partial y}{\partial x} = f(x, y),$$

which defines a function y of the three variables x, x_0, y_0, and we propose to determine an integral of that equation which is analytic in the neighborhood of the point $x = \alpha$, $x_0 = \alpha$, $y_0 = \beta$ and which reduces to y_0 for $x = x_0$. This last condition is not in the same form as that of the preceding paragraph, but it suffices, in order to overcome the difficulty, to take instead of x and of x_0 two new independent variables $u = x + x_0$ and $v = x - x_0$. Then the equation (28) becomes

$$(29) \qquad \frac{\partial y}{\partial u} + \frac{\partial y}{\partial v} = f\left(\frac{u+v}{2}, y\right),$$

and we are led to seek an integral of this new equation which is analytic in the neighborhood of the values $u = 2\alpha$, $v = 0$, $y_0 = \beta$ and which reduces to y_0 for $v = 0$. By the general theorem, there exists an analytic integral, and only one, which satisfies these conditions; we shall denote it by $\phi(x, x_0, y_0)$, supposing that we have replaced u and v by their values in terms of x and y. Let D be a region defined by the conditions $|x - \alpha| \leqq r$, $|x_0 - \alpha| \leqq r$, $|y_0 - \beta| \leqq \rho$, in which the function $\phi(x, x_0, y_0)$ is regular. The function ϕ has the following properties in this region. In the first place, from the very way in which we have obtained it, if x_0 and y_0 are constants, it represents the integral of the differential equation $y' = f(x, y)$, which takes on the value y_0 for $x = x_0$. This integral is surely analytic whenever $|x - \alpha|$ is less than r, for any point (x_0, y_0) in the region D.

The development of $\phi(x, x_0, y_0)$ is of the form

$$y = y_0 + (x - x_0) P(x, x_0, y_0),$$

where P also denotes a regular function. By the general theory of implicit functions, we can solve the above relation, obtaining $y_0 = \psi(x, x_0, y)$, in which the right-hand side is also a power series. *The function $\psi(x, x_0, y)$ is identical with $\phi(x_0, x, y)$.* In fact, let x_0 and x_1 be two values of x in the region D; then the integral which is equal to y_0 for $x = x_0$ takes on at the point x_1 a certain value y_1, and we have $y_1 = \phi(x_1, x_0, y_0)$. But it is evident that the relation between the two pairs of values (x_0, y_0), (x_1, y_1) is a reciprocal one; hence we have also $y_0 = \phi(x_0, x_1, y_1)$.

Let x_0' be any value of x such that we have $|x_0' - \alpha| < r$. Every analytic integral of the equation (28), passing through any point (x_0, y_0) of the region D, satisfies a relation of the form

(30) $$\phi(x_0', x, y) = C.$$

For, let us consider the analytic integral equal to y_0 for $x = x_0$. That integral takes on a value y_0' when x has the value x_0', and we have, from the definition of the function ϕ, $\phi(x_0', x_0, y_0) = y_0'$. Let x be another value of the variable in the same region and y the corresponding value of the integral. We have also $\phi(x_0', x, y) = y_0'$, and therefore the analytic integral considered does satisfy a relation of the form (30). By differentiating it with respect to x and replacing y' by its value $f(x, y)$ we find that the function $\phi(x_0', x, y)$ satisfies the relation

(31) $$\frac{\partial \phi}{\partial x} + \frac{\partial \phi}{\partial y} f(x, y) = 0.$$

This relation reduces necessarily to an identity, for it must be true for $x = x_0$, $y = y_0$, and the point (x_0, y_0) is any point of the region D.

This enables us to answer a question left undecided in § 22. In the plane of the variable x let any curve Γ approach the point x_0 as a limit. We shall say that a function y of the variable x which can be continued analytically along the whole length of Γ approaches y_0 as x approaches x_0 on Γ if for every positive number ϵ we can find a corresponding positive number η such that $|y - y_0|$ remains less than ϵ for all the values of x lying on Γ in the interior of a circle with a radius η and with the center x_0.

The reasoning of Briot and Bouquet does not prove that there do not exist other integrals than the analytic integral, approaching y_0 as x approaches x_0 in the manner which has just been defined. This, however, is the fact. For let us consider a definite point (x_0, y_0) of the region D, and let us take for the new dependent variable in the equation (28) the function $Y = \phi(x_0, x, y)$ defined above. Then we have

$$\frac{dY}{dx} = \frac{\partial \phi}{\partial x} + \frac{\partial \phi}{\partial y}\frac{dy}{dx},$$

and, by the relation (31), the given differential equation reduces to $dY/dx = 0$. If, now, y approaches y_0 when x approaches x_0, the same thing is true of Y, and the only integral of the new equation $dY/dx = 0$ which satisfies this condition is evidently $Y = y_0$. The integral sought must therefore satisfy the relation

$$\phi(x_0, x, y) = y_0,$$

or

(32) $\qquad y_0 = y + (x - x_0) P(x, y, x_0),$

and, by the theorem on implicit functions (I, § 193, 2d ed.; § 187, 1st ed.), there is only one root of the equation (32) approaching y_0 as x approaches x_0, and that root is an analytic function.*

It follows that every integral of the equation (28) which passes through a point of the region D satisfies a relation of the form (30). On that account we say that that equation represents the *general integral* of the differential equation in this region. The number C is the constant of integration which remains arbitrary at least between certain limits. We have seen that we could also put the equation (30) in the equivalent form $y = \phi(x, x_0', y_0')$, where the constant of integration is y_0'.

---

* PICARD, *Traité d'Analyse*, Vol. II, pp. 315–317. PAINLEVÉ, *Leçons de Stockholm*, p. 394.

All these properties can be extended to a system of differential equations of the form

(33) $\quad \dfrac{dy_1}{dx} = f_1(x, y_1, y_2, \cdots, y_n), \qquad \dfrac{dy_2}{dx} = f_2, \qquad \cdots, \qquad \dfrac{dy_n}{dx} = f_n.$

Let us suppose that the right-hand sides are analytic in the neighborhood of the system $x = \alpha$, $y_1 = \beta_1, \cdots, y_n = \beta_n$. We may again regard the preceding equations as a system of partial differential equations involving the n dependent variables $y_1, y_2, \cdots, y_n$ and the $n + 2$ independent variables x, x_0, $(y_1)_0$, $(y_2)_0, \cdots, (y_n)_0$, and we may seek the integrals of this system which are regular in the neighborhood of the values $x = \alpha$, $x_0 = \alpha$, $(y_1)_0 = \beta_1, \cdots, (y_n)_0 = \beta_n$ and which reduce to $(y_1)_0$, $(y_2)_0, \cdots, (y_n)_0$ respectively for $x = x_0$.

Let

(34) $\quad \begin{cases} y_1 = \phi_1[x, x_0, (y_1)_0, \cdots, (y_n)_0], & y_2 = \phi_2, \quad \cdots, \\ y_n = \phi_n[x, x_0, (y_1)_0, \cdots, (y_n)_0] \end{cases}$

be the n functions thus defined, which we suppose to be analytic in the region D defined by the conditions $|x - \alpha| \leqq r$, $|x_0 - \alpha| \leqq r$, $|(y_i)_0 - \beta_i| \leqq \rho$. From the equations (34) we derive, conversely,

(35) $\quad (y_1)_0 = \Phi_1(x_0, x, y_1, \cdots, y_n), \quad \cdots, \quad (y_n)_0 = \Phi_n(x_0, x, y_1, \cdots, y_n),$

and each of these functions Φ_i satisfies, for any value of x_0, the relation

(36) $\quad \dfrac{\partial \Phi_i}{\partial x} + \dfrac{\partial \Phi_i}{\partial y_1} f_1 + \cdots + \dfrac{\partial \Phi_i}{\partial y_n} f_n = 0.$

We prove this just as before by observing that the analytic integrals which take the values $(y_1)_0, \cdots, (y_n)_0$ for $x = x_0$ satisfy the relations (35), and therefore the relations (36), which we deduce from them by differentiating with respect to the independent variable x and by replacing the derivative dy_i/dx by f_i. These relations (36) must reduce to identities; for if x_0 is supposed fixed, we can show as above that we can choose $(y_1)_0, \cdots, (y_n)_0$ in such a way that the *integral curve* [*] passes through any given point of the region D. The left-hand side of the equation (36) must therefore be zero for the coördinates of any point whatever of this region.

If in the equations (33) we take for new dependent variables the n functions $Y_i = \Phi_i(x_0, x, y_1, \cdots, y_n)$, where x_0 is constant, these

---

[*] As a generalization we shall say that every system of integrals of the equations (33) defines an *integral curve*.

equations become, by the conditions (36),

(37) $$\frac{dY_1}{dx} = 0, \quad \frac{dY_2}{dx} = 0, \quad \cdots, \quad \frac{dY_n}{dx} = 0.$$

It follows that all the integrals of the system (33) satisfy relations of the form (35), where $(y_1)_0, \cdots, (y_n)_0$ are constants — at least all of those integrals which have a point in the interior of the region D where the functions ϕ are regular. We shall say, then, that the equations (35) represent the general integral of the system (33) in this region.

From these equations it follows also that there are no other systems of integrals than the analytic integrals which approach $(y_1)_0, \cdots, (y_n)_0$ when x approaches x_0. We have, in fact,

$$\phi_i = y_i + (x - x_0) P_i(x_0, x, y_1, \cdots, y_n),$$

and the Jacobian $D(\phi_1, \phi_2, \cdots, \phi_n)/D(y_1, y_2, \cdots, y_n)$ reduces to unity for $x = x_0$. According to the general theory of implicit functions, the equations (35) have only a single system of solutions for $y_1, y_2, \cdots, y_n$, which approach $(y_1)_0, \cdots, (y_n)_0$ when x approaches x_0, and these solutions are analytic.

To sum up, through every point of the region D there passes an integral curve, and only one, represented by n equations $y_i = \psi_i(x)$, where the functions ψ_i are analytic so long as $|x - a| \leq r$.

II. THE METHOD OF SUCCESSIVE APPROXIMATIONS. THE CAUCHY-LIPSCHITZ METHOD

27. Successive approximations. The method of successive approximations has been applied with success by E. Picard to ordinary differential equations and to a great number of cases of partial differential equations. We shall apply it to the treatment of differential equations with an important addition due to Ernst Lindelöf.

Let $y(x)$ be an integral of the differential equation $dy/dx = f(x, y)$ taking on the value y_0 for $x = x_0$. The function $y(x)$ satisfies the relation

(38) $$y(x) = y_0 + \int_{x_0}^{x} f[t, y(t)] dt,$$

and conversely. The equation (38) is an *integral equation* which is equivalent to the two conditions $y'(x) = f[x, y(x)], y(x_0) = y_0$ and which lends itself readily to the method of successive approximations. We shall develop the method on a system of two equations of the first order

(39) $$\frac{dy}{dx} = f(x, y, z), \quad \frac{dz}{dx} = \phi(x, y, z),$$

supposing first that the variables are real. We shall assume that the two functions f and ϕ are continuous when x varies from x_0 to $x_0 + a$ and when y and z vary respectively between the limits $(y_0 - b, y_0 + b)$ and $(z_0 - c, z_0 + c)$; that the absolute value of each of these functions f and ϕ remains less than a positive number M when the variables x, y, z remain within the preceding limits; and, finally, that there exist two positive numbers A and B such that we have

(40) $$\begin{cases} |f(x, y, z) - f(x, y', z')| < A|y - y'| + B|z - z'|, \\ |\phi(x, y, z) - \phi(x, y', z')| < A|y - y'| + B|z - z'| \end{cases}$$

for any positions of the points (x, y, z) and (x, y', z') in the preceding region.

Let us suppose, for ease in the reasoning, $a > 0$, and let h be the smallest of the three positive numbers a, b/M, c/M. We shall prove that *the equations* (39) *have a system of integrals which are continuous in the interval* $(x_0, x_0 + h)$ *and which take on the values* y_0 *and* z_0 *for* $x = x_0$. For this purpose we shall write the equations (39) in the form of integral equations:

(41) $$y(x) = y_0 + \int_{x_0}^{x} f[t, y(t), z(t)] dt, \quad z(x) = z_0 + \int_{x_0}^{x} \phi[t, y(t), z(t)] dt,$$

and we shall solve these equations by successive approximations in the same way as for a system of simultaneous equations (I, § 34, 2d ed.; § 25 ftn., 1st ed.), taking for the first approximation values the initial values y_0 and z_0 themselves. We are thus led to write

(42) $$\begin{cases} y_1(x) = y_0 + \int_{x_0}^{x} f(t, y_0, z_0) dt, \\ z_1(x) = z_0 + \int_{x_0}^{x} \phi(t, y_0, z_0) dt, \\ y_2(x) = y_0 + \int_{x_0}^{x} f[t, y_1(t), z_1(t)] dt, \\ z_2(x) = z_0 + \int_{x_0}^{x} \phi[t, y_1(t), z_1(t)] dt \end{cases}$$

and, in general,

(43) $$\begin{cases} y_n(x) = y_0 + \int_{x_0}^{x} f[t, y_{n-1}(t), z_{n-1}(t)] dt, \\ z_n(x) = z_0 + \int_{x_0}^{x} \phi[t, y_{n-1}(t), z_{n-1}(t)] dt. \end{cases}$$

Let us prove first that this process of approximation can be continued indefinitely if x is contained in the interval $(x_0, x_0 + h)$. We have, in the first place, if x is within that interval,

$$|y_1 - y_0| < Mh < b,$$

and, similarly, $|z_1 - z_0| < c$. If we replace y and z by y_1 and z_1 in the functions f and ϕ, the functions of x thus obtained are therefore continuous between x_0 and $x_0 + h$, and their absolute values remain less than M. For the same reason as before, y_2 and z_2 are continuous functions of x in the interval $(x_0, x_0 + h)$, and we have in this interval $|y_2 - y_0| < b$, $|z_2 - z_0| < c$. The reasoning can be continued indefinitely; all the functions y_n and z_n are continuous between x_0 and $x_0 + h$, and we always have in this interval $|y_n - y_0| < b$, $|z_n - z_0| < c$.

SUCCESSIVE APPROXIMATIONS

In order to prove that y_n and z_n approach limits when n becomes infinite, let us notice that we derive first, from the first of the relations (42),

(44) $\qquad |y_1(x) - y_0| < M(x - x_0), \qquad |z_1(x) - z_0| < M(x - x_0),$

where x is any value whatever except x_0 in the interval $(x_0, x_0 + h)$. We have next

$$y_2(x) - y_1(x) = \int_{x_0}^{x} \{f[t, y_1(t), z_1(t)] - f(t, y_0, z_0)\} dt,$$

and, by taking account of the first of the inequalities (40),

$$|y_2(x) - y_1(x)| < \int_{x_0}^{x} A |y_1(t) - y_0| dt + \int_{x_0}^{x} B |z_1(t) - z_0| dt ;$$

and therefore, by the inequalities (44),

$$|y_2(x) - y_1(x)| < (A + B) M \frac{(x - x_0)^2}{2!}.$$

We have an analogous result for $|z_2(x) - z_1(x)|$, and, continuing in this way, we see that we have in general

(45) $\qquad \begin{cases} |y_n(x) - y_{n-1}(x)| < M(A + B)^{n-1} \dfrac{(x - x_0)^n}{n!}, \\ |z_n(x) - z_{n-1}(x)| < M(A + B)^{n-1} \dfrac{(x - x_0)^n}{n!}. \end{cases}$

The two series

(46) $\qquad \begin{cases} y_0 + (y_1 - y_0) + (y_2 - y_1) + \cdots + (y_n - y_{n-1}) + \cdots, \\ z_0 + (z_1 - z_0) + (z_2 - z_1) + \cdots + (z_n - z_{n-1}) + \cdots, \end{cases}$

whose terms are all continuous functions of x in the interval $(x_0, x_0 + h)$, are therefore uniformly convergent in that interval. The values of these two series, $Y(x)$ and $Z(x)$, are consequently continuous functions of x between x_0 and $x_0 + h$. As the number n becomes infinite, the relations (43) become, at the limit,

$$Y(x) = y_0 + \int_{x_0}^{x} f[t, Y(t), Z(t)] dt, \qquad Z(x) = z_0 + \int_{x_0}^{x} \phi[t, Y(t), Z(t)] dt.$$

For we have just seen that the differences $Y(x) - y_{n-1}(x)$, $Z(x) - z_{n-1}(x)$ approach zero uniformly in the interval $(x_0, x_0 + h)$, and therefore, by virtue of the relations (40), the integrals

$$\int_{x_0}^{x} \{f[t, Y(t), Z(t)] - f[t, y_{n-1}(t), z_{n-1}(t)]\} dt,$$

$$\int_{x_0}^{x} \{\phi[t, Y(t), Z(t)] - \phi[t, y_{n-1}(t), z_{n-1}(t)]\} dt$$

approach zero when n becomes infinite. The functions $Y(x)$ and $Z(x)$ therefore satisfy all the given conditions.

The preceding method is evidently applicable, whatever may be the number of the equations in the system. The inequalities (40), which play an essential part in the demonstration, are certainly satisfied for suitable values of A and B whenever the functions f and ϕ have continuous partial derivatives with respect to y and z within the limits indicated for the variables; this is an easy consequence of the law of the mean (I, § 20, 2d ed.; § 11, 1st ed.). Let us also notice that if the functions f and ϕ remain continuous when x varies between $x_0 - a$ and $x_0 + a$, and the variables y and z between the same limits as above, the same reasoning proves the existence of a system of integrals, $Y(x)$ and $Z(x)$, which take on the values y_0 and z_0 for $x = x_0$ and are continuous in the interval $(x_0 - h, x_0 + h)$, where h has the same meaning as before.

There are no other systems of integrals than $Y(x)$ and $Z(x)$ taking on the values y_0 and z_0 for $x = x_0$. The reasoning being always the same, let us take for simplicity a single equation $dy/dx = f(x, y)$, and let us put, as before,

$$y_1 = y_0 + \int_{x_0}^{x} f(t, y_0)\,dt, \quad \cdots, \quad y_n = y_0 + \int_{x_0}^{x} f[t, y_{n-1}(t)]\,dt.$$

Let $Y_1(x)$ be an integral of that equation which takes on the value y_0 for $x = x_0$ and which is continuous in the interval $(x_0, x_0 + a')$, where a' is less than the smaller of the numbers a and b/M and such that we have $|Y_1(x) - y_0| < b$ in this interval. Since Y_1 satisfies the given equation, we can write

$$Y_1(x) - y_0 = \int_{x_0}^{x} f[t, Y_1(t)]\,dt,$$

and, consequently,

$$Y_1(x) - y_n(x) = \int_{x_0}^{x} \{f[t, Y_1(t)] - f[t, y_{n-1}(t)]\}\,dt.$$

Let us put successively in that relation $n = 1, 2, 3, \cdots$; we have first

$$|Y_1(x) - y_1(x)| < Ab(x - x_0),$$

then

$$|Y_1(x) - y_2(x)| < A\int_{x_0}^{x} Ab(t - x_0)\,dt = A^2 b \frac{(x - x_0)^2}{2!},$$

and, in general,

$$|Y_1(x) - y_n(x)| < A^n b \frac{(x - x_0)^n}{n!}.$$

The right-hand side of that inequality approaches zero when n becomes infinite; the integral Y_1 is therefore identical with the limit of y_n, that is, with Y.*

28. The case of linear equations. The general reasoning proves that the integrals are certainly continuous in the interval $(x_0, x_0 + h)$ defined above; but in quite a number of cases we can state the existence of a more extended interval in which the integrals are continuous. If, in fact, we go over the proof again, we see that the conditions $h < b/M$, $h < c/M$ are needed only to make sure that the intermediate functions $y_1, z_1, y_2, z_2, \cdots$ do not get out of the intervals $(y_0 - b, y_0 + b)$, $(z_0 - c, z_0 + c)$, so that the functions $f(x, y_i, z_i)$, $\phi(x, y_i, z_i)$ shall be continuous functions of x between x_0 and $x_0 + h$. If the functions $f(x, y, z)$, $\phi(x, y, z)$ remain continuous when x varies from x_0 to $x_0 + a$, and when y and z vary from $-\infty$ to $+\infty$, it is unnecessary to make these requirements. All the functions y_i and z_i are continuous in the interval $(x_0, x_0 + a)$.

---

*Regarding questions concerning the approximate integration of differential equations, the reader is referred to the articles of E. Cotton (*Acta mathematica*, Vol. XXXI; *Bulletin de la Société mathématique de France*, Vols. XXXVI, XXXVII, and XXXVIII; *Annales de l'Université de Grenoble*, Vol. XXI).

Again, in order to prove the convergence of the two series (46) it is sufficient that there exist two positive numbers A and B such that the two inequalities (40) are satisfied for any values of y, y', z, z' if x remains in the interval $(x_0, x_0 + a)$. We recognize, in fact, on going over the calculations made above, that the inequalities (45) still hold, provided that we indicate by M an upper bound of $|f(x, y_0, z_0)|$ and of $|\phi(x, y_0, z_0)|$ in the interval $(x_0, x_0 + a)$.

These conditions are satisfied, according to the law of the mean, if the functions $f(x, y, z)$, $\phi(x, y, z)$ have partial derivatives with respect to the variables y and z which remain finite for all values of y and z when x varies from x_0 to $x_0 + a$. Such, for example, is the case for the equation

$$\frac{dy}{dx} = x + \sin y;$$

the right-hand side is a continuous function, whatever x and y may be, and the partial derivative $\partial f / \partial y$ is at most equal to unity in absolute value. All the integrals of that equation are therefore continuous functions when x varies from $-\infty$ to $+\infty$.*

The preceding conclusions apply in particular to systems of linear equations

(47) $\qquad \dfrac{dy_i}{dx} = a_{i1} y_1 + a_{i2} y_2 + \cdots + a_{in} y_n + b_i, \qquad (i = 1, 2, \cdots, n)$

where the coefficients a_{ik}, b_i are functions of x. If all these functions are continuous in an interval (x_0, x_1), all the integrals of this system are likewise continuous in this interval; if the coefficients are polynomials, all the integrals are then continuous when x varies from $-\infty$ to $+\infty$.

Limiting ourselves to real variables, we see that the integrals of *linear* equations can have no other singular points than those of the coefficients. This very important property cannot be extended to many other equations, even though they are apparently just as simple — for example, to the equation $y' = y^2$.

Note. We often have occasion to study systems of linear equations whose coefficients are analytic functions of certain parameters. Let us suppose, for definiteness, that the coefficients a_{ik} and b_i of the equations (47) are continuous functions of x in an interval (a, b), and that they depend also upon a parameter λ of which they are analytic functions in a region D.

The integrals of this system which take on given initial values for a value x_0 of x included between a and b are represented in the whole interval (a, b) by uniformly convergent series, and from the very manner in which we obtain them it is clear that all the terms of this series are analytic functions of the

---

* We can deduce an analogous theorem from the calculus of limits. Let $f(x, y)$ be a function which is *real* for every system of real values of x and y and analytic in their neighborhood. Suppose, besides, that $|f(x, y)|$ remains less than a fixed number M when we have respectively $|\mathcal{R}(x/i)| \leqq a$ and $|\mathcal{R}(y/i)| \leqq b$. If x_0, y_0 are a pair of any *real* values of x and y, the function $f(x, y)$ is analytic in the region defined by the inequalities $|x - x_0| \leqq a$, $|y - y_0| \leqq b$, and its absolute value is less than M. Then, by the calculus of limits, the integral of the equation $y' = f(x, y)$, which is equal to y_0 for $x = x_0$, is surely analytic in a circle C whose radius r is *independent* of x_0, y_0. We can follow the analytic extension of that integral along the real axis by means of circles of radius r, and we see that it is analytic in the interior of the strip bounded by two parallels to the real axis at a distance r from it.

parameter λ in D. *These integrals are therefore themselves analytic functions of λ in the region D* (Part I, § 89).

Most frequently the coefficients a_{ik} and b_i are integral functions of the parameter λ; the integrals are therefore themselves integral functions of λ. We can obtain directly the developments, according to powers of λ, of the integrals which take on given initial values, by first substituting in the two sides of the equations (47) developments of the form

$$y_i = u_{i0} + u_{i1}\lambda + \cdots + u_{ip}\lambda^p + \cdots, \qquad (i = 1, 2, \cdots, n)$$

where the variables u_{ik} are functions of x, and by then equating coefficients. The functions u_{i0} must take on the given initial values for $x = x_0$, while the other functions u_{ik}, where $k \geqq 1$, must be zero for $x = x_0$.

Proceeding in this way, we find, step by step, systems of linear differential equations for determining these coefficients. We shall return to this subject later.

29. Extension to analytic functions. The method can be extended to complex variables. To do so it suffices to observe that we have for analytic functions of one or several variables inequalities analogous to the inequalities (40). First, let $f(x)$ be an analytic function of a complex variable x, in a region bounded by a convex curve C and also on the boundary, and let A be the maximum value of $|f'(x)|$ in this region. The difference $f(x_2) - f(x_1)$, where x_1 and x_2 are any two points of that region, is equal to the definite integral $\int f'(x)\,dx$ taken along the straight line joining these two points. We have, therefore,

$$|f(x_2) - f(x_1)| < A\,|x_2 - x_1|.$$

Similarly, let $f(x, y)$ be an analytic function of the two variables x and y when these variables remain respectively in two regions Ω and Ω' bounded by two closed convex curves C and C', and let A and B be the maximum values of $|f'_x|$ and of $|f'_y|$ in this region. If x_1 and x_2 are any two values of x in Ω, and y_1 and y_2 any two values of y in Ω', we can write

$$f(x_2, y_2) - f(x_1, y_1) = [f(x_2, y_2) - f(x_1, y_2)] + [f(x_1, y_2) - f(x_1, y_1)],$$

and, consequently, from what we have just shown, we have

$$|f(x_2, y_2) - f(x_1, y_1)| < A\,|x_2 - x_1| + B\,|y_2 - y_1|.$$

The proof is the same whatever the number of the independent variables.

Having seen this, let us limit ourselves, for simplicity, to the case of a single equation,

$$(48) \qquad \frac{dy}{dx} = f(x, y),$$

the right-hand side of which we shall suppose to be analytic in the region defined by the inequalities $|x - x_0| \leqq a$, $|y - y_0| \leqq b$. Let M be the maximum value of $|f(x, y)|$ in this region, and h the smaller of the two numbers a and b/M. In the plane of the variable x let us describe a circle C_h of radius h about the point x_0 as center, and let us put, as above,

$$y_1 = y_0 + \int_{x_0}^{x} f(t, y_0)\,dt, \qquad y_2 = y_0 + \int_{x_0}^{x} f[t, y_1(t)]\,dt, \qquad \cdots,$$

$$y_n = y_0 + \int_{x_0}^{x} f[t, y_{n-1}(t)]\,dt,$$

where the upper limit x is a point within C_h. We prove first, step by step, that we have

$$|y_1 - y_0| < b, \qquad |y_2 - y_0| < b, \qquad \cdots, \qquad |y_n - y_0| < b, \qquad \cdots.$$

All the functions $y_1, y_2, \cdots, y_n, \cdots$ are therefore analytic functions of x in the circle C_h, and the process can be continued indefinitely. Moreover, we have

$$(49) \qquad y_n(x) - y_{n-1}(x) = \int_{x_0}^{x} \{f[t, y_{n-1}(t)] - f[t, y_{n-2}(t)]\} dt,$$

where the integral is taken along the straight line joining the two points x_0, x. Let A be the maximum value of $|\partial f/\partial y|$ in the region $|x - x_0| \leqq h$, $|y - y_0| \leqq b$; then, according to the observations made just above, we have always

$$|f[t, y_{n-1}(t)] - f[t, y_{n-2}(t)]| < A |y_{n-1}(t) - y_{n-2}(t)|.$$

In order to prove that we have an inequality analogous to the inequalities (45), let us suppose that we have

$$|y_{n-1}(t) - y_{n-2}(t)| < M A^{n-2} \frac{|t - x_0|^{n-1}}{(n-1)!},$$

which is evidently the case for $n = 2$. Let $x = x_0 + r e^{\theta i}$; the change of variable $t - x_0 = \rho e^{\theta i}$ reduces the integral (49) to an integral taken along the real axis from 0 to r, and we have (Part I, § 44)

$$|y_n(x) - y_{n-1}(x)| < \int_0^r M A^{n-1} \frac{\rho^{n-1}}{(n-1)!} d\rho = M A^{n-1} \frac{r^n}{n!},$$

or

$$|y_n(x) - y_{n-1}(x)| < M A^{n-1} \frac{|x - x_0|^n}{n!}.$$

The proof can be completed as before. The series whose general term is $y_n - y_{n-1}$ is uniformly convergent in the circle C_h, and, since all the terms are analytic functions, the sum of that series is an analytic function in the same circle (Part I, § 39), which satisfies the equation (48) and which takes on the value y_0 for $x = x_0$. The development in power series of this integral is necessarily identical with that furnished by the calculus of limits, but the limit obtained for the radius of convergence is greater than that given by the first method.

The remark relative to linear equations applies also to analytic functions. Let us suppose that the coefficients a_{ik} and b_i of the equations (47) are analytic functions of the complex variable x. Let us mark in the plane the singular points of these functions, and let us suppose that from each of these singular points a ray is drawn following the prolongation of the segments from x_0 to the singular point. The set of points of the plane which are not situated upon any of the preceding lines is called the *star* corresponding to the system of singular points. The straight line which joins the point x_0 to a point x of the star does not pass through any of the singular points, and the method of § 28 proves that all the integrals of the system (47) are analytic functions along that straight line. The point x being any point of the star, it follows that all the integrals of the linear system (47) are analytic functions in the whole star — a result which will be established later in another manner (§ 37).

The method of successive approximations enables us also to obtain for the integrals developments in series converging in the whole star. Let A be a region of the plane bounded by a closed curve C lying entirely in the star; the series furnished by the method of successive approximations are *uniformly convergent in A*. The remaining details of the proof are left to the reader, since they do not differ essentially from the details of the proof given before.

30. The Cauchy-Lipschitz method. The first proof given by Cauchy of the existence of integrals of a system of differential equations has been preserved in the lectures by Moigno published in 1844. It was considerably simplified by Lipschitz, who made clear just what hypotheses were necessary for the validity of the proof.

In order to gain a clear grasp of the whole process, let us take the simple equation
$$\frac{dy}{dx} = f(x).$$

We have shown (I, § 78, 2d ed.; § 76, 1st ed.) that the integral of this equation which takes on the value y_0 for $x = x_0$ is the limit of the sum

(50) $\quad y_0 + f(x_0)(x_1 - x_0) + f(x_1)(x_2 - x_1) + \cdots + f(x_{n-1})(x - x_{n-1}),$

where $x_1, x_2, \cdots, x_{n-1}$ are $n-1$ points of the interval (x_0, x), as the number n becomes infinite in such a way that all the intervals $(x_i - x_{i-1})$ approach zero. It is this process, suitably generalized, which leads to Cauchy's first method. In order to simplify the exposition, we shall take the case of a single equation,

(51) $$\frac{dy}{dx} = f(x, y).$$

We shall suppose that the function $f(x, y)$ of the real variables x, y is continuous when x varies from x_0 to $x_0 + a$ and when y varies from $y_0 - b$ to $y_0 + b$, and that there exists a positive number K such that

(52) $\quad |f(x, y') - f(x, y)| < K |y' - y|,$

where y and y' are any two numbers included between $y_0 - b$ and $y_0 + b$, and where x lies between x_0 and $x_0 + a$.

This condition, the importance of which was brought out by Lipschitz, will be called, for brevity, *the Lipschitz condition*. It has already been used in the method of successive approximations (§ 27; and I, § 34, 2d ed.; § 25 ftn., 1st ed.).

Let M be the upper limit of $|f(x, y)|$ in the preceding region, and h the smaller of the two numbers a and b/M (we suppose $a > 0$, $b > 0$). In order to prove that the equation (51) has an integral which takes on the value y_0 for $x = x_0$ and which is continuous in the interval $(x_0, x_0 + h)$, we shall imitate so far as possible the procedure followed in establishing the existence of a primitive function for $f(x)$. Let x be a value of the variable belonging to the interval. Let us take between x_0 and x a certain number of intermediate values, $x_1, x_2, \cdots, x_{i-1}, x_i, \cdots, x_{n-1}$, proceeding in increasing order from x_0 to x. We shall put successively

(53) $\quad y_1 = y_0 + f(x_0, y_0)(x_1 - x_0), \quad y_2 = y_1 + f(x_1, y_1)(x_2 - x_1), \quad \cdots,$

and, in general,

(54) $\quad y_i = y_{i-1} + f(x_{i-1}, y_{i-1})(x_i - x_{i-1}). \quad (i = 1, 2, \cdots, n-1)$

The sum

(55) $\quad \begin{cases} y_n = y_0 + f(x_0, y_0)(x_1 - x_0) + f(x_1, y_1)(x_2 - x_1) + \cdots \\ \qquad + f(x_{n-1}, y_{n-1})(x - x_{n-1}) \end{cases}$

presents an evident analogy with the sum (50), to which it reduces when the function $f(x, y)$ does not depend upon y. We are thus led to investigate whether or not that sum approaches a limit when the number n becomes infinite. We shall generalize the question by defining first two sums analogous to the quantities S and s (I, § 72, 2d ed.; § 71, 1st ed.).

Let us consider the triangle ABC formed by the straight lines defined by the equations

$$X = x_0 + h, \qquad Y = y_0 + M(X - x_0), \qquad Y = y_0 - M(X - x_0).$$

From the way in which we have defined h, the function $f(x, y)$ is continuous when the point (x, y) remains in the interior or on the sides of this triangle, and its absolute value is at most equal to M.

The parallels to the y-axis, $X = x_1$, $X = x_2$, $\cdots$, $X = x$, divide the triangle ABC into a certain number of isosceles trapezoids of which the first reduces

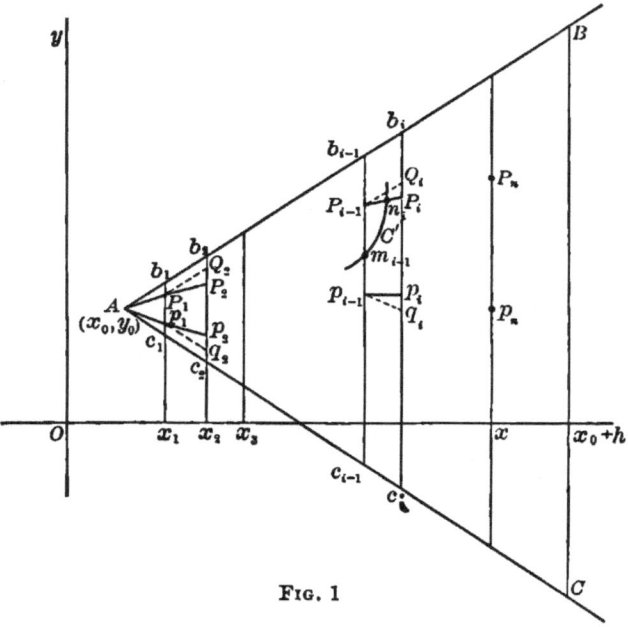

FIG. 1

to a triangle. Let M_1 and m_1 denote respectively the maximum and minimum values of $f(x, y)$ in the triangle Ab_1c_1; then we have $-M \leqq m_1 < M_1 \leqq M$. Through the point A let us draw the straight lines with slopes equal to M_1 and m_1, meeting the straight line $X = x_1$ in two points, P_1 and p_1, whose ordinates are respectively $Y_1 = y_0 + M_1(x_1 - x_0)$ and $y_1 = y_0 + m_1(x_1 - x_0)$. The letter y_1 no longer denotes the same thing as in the expressions (53) to (55). These points, P_1 and p_1, are evidently in the interior of the triangle ABC or on its sides, and we have $Y_1 > y_1$. Through the point P_1 let us draw the straight line with the slope M up to its intersection with the straight line b_2c_2 in Q_2, and through p_1 let us draw, similarly, the straight line with slope $-M$ up to its intersection q_2

with the same straight line b_2c_2. Let M_2 and m_2 be the maximum and minimum values of $f(x, y)$ in the trapezoid $P_1Q_2q_2p_1$; the straight line with the slope M_2 drawn through P_1 meets the straight line b_2c_2 in a point P_2 whose ordinate is

$$Y_2 = Y_1 + M_2(x_2 - x_1),$$

and the straight line with the slope m_2 drawn through p_1 meets b_2c_2 in a point p_2 with the ordinate $y_2 = y_1 + m_2(x_2 - x_1)$. We have evidently $Y_2 > y_2$ and $Y_2 - y_2 \geqq Y_1 - y_1$, the equality holding only if the function $f(x, y)$ is constant in the trapezoid $P_1Q_2q_2p_1$. This process can be continued. Having obtained two points, P_{i-1} and p_{i-1}, on the straight line $c_{i-1}b_{i-1}$, let us draw through P_{i-1} a parallel to AB, and through p_{i-1} a parallel to AC. We thus form an isosceles trapezoid $P_{i-1}Q_iq_ip_{i-1}$. Let M_i be the maximum value of $f(x, y)$ in this trapezoid, and m_i the minimum value; the straight line with the slope M_i drawn through P_{i-1} meets the straight line c_ib_i in a point P_i, and the straight line with the slope m_i drawn through p_{i-1} meets c_ib_i in a point p_i. We thus form two broken lines starting from the point A, namely, $AP_1P_2 \cdots P_{i-1}P_i \cdots P_n$, or L, and $Ap_1p_2 \cdots p_{i-1}p_i \cdots p_n$, or l, ending in the two points P_n and p_n of the straight line $X = x$. From the manner in which these two lines were constructed it is evident that they both lie in the triangle ABC, that the line L is never below l, and that the distance between these two lines, measured on a parallel to the axis Oy, cannot diminish when the abscissa increases from x_0 to x. The ordinates Y_n and y_n of the two extreme points are entirely analogous to the sums S and s (I, § 72, 2d ed.; § 71, 1st ed.). We shall put $S = Y_n$, $s = y_n$.

To each method of subdivision of the interval (x_0, x) corresponds a sum S and a sum s. If we subdivide each of the partial intervals (x_{i-1}, x_i) into still smaller intervals in an arbitrary manner, the preceding geometric construction shows immediately that the line L' corresponding to this new division is never above L, and the line l' is never below l. We have, therefore, $S' \leqq S$, $s' \geqq s$, where the accented letters denote the sums relative to the second division. We conclude from this (as in § 72, 2d ed.; § 71, 1st ed.) that if S, s, S_1, s_1 represent respectively the sums relative to any two methods of division *whatever* of the interval (x_0, x), we have $s \leqq S_1$, $s_1 \leqq S$. Indicating by I the lower limit of the sums S, and by I' the upper limit of the sums s, we have, therefore, $I' \leqq I$.

In order that the sums S and s shall have a common limit when the maximum length of the partial intervals approaches zero, it is necessary and sufficient that $S - s$ approach zero. In fact, we may write

$$S - s = S - I + I - I' + I' - s,$$

and the difference $S - s$ cannot be less than a number ϵ unless each of the numbers $S - I$, $I - I'$, $I' - s$ (no one of which can be negative) is itself less than ϵ. Since ϵ is an arbitrary positive number, this cannot happen unless we have $I' = I$, and it is, moreover, necessary that S and s shall have the same limit I. In order to prove that $S - s$ has zero for its limit, it is not sufficient to suppose that the function $f(x, y)$ is continuous, and it is here that the Lipschitz condition plays a part.

Let Y_i and y_i be the ordinates of the points P_i and p_i, and δ_i the difference $Y_i - y_i$. Since the function $f(x, y)$ is continuous in the triangle ABC, corresponding to every positive number λ we can find another positive number σ such that

$$|f(x, y) - f(x', y')| < \lambda,$$

provided that the distance between the two points (x, y) and (x', y') of the triangle ABC is less than σ. We shall suppose that all the differences $x_i - x_{i-1}$ are less than σ. From the construction by which the points P_i, p_i are obtained from the points P_{i-1}, p_{i-1}, we have

$$\delta_i = \delta_{i-1} + (M_i - m_i)(x_i - x_{i-1}).$$

On the other hand, we can write

$$M_i - m_i = f(x_i', y_i') - f(x_i'', y_i'')$$
$$= f(x_i', y_i') - f(x_i'', y_i') + [f(x_i'', y_i') - f(x_i'', y_i'')],$$

where (x_i', y_i') and (x_i'', y_i'') are the coördinates of two points of the trapezoid $P_{i-1}Q_iq_ip_{i-1}$. We have, therefore, by the condition (52),

$$M_i - m_i < \lambda + K|y_i'' - y_i'|.$$

But the difference $|y_i'' - y_i'|$ is at most equal to $\delta_{i-1} + 2M(x_i - x_{i-1})$, and we have

$$M_i - m_i < \lambda + 2MK(x_i - x_{i-1}) + K\delta_{i-1}.$$

If we take all the intervals so small that each of the products $2MK(x_i - x_{i-1})$ is less than λ, the difference $M_i - m_i$ will be less than $2\lambda + K\delta_{i-1}$, and consequently we shall have the inequality

(56) $$\delta_i < \delta_{i-1}[1 + K(x_i - x_{i-1})] + 2\lambda(x_i - x_{i-1}),$$

which can be written in the form

$$\delta_i + \frac{2\lambda}{K} < \left(\delta_{i-1} + \frac{2\lambda}{K}\right)[1 + K(x_i - x_{i-1})].$$

We have, therefore, a fortiori,

$$\delta_i + \frac{2\lambda}{K} < e^{K(x_i - x_{i-1})}\left(\delta_{i-1} + \frac{2\lambda}{K}\right).$$

Putting $i = 1, 2, \cdots, n$ successively in this last inequality and multiplying the two sides of the inequalities obtained, we find

$$\delta_n + \frac{2\lambda}{K} < \frac{2\lambda}{K} e^{K(x - x_0)},$$

or

$$S - s = \delta_n < \frac{2\lambda}{K}[e^{K(x - x_0)} - 1].$$

Since it is possible to take the positive number λ as small as we wish, provided that all the partial intervals are themselves less than another suitably chosen positive number, we see that the sums S and s have the same limit. That limit is a function of x, say $F(x)$, defined in the interval $(x_0, x_0 + h)$. We shall now show that this function $F(x)$ is an integral of the given equation (51), and that it reduces to y_0 for $x = x_0$. In showing this we shall continue to make use of the geometric representation.

If all the partial intervals approach zero, not only the extremities of the two broken lines L and l approach a *limit point*, but the lines themselves approach a limiting curve. Any straight line parallel to BC meets the line L in a point P, and the line l in a point p, and the distance Pp is less than $S - s$. From the properties of these broken lines, all the points P have their ordinates greater than the ordinates of the corresponding points p; and since the distance Pp approaches zero, it follows that the points P and p approach a single limit point π lying on the line considered. The locus of these points, π, is evidently a curve C lying between the two broken lines L and l and passing through the point A. The ordinate of a point of that curve with the abscissa x is equal to the function $F(x)$ just defined, for in order to obtain the position of the point π on the line $X = x$, we make use of only the portions of the broken lines which are on the left of that line. Let us suppose the two broken lines L and l produced up to the side BC, all the partial intervals being less than the smaller of the two numbers σ, $\lambda/(2MK)$, and let $P(x)$ and $Q(x)$ be two continuous functions which represent the ordinates of a point of the line L and of the line l in the interval $(x_0, x_0 + h)$. The difference $P(x) - Q(x)$ is less than $2\lambda(e^{Kh} - 1)/K$, and each of the functions $P(x)$, $Q(x)$ differs from $F(x)$ by a still smaller quantity. Since λ can be made as small as we wish, we see that we can construct a uniformly convergent series of continuous functions in the interval $(x_0, x_0 + h)$ which has $F(x)$ for its sum; this function is therefore itself continuous (see Vol. I, § 31, 2d ed.; § 173, 1st ed.).

Every broken line included between L and l has evidently the same curve C for its limit. Such would be the broken line Λ, whose successive vertices have the coördinates obtained by the recurrent formula

$$z_i = z_{i-1} + f(x_{i-1}, z_{i-1})(x_i - x_{i-1}),$$

the first vertex being the point (x_0, y_0). Thus we find again the expressions (54) which served as our starting point. Let us notice also that if we apply the construction starting with a point $M'(x', y')$ on the curve C, we obtain two broken lines L' and l' lying between L and l, which also approach more and more the portion of C included between M' and the straight line BC. Let now $M'(x', y')$ and $M''(x'', y'')$ be two neighboring points of C ($x'' > x'$). The slope of the straight line $M'M''$ lies between the maximum and minimum values of $f(x, y)$ when the point (x, y) moves over the triangle formed by the straight lines

$$X = x'', \qquad Y - y' = M(X - x'), \qquad Y - y' = -M(X - x');$$

if the difference $x'' - x'$ is less than a suitably chosen positive number, these two values of $f(x, y)$ will differ from $f(x', y')$ and from $f(x'', y'')$ by as little as we wish. If one of the two points, M'' for example, approaches the first one as a limit, the slope of $M'M''$ will therefore have for its limit $f(x', y')$. The function $F(x)$ consequently satisfies the given differential equation (51). It is, moreover, evident that the curve C passes through the point A, that is, that we have $F(x_0) = y_0$.

The curve C is the only solution of the problem. If there existed a second solution C', this curve C' could not be at the same time below all the lines L and above all the lines l, since these lines approach the curve C. We can therefore find a line — for example, L — which will be cut by this curve C'. Since C' is below the line L in the neighborhood of the point A, let us suppose that it passes above L, crossing that line in a point n_i of the side $P_{i-1}P_i$, and let m_{i-1} be the point of C' with the abscissa x_{i-1}. The slope of the chord $m_{i-1}n_i$ is equal, by the law of the mean, to the value of the function $f(x, y)$ at a point of the arc $m_{i-1}n_i$; hence this slope cannot be greater than the slope of the side $P_{i-1}P_i$, since the arc $m_{i-1}n_i$ is in the trapezoid $P_{i-1}Q_iq_ip_{i-1}$. But the figure shows that the slope of the chord must be the greater.

Cauchy's first method and that of the successive approximations give, as we see, the same limit for the interval in which the integral surely exists. But from a theoretical point of view Cauchy's method is unquestionably superior: we shall show, in fact, that this method enables us to find the integral in every finite interval in which the integral is continuous. More precisely, let us suppose that the equation (51) has an integral $y = F(x)$ continuous in the interval $(x_0, x_0 + l)$, that the function $f(x, y)$ is itself continuous in the region (E) of the xy-plane bounded by the two straight lines $x = x_0$, $x = x_0 + l$ and by the two curves $Y = F(x) \pm \eta$, where η is a positive number taken at pleasure, and that $f(x, y)$ satisfies the condition (52) in this region. Let us suppose that we divide the interval $(x_0, x_0 + l)$ into smaller partial intervals and that we construct the broken line Λ by the method which has just been explained, relative to this manner of division and starting from the point (x_0, y_0). *If all the partial intervals are less than a suitable positive number σ, this broken line will lie entirely in the region (E), and the difference of the ordinates of two points having the same abscissa, taken on the integral curve C and on the line Λ, will be less than any positive number ϵ given in advance.*

Let $x_0, x_1, x_2, \cdots, x_{i-1}, x_i \cdots, x_{n-1}, x_0 + l$ be the abscissas of the points of division, let $y_0, y_1, \cdots, Y$ be the corresponding ordinates of the curve C, and let $y_0, z_1, z_2, \cdots, z_i, \cdots, z_n$ be the ordinates of the vertices of the line Λ. Let us first suppose that all the vertices to the left of the vertex (x_i, z_i) are in the region (E), and let us consider the problem of calculating an upper bound for the difference $d_i = |z_i - y_i|$.

We have, on the one hand, from the very definition of Λ,

$$z_i = z_{i-1} + f(x_{i-1}, z_{i-1})(x_i - x_{i-1}).$$

On the other hand, from the law of the mean, we have also

$$y_i = y_{i-1} + f(x_i', y_i')(x_i - x_{i-1}),$$

where (x_i', y_i') are the coördinates of a point of C, and where x_i' lies between x_{i-1} and x_i. We derive from these equations

(57) $\quad z_i - y_i = z_{i-1} - y_{i-1} + (x_i - x_{i-1})[f(x_{i-1}, z_{i-1}) - f(x_i', y_i')];$

and the coefficient of $(x_i - x_{i-1})$ can be written in the form

$$[f(x_{i-1}, z_{i-1}) - f(x_{i-1}, y_{i-1})] + [f(x_{i-1}, y_{i-1}) - f(x_i', y_i')].$$

The absolute value of the first difference is, by the condition (52), less than $K d_{i-1}$. On the other hand, since the function $f(x, y)$ is continuous in the region (E), it is a continuous function of x along C, and we can find a positive number σ so small that $|f(x, y) - f(x', y')|$ is less than a given positive number 2λ for any two points of the curve C, provided that $|x - x'|$ is less than σ. Having chosen the number σ in this way, we have

(58) $\quad\quad d_i < d_{i-1} + (x_i - x_{i-1})(2\lambda + K d_{i-1}),$

a relation which is very similar to the relation (56), and from which we obtain, as before, the inequality

$$d_i < \frac{2\lambda}{K}[e^{K(x_i - x_0)} - 1].$$

Let us suppose that the number λ is so small that we have $2\lambda(e^{Kl} - 1) < K\eta$. We may then establish, step by step, that each of the differences $d_1, d_2, \cdots, d_n$ is less than η. All the vertices of the broken line Λ are therefore in the region (E).

Let $P(x)$ be the ordinate of a point of the line Λ; similarly, let $Q(x)$ be the ordinate of a point of the auxiliary broken line Λ' obtained by joining the points of C having the abscissas $x_0, x_1, x_2, \cdots, x_{n-1}, x_0 + l$. Then we have

$$P(x) - F(x) = P(x) - Q(x) + Q(x) - F(x).$$

If the oscillation of the function $F(x)$ in each of the partial intervals is less than $\epsilon/2$, we have always $|Q(x) - F(x)| < \epsilon/2$ (see Vol. I, § 206, 2d ed.; § 199, 1st ed.). If also the number η is less than $\epsilon/2$, we have $|P(x) - Q(x)| < \epsilon/2$, and therefore $|P(x) - F(x)| < \epsilon$. Then the continuous function $P(x)$ represents the function $F(x)$ with an error less than ϵ in the whole interval $(x_0, x_0 + l)$.

The Cauchy-Lipschitz method can be extended to systems of differential equations without any other difficulty than some complications in the formulæ. It applies also to complex variables. The investigations of E. Picard and of Painlevé have shown that the method leads to developments of the integrals in convergent series in the whole region of their existence if the right-hand sides of the given equations remain analytic in this region.

III. FIRST INTEGRALS. MULTIPLIERS

31. First integrals. Given a system of $n - 1$ *analytic* differential equations of the first order, we shall write these equations in the symmetric form

(59) $$\frac{dx_1}{X_1} = \frac{dx_2}{X_2} = \cdots = \frac{dx_n}{X_n},$$

where the denominators $X_1, X_2, \cdots, X_n$ are functions of the n variables $x_1, x_2, \cdots, x_n$. This form of the equations does not involve a choice of the independent variable, which may be any one of the variables or may be chosen arbitrarily. We have seen above that, under certain conditions which have been defined, all the integrals of this system which pass through any point of a region D are represented by a system of equations of the form

(60) $$\begin{cases} f_1(x_1, x_2, \cdots, x_n) = C_1, & f_2(x_1, x_2, \cdots, x_n) = C_2, \quad \cdots, \\ f_{n-1}(x_1, x_2, \cdots, x_n) = C_{n-1}, \end{cases}$$

where $f_1, f_2, \cdots, f_{n-1}$ are $(n-1)$ functions analytic in D, and where $C_1, C_2, \cdots, C_{n-1}$ are constants which may be arbitrarily chosen, at least within certain limits (§ 26). The formulæ (60) represent the *general integral* of the system (59) in the region D; but there may be other values of the variables also, for which (60) represents the solution. It may happen that we obtain several different systems of formulæ representing the general integral in different regions. It is also clear that, in the same region D, the system of equations (60) is not the only possible representation. We can replace the $(n-1)$ functions f_i by $(n-1)$ functions F_i which depend only upon the functions f_i, provided that these $(n-1)$ functions F_i are independent functions of the variables f_i.

However the functions f_i have been taken, if the formulæ (60) represent the general integral of the system (59), the functions f_i satisfy the same partial differential equation of the first order. For, let us suppose the coördinates of a point $x_1, x_2, \cdots, x_n$ of an integral curve expressed as functions of a variable parameter. If we replace the coördinates $x_1, x_2, \cdots, x_n$ in f_i by their expressions as functions of this parameter, the result reduces to a constant. We have, therefore, $df_i = 0$, and, replacing the differentials $dx_1, dx_2, \cdots$ in df_i by the proportional quantities $X_1, X_2, \cdots$, we find that f_i satisfies the relation

$$(61) \quad X(f) = X_1 \frac{\partial f}{\partial x_1} + X_2 \frac{\partial f}{\partial x_2} + \cdots + X_n \frac{\partial f}{\partial x_n} = 0.$$

This relation must reduce to an identity, when f is replaced by f_i, since we can choose the constants C_i in such a way that the integral curve passes through any point of D. The $(n-1)$ functions $f_1, f_2, \cdots, f_{n-1}$ are therefore $(n-1)$ integrals of the equation $X(f) = 0$. Every function $\Pi(f_1, f_2, \cdots, f_{n-1})$ is also an integral of the same equation, whatever may be the function Π, by the relation

$$X(\Pi) = \frac{\partial \Pi}{\partial f_1} X(f_1) + \frac{\partial \Pi}{\partial f_2} X(f_2) + \cdots + \frac{\partial \Pi}{\partial f_{n-1}} X(f_{n-1}),$$

which is easily verified.

Conversely, we obtain in this way all the integrals of the equation $X(f) = 0$. For, eliminating the coefficients X_i from the n relations

$$X(f) = 0, \quad X(f_1) = 0, \quad \cdots, \quad X(f_{n-1}) = 0,$$

we obtain

$$\frac{D(f, f_1, f_2, \cdots, f_{n-1})}{D(x_1, x_2, \cdots, x_n)} = 0,$$

which shows that f is a function $\Pi(f_1, f_2, \cdots, f_{n-1})$ of the $(n-1)$ particular integrals $f_1, f_2, \cdots, f_{n-1}$ (I, § 55, 2d ed.; § 28, 1st ed.). We can also verify this by a change of variables. Let us suppose, in fact, that we take a new system of independent variables $y_1, y_2, \cdots, y_n$, where the $n-1$ variables $y_1, y_2, \cdots, y_{n-1}$ are precisely the functions $f_1, f_2, \cdots, f_{n-1}$ themselves, and where the variable y_n is chosen in such a way as to form with $y_1, y_2, \cdots, y_{n-1}$ a system of n independent functions of the original variables $x_1, x_2, \cdots, x_n$. Then the equation $X(f) = 0$ is replaced by an equation of the same form

$$(62) \qquad Y(f) = Y_1 \frac{\partial f}{\partial y_1} + Y_2 \frac{\partial f}{\partial y_2} + \cdots + Y_n \frac{\partial f}{\partial y_n} = 0,$$

which must have the $(n-1)$ particular integrals

$$f = y_1, \quad \cdots, \quad f = y_{n-1}.$$

We have, therefore,

$$Y_1 = Y_2 = \cdots = Y_{n-1} = 0,$$

and the equation (62) reduces to $\partial f/\partial y_n = 0$. The general integral is therefore an arbitrary function of $y_1, y_2, \cdots, y_{n-1}$.*

The integration of the partial differential equation $X(f) = 0$ is therefore reduced to the integration of the proposed system of differential equations (59). Conversely, let us suppose that we have obtained an integral f of the equation $X(f) = 0$ in any manner whatever. If we replace $x_1, x_2, \cdots, x_n$ in that function by the coördinates of a point of an integral curve, supposed to be expressed as functions of a variable parameter which may be one of the coördinates themselves, *the result obtained reduces to a constant*. In fact, if we suppose that $x_1, x_2, \cdots, x_n$ are functions of a variable parameter satisfying the relations (59), the total differential df of the preceding function reduces to $KX(f)$, where K denotes the common value of the ratios dx_i/X_i. The equation $f = C$ is therefore a consequence of the given system of differential equations. For this reason we say that the function f is a *first integral* of that system.†

If we know $n-1$ independent first integrals, we can write immediately the general integral of the system (59); if we know only p independent first integrals ($p < n-1$), we can reduce the integration of the given system to the integration of a system of $n-p-1$ differential equations. For, let $f_1, f_2, \cdots, f_p$ be these p first integrals. From the p relations

$$f_1 = C_1, \quad f_2 = C_2, \quad \cdots, \quad f_p = C_p$$

---

* The two modes of reasoning do not require that the function f should be analytic. The only necessary conditions are those which are required in order that we may apply the formulæ for change of variables, that is, the existence and the continuity of the partial derivatives of the desired function f.

† The reasoning would no longer apply if the factor K were infinite for all the points of the integral curve, which would be the case if the coördinates of all the points of that curve were to make the n functions X_i vanish. It is also necessary to make an exception of the integrals which are such that at least one of the functions $X_1, X_2, \cdots, X_n$ is not analytic in the neighborhood of any point of that curve. This case arises when there are singular integrals.

we can obtain p of the variables $x_1, x_2, \cdots, x_n$, for example, $x_1, x_2, \cdots, x_p$, as functions of the remaining $n-p$ variables $x_{p+1}, \cdots, x_n$ and the p arbitrary constants $C_1, C_2, \cdots, C_p$. It will suffice, then, to determine $x_{p+1}, x_{p+2}, \cdots, x_n$ as functions of a single independent variable. If we denote by $\overline{X_{p+1}}, \overline{X_{p+2}}, \cdots, \overline{X_n}$ the new functions resulting from $X_{p+1}, X_{p+2}, \cdots, X_n$ after we have replaced $x_1, x_2, \cdots, x_p$ in them by their expressions, it will suffice, therefore, to integrate the new system,

$$(63) \qquad \frac{dx_{p+1}}{\overline{X}_{p+1}} = \frac{dx_{p+2}}{\overline{X}_{p+2}} = \cdots = \frac{dx_n}{\overline{X}_n},$$

in which the new denominators depend upon p arbitrary constants.

We can also reason in another way. If we take a new system of independent variables, $y_1, y_2, \cdots, y_n$, where the p variables $y_1, y_2, \cdots, y_p$ are identical with the p known first integrals $f_1, f_2, \cdots, f_p$, the equation $X(f) = 0$ is replaced by an equation of the same form, $Y(f) = 0$, which must have for integrals $f = y_1, \cdots, f = y_p$. That equation is therefore of the form

$$Y_{p+1}\frac{\partial f}{\partial y_{p+1}} + \cdots + Y_n \frac{\partial f}{\partial y_n} = 0,$$

and its integration reduces to that of a system of $n - p - 1$ differential equations of the first order,

$$\frac{dy_{p+1}}{Y_{p+1}} = \cdots = \frac{dy_n}{Y_n}.$$

We see from this the importance of looking for first integrals. In each particular case the discovery of a new first integral constitutes a step farther toward the complete solution. It would not be possible to give a very definite rule of procedure for this purpose. Let us merely notice that the problem amounts to forming an *integrable combination* of the equations (59), that is, to determining n factors, $\mu_1, \mu_2, \cdots, \mu_n$, so that

and that
$$\mu_1 X_1 + \mu_2 X_2 + \cdots + \mu_n X_n = 0,$$

$$\mu_1 dx_1 + \mu_2 dx_2 + \cdots + \mu_n dx_n$$

is an exact differential $d\phi$. For it is clear that we can deduce from the equations (59) a new ratio equal to the first

$$\frac{dx_i}{X_i} = \frac{\mu_1 dx_1 + \cdots + \mu_n dx_n}{\mu_1 X_1 + \cdots + \mu_n X_n};$$

hence the relation
$$d\phi = \mu_1 dx_1 + \cdots + \mu_n dx_n = 0$$

is a consequence of the equations (59) if

$$\mu_1 X_1 + \cdots + \mu_n X_n = 0.$$

It follows that we can find a first integral by quadratures if we know the factors μ_i. This is the case in particular whenever we can find n factors, $\mu_1, \mu_2, \cdots, \mu_n$, such that the factor μ_i depends only upon the variable x_i, and such that

$$\Sigma \mu_i X_i = 0.$$

Let us also observe that, if we have obtained p first integrals of the system (59), it may happen that the new system (63) can be integrated completely for particular numerical values of the constants $C_1, C_2, \cdots, C_p$, while the actual integration is impossible for arbitrary values of these constants.

Example 1. Let it be required to integrate the system

(64) $$\frac{du}{dx} = vw, \qquad \frac{dv}{dx} = wu, \qquad \frac{dw}{dx} = uv.$$

We easily see two integrable combinations $u\,du = v\,dv = w\,dw$. We have, therefore, two first integrals, $u^2 - v^2 = C_1$, $u^2 - w^2 = C_2$. Hence, putting the values of v and of w obtained from these relations in the first of the equations (64), we have for the determination of u the differential equation

(65) $$\frac{du}{dx} = \sqrt{(u^2 - C_1)(u^2 - C_2)},$$

the general integral of which is an elliptic function (§ 11), reducing in special cases to a simply periodic function or even to a rational function. Since the given system is symmetric in u, v, w, we conclude that v and w are also elliptic functions.

Example 2. Let us consider the system

(66) $$\frac{du}{dx} = rv - qw, \qquad \frac{dv}{dx} = pw - ru, \qquad \frac{dw}{dx} = qu - pv,$$

where p, q, r are given functions of x. We have again an integrable combination, $u\,du + v\,dv + w\,dw = 0$, from which we derive the first integral, $u^2 + v^2 + w^2 = C$. Discarding the case where C is zero, we may suppose $C = 1$, for the system (66) is not changed by multiplying u, v, w by the same constant factor. Instead of solving the relation $u^2 + v^2 + w^2 = 1$ for one of the unknowns, we can proceed in a more symmetric manner by considering u, v, w as the coördinates of a point of a sphere of radius *unity* and expressing them as functions of two variable parameters — for example, in terms of the parameters which determine the rectilinear generators of the sphere. Let us put for that purpose

$$\frac{u + iv}{1 - w} = \frac{1 + w}{u - iv} = \lambda, \qquad \frac{u + iv}{1 + w} = \frac{1 - w}{u - iv} = -\mu,$$

which gives

$$u = \frac{1 - \lambda\mu}{\lambda - \mu}, \qquad v = i\frac{1 + \lambda\mu}{\lambda - \mu}, \qquad w = \frac{\lambda + \mu}{\lambda - \mu}.$$

Substituting these values of u, v, w in the system (66), we find after some easy calculations that λ and μ must satisfy the same Riccati equation,

(67) $$\frac{d\sigma}{dz} = -ir\sigma + \frac{q-ip}{2} + \frac{q+ip}{2}\sigma^2.$$

Hence the integration of the given system is reduced to the integration of a Riccati equation.*

Example 3. Let us consider the equation integrated by Liouville,

$$y'' + \phi(x)y' + f(y)y'^2 = 0.$$

Putting $y' = z$, we may replace the given equation by the system

$$\frac{dx}{1} = \frac{dy}{z} = \frac{-dz}{\phi(x)z + f(y)z^2},$$

from which we derive the integrable combination $dz/z + \phi(x)dx + f(y)dy = 0$. The given equation of the second order has therefore the first integral,

$$y' e^{\int_{x_0}^{x} \phi(x)dx} e^{\int_{y_0}^{y} f(y)dy} = C,$$

which we could also have obtained directly by dividing all the terms of the equation of the second order by y'. The preceding equation of the first order is of the form $y' = CXY$; hence, by separating the variables, the integration may be completed by two quadratures.

Note 1. We sometimes replace the system (59) by the system

(68) $$\frac{dx_1}{X_1} = \frac{dx_2}{X_2} = \cdots = \frac{dx_n}{X_n} = dt,$$

where t is an auxiliary variable which is introduced in many cases only for the sake of greater symmetry in the reasoning. If the original system (59) has been integrated, we can obtain t by a quadrature, for if we replace $x_2, x_3, \cdots, x_n$, for example, by their expressions in terms of x_1 and of the constants $C_1, C_2, \cdots, C_{n-1}$ in X_1, we are led to a relation,

$$dt = P(x_1, C_1, C_2, \cdots, C_{n-1})dx_1,$$

from which we can find t by a quadrature. It follows from this that the general integral of the new system (68) will be represented by the n equations of the form

(69) $$\begin{cases} f_1 = C_1, & f_2 = C_2, & \cdots, & f_{n-1} = C_{n-1}, \\ f_n(x_1, x_2, \cdots, x_n) = t - t_0, \end{cases}$$

where $f_1, f_2, \cdots, f_{n-1}$ are $(n-1)$ independent integrals of $X(f) = 0$, and where t_0 is a new arbitrary constant.

Conversely, in order to obtain the integral curve of the system (59) that passes through the given point $x_1^0, x_2^0, \cdots, x_n^0$, we can look for the integrals of

---

* See DARBOUX, *Théorie des surfaces*, Vol. I, chap. ii.

the system (68), where t is considered as the independent variable, which for $t=0$ take on the values $x_1^0, x_2^0, \cdots, x_n^0$ respectively. Let

(70) $$\begin{cases} x_1 = \phi_1(t; x_1^0, \cdots, x_n^0), \quad x_2 = \phi_2(t; x_1^0, \cdots, x_n^0), \quad \cdots, \\ x_n = \phi_n(t; x_1^0, \cdots, x_n^0) \end{cases}$$

be these integrals; it is clear that the preceding expressions represent the integral curve sought. We should have to make an exception only if all the functions X_i were zero for the initial values x_i^0 and analytic in the neighborhood. In this case the expressions (70) should reduce to $x_i = x_i^0$. But, since the ratios $dx_2/dx_1, \cdots, dx_n/dx_1$ appear in an indeterminate form, nothing justifies us so far in saying that there is no integral curve passing through the given point. This is a case which will be examined later (§ 75).

Note 2. The relation which exists between the system of differential equations (59) and the linear equation (61) proves that $X(f)$ is a *covariant* of the system (59). The meaning of this statement is as follows: Let us suppose that we take a new system of independent variables, $y_1, y_2, \cdots, y_n$, connected with the variables $x_1, x_2, \cdots, x_n$ by the relations

(71) $$x_i = \phi_i(y_1, y_2, \cdots, y_n). \qquad (i = 1, 2, \cdots, n)$$

By the formulæ for change of variables, $\partial f/\partial x_i$ is a linear homogeneous function of the derivatives $\partial f/\partial y_i$, and $X(f)$ changes into an expression of the same form,

(72) $$Y(f) = Y_1 \frac{\partial f}{\partial y_1} + Y_2 \frac{\partial f}{\partial y_2} + \cdots + Y_n \frac{\partial f}{\partial y_n} = 0,$$

where $Y_1, Y_2, \cdots, Y_n$ are functions of $y_1, y_2, \cdots, y_n$. This being true, we may now assert that the same change of variables applied to the system (59) leads to the new system of differential equations,

(73) $$\frac{dy_1}{Y_1} = \frac{dy_2}{Y_2} = \cdots = \frac{dy_n}{Y_n}.$$

We could establish this by a direct calculation, but it results also from the preceding properties. In fact, let

(74) $$\frac{dy_1}{Z_1} = \frac{dy_2}{Z_2} = \cdots = \frac{dy_n}{Z_n}$$

be the system to which we are led by applying to the original system (59) the change of variables (71); it suffices to show that $Z_1, Z_2, \cdots, Z_n$ are proportional to $Y_1, Y_2, \cdots, Y_n$. Now let $f(x_1, x_2, \cdots, x_n)$ be a first integral of the system (59) and

$$F(y_1, y_2, \cdots, y_n)$$

the function derived from $f(x_1, x_2, \cdots, x_n)$ by the change of variables. Since we have $X(f) = 0$, we have also $Y(F) = 0$. Besides, $F(y_1, y_2, \cdots, y_n)$ is evidently a first integral of the new system (74), that is, an integral of the linear equation

$$Z(F) = Z_1 \frac{\partial F}{\partial y_1} + \cdots + Z_n \frac{\partial F}{\partial y_n} = 0.$$

Since the linear equations $Y(F) = 0$, $Z(F) = 0$ have the same integrals, their coefficients are proportional, which proves the theorem.

This last point in the proof results from the fact that a linear equation $X(f) = 0$ is completely determined, except for a factor, when we know $(n-1)$ *independent* integrals, $f_1, f_2, \cdots, f_{n-1}$, of it. In fact, the $(n-1)$ equations, $X(f_i) = 0$, linear and homogeneous in $X_1, X_2, \cdots, X_n$, determine the ratios of these coefficients as unknowns, for the determinants of order $(n-1)$ formed from the partial derivatives of the functions f_i cannot all be zero at the same time (I, § 55, 2d ed.; § 28, 1st ed.). It may be noticed that the most general linear equation having the $(n-1)$ integrals f_i can be written in the form

$$\Pi(x_1, x_2, \cdots, x_n) \frac{D(f, f_1, f_2, \cdots, f_{n-1})}{D(x_1, x_2, \cdots, x_n)} = 0,$$

where $\Pi(x_1, x_2, \cdots, x_n)$ is an arbitrary function.

32. Multipliers. The theory of integrating factors has been extended by Jacobi to simultaneous differential equations. Let $f_1, f_2, \cdots, f_{n-1}$ be independent first integrals of the system (59). The equation $X(f) = 0$ is, as we have already remarked, identical with the equation

$$\Delta = \frac{D(f, f_1, f_2, \cdots, f_{n-1})}{D(x_1, x_2, \cdots, x_n)} = 0.$$

Writing the condition that the coefficients of the derivatives $\partial f/\partial x_i$ in the two equations are proportional, we are led to n relations which may be written in the form

(75) $$\Delta_i = MX_i, \qquad (i = 1, 2, \cdots, n)$$

where Δ_i denotes the coefficient of $\partial f/\partial x_i$ in the determinant Δ. This factor M is called a *multiplier*.

Whatever the first integrals $f_1, f_2, \cdots, f_{n-1}$ may be, this function M satisfies the linear partial differential equation

(76) $$\frac{\partial (MX_1)}{\partial x_1} + \frac{\partial (MX_2)}{\partial x_2} + \cdots + \frac{\partial (MX_n)}{\partial x_n} = 0.$$

Substituting for each of the products $MX_i = \Delta_i$ its equivalent expression as a determinant of order $n-1$, and carrying out the indicated differentiations, each term of the left-hand side is, in fact, the product of a derivative of the second order, such as $\partial^2 f_h/\partial x_i \partial x_k (i \neq k)$, and $(n-2)$ partial derivatives of the first order. To prove that the result is zero, it suffices to show that it does not contain any derivatives of the second order. Let us take, for example, the derivative $\partial^2 f_1/\partial x_1 \partial x_2$. This derivative appears in two terms; in one it is multiplied by $D(f_2, f_3, \cdots, f_{n-1})/D(x_3, x_4, \cdots, x_n)$, and in the other by the same coefficient but with the opposite sign. The sum of these two terms is therefore zero, and similarly for all the others.

If M_1 is a particular integral of the equation (76), the substitution $M = M_1 \mu$ reduces that equation to the form $X(\mu) = 0$. If we know a multiplier M of the system (59), the general integral of the equation (76) is accordingly $M\Pi(f_1, f_2, \cdots, f_{n-1})$, where Π is an arbitrary function. Every function of this form is also a multiplier; in other words, there exist $(n-1)$ first integrals $F_1, \cdots, F_{n-1}$, such that $M\Pi(f_1, f_2, \cdots, f_{n-1})$ can be deduced from $F_1, F_2, \cdots, F_{n-1}$ in the same way that M was deduced from $f_1, f_2, \cdots, f_{n-1}$. For this purpose it is sufficient that we have, supposing $X_1 \neq 0$,

$$\frac{1}{X_1} \frac{D(F_1, F_2, \cdots, F_{n-1})}{D(x_2, x_3, \cdots, x_n)} = \frac{1}{X_1} \frac{D(F_1, F_2, \cdots, F_{n-1})}{D(f_1, f_2, \cdots, f_{n-1})} \frac{D(f_1, f_2, \cdots, f_{n-1})}{D(x_2, \cdots, x_n)} = M\Pi,$$

or

$$\frac{D(F_1, F_2, \cdots, F_{n-1})}{D(f_1, f_2, \cdots, f_{n-1})} = \Pi(f_1, f_2, \cdots, f_{n-1}).$$

This condition can be satisfied in an infinite number of ways. Indeed, $n-2$ of the first integrals F_i may be assigned arbitrarily in advance.

Let us consider the system

(77) $$\frac{dx_1}{X_1} = \frac{dx_2}{X_2} = \cdots = \frac{dx_n}{X_n} = dt,$$

with the auxiliary variable t. This system can be reduced to the simple form

(78) $$dy_1 = dy_2 = \cdots = dy_{n-1} = 0, \quad dy_n = dt$$

by taking for the variables the $n-1$ first integrals $f_1, f_2, \cdots, f_{n-1}$ and the function f_n, which appears in the preceding formulæ (69). It is easy to obtain the general expression for the multipliers in terms of the variables y_i, for every multiplier is of the form

$$M = \frac{1}{X_1} \frac{D(y_1, y_2, \cdots, y_{n-1})}{D(x_2, x_3, \cdots, x_n)} \Pi(y_1, y_2, \cdots, y_{n-1}).$$

On the other hand, we have

$$X_1 = \frac{dx_1}{dt} = \frac{\partial x_1}{\partial y_1} \frac{dy_1}{dt} + \cdots + \frac{\partial x_1}{\partial y_n} \frac{dy_n}{dt} = \frac{\partial x_1}{\partial y_n}.$$

From the relations $y_1 = f_1, \cdots, y_n = f_n$, which define the change of the variables, we derive, by differentiating with respect to y_n and solving,

$$\frac{\partial x_1}{\partial y_n} = (-1)^{n-1} \frac{\dfrac{D(y_1, y_2, \cdots, y_{n-1})}{D(x_2, x_3, \cdots, x_n)}}{\dfrac{D(y_1, y_2, \cdots, y_n)}{D(x_1, x_2, \cdots, x_n)}},$$

and the general expression for the multiplier can be written in the form

(79) $$\frac{1}{M} = \frac{D(x_1, x_2, \cdots, x_n)}{D(y_1, y_2, \cdots, y_n)} \Phi(y_1, y_2, \cdots, y_{n-1}),$$

where Φ is an arbitrary function of $y_1, y_2, \cdots, y_{n-1}$.

Let us suppose, now, that after carrying out any change of variables affecting only the x_i's without changing the variable t, we have reduced the system (77) to the form

(80) $$\frac{dx_1'}{X_1'} = \frac{dx_2'}{X_2'} = \cdots = \frac{dx_n'}{X_n'} = dt,$$

where the X_i''s are functions of the new variables x_i' independent of t. If M' is a multiplier of this new system, we have

(81) $$\frac{1}{M'} = \frac{D(x_1', x_2', \cdots, x_n')}{D(y_1, y_2, \cdots, y_n)} \Phi(y_1, y_2, \cdots, y_{n-1}).$$

Taking the same function Φ in the two expressions, we derive from them, by dividing their corresponding sides, the relation

(82) $$M' = M \frac{D(x_1, x_2, \cdots, x_n)}{D(x_1', x_2', \cdots, x_n')}.$$

Hence, *if we know a multiplier M for the system* (77), *we can derive from it a multiplier M' for the transformed system.*

This property explains the practical importance of multipliers. Let us suppose that we know $n-2$ first integrals of the system (59), and also a multiplier. We can then reduce this system to the form

$$\frac{dx_1'}{0} = \cdots = \frac{dx_{n-2}'}{0} = \frac{dx_{n-1}'}{X_{n-1}'} = \frac{dx_n'}{X_n'} = dt$$

by a change of variables, and we can then find a multiplier M' for this new system, that is, a solution of the equation

$$\frac{\partial(M'X_{n-1}')}{\partial x_{n-1}'} + \frac{\partial(M'X_n')}{\partial x_n'} = 0.$$

It follows that M' is an integrating factor for $X_n' dx_{n-1}' - X_{n-1}' dx_n'$, and the integration can be finished by quadratures.

A particular case which presents itself frequently in mechanics is the one for which we have $\Sigma \partial X_i / \partial x_i = 0$. The equation (76) reduces then to $X(M) = 0$, and we know at once a multiplier $M=1$.

This remark applies also to the equation of the second order, $y'' = f(x, y)$, the integration of which leads to that of the system

$$\frac{dx}{1} = \frac{dy}{y'} = \frac{dy'}{f(x,y)}.$$

If we know a first integral of it, $\psi(x, y, y') = C$, we can, from what precedes, finish the integration by quadratures. This is easily verified as follows: Let us suppose that the equation $\psi(x, y, y') = C$ has been solved for y':

$$y' = \phi(x, y, C).$$

Since all the integrals of this equation of the first order must satisfy the equation $y''=f(x,y)$, whatever may be the constant C, we must have $\partial \phi/\partial x + (\partial \phi/\partial y)\phi = f$. Hence, since f does not contain C,

$$\frac{\partial^2 \phi}{\partial C \partial x} + \frac{\partial^2 \phi}{\partial C \partial y}\phi + \frac{\partial \phi}{\partial y}\frac{\partial \phi}{\partial C} = 0,$$

which states that $\partial \phi / \partial C$ is an integrating factor for $dy - \phi dx$.

33. Invariant integrals. The invariant property of the multipliers relative to every change of variables can be brought into relation with the general theory of *invariant integrals*, due to Poincaré,[*] and about which we shall say a few

---

[*] *Les méthodes nouvelles de la Mécanique céleste*, Vol. III, chap. xxii, and the following chapters. See also GOURSAT, *Sur les invariants intégraux*, in *Journal de Mathématiques*, 6th series, Vol. IV.

words. Let us consider in particular a system of three differential equations,

(83) $$\frac{dx}{X} = \frac{dy}{Y} = \frac{dz}{Z} = dt,$$

where X, Y, Z are functions of x, y, z. In order to simplify the statements, we shall regard these equations as defining the movement of a particle in space, where the variable t represents the time. The particle which, at the time $t = 0$, is at a point $M_0(x_0, y_0, z_0)$ has arrived at the time t at a point M_t whose coördinates are (x, y, z). If the point M_0 describes a certain region D_0 of space, the point M_t describes a corresponding region D_t. Now let $M(x, y, z)$ be a function of the variables x, y, z; we shall say that the triple integral

$$I = \iiint M(x, y, z)\, dx\, dy\, dz$$

is an *invariant integral* of the system (83) if the value of that triple integral,

$$\iiint_{D_t} M(x, y, z)\, dx\, dy\, dz,$$

extended over the region D_t, is independent of t and equal to the same integral extended over the region D_0. For example, if the equations (83) define the movement of an *incompressible* fluid, the volume of the region D_t is constant and the integral $\iiint dx\, dy\, dz$ is an invariant integral.

Invariant line and surface integrals are defined in a similar way. If the point M_0 describes a curve L_0 or a surface Σ_0, the point M_t describes a curve L_t or a surface Σ_t. A line integral

$$\int \alpha\, dx + \beta\, dy + \gamma\, dz$$

is an invariant integral if the value of that integral along the curve L_t is independent of t and equal to the same line integral taken along L_0. Similarly, a surface integral

$$\iint P\, dy\, dz + Q\, dz\, dx + R\, dx\, dy$$

is an invariant integral if the value of that integral extended over the surface Σ_t is independent of t.

These notions can be extended without difficulty to the most general systems of differential equations of the form (68). For such a system there are n classes of invariant integrals, of the 1st order, of the 2d order, $\cdots$, of the nth order, according to the order of multiplicity of the integral considered. The conditions that a multiple integral of order p shall be an invariant integral are easily obtained by means of the formulæ for the change of variables in multiple integrals. We shall develop the calculations for a multiple integral of order n. Let

$$I(t) = \iint \cdots \int_{D_t} M(x_1, x_2, \cdots, x_n)\, dx_1\, dx_2 \cdots dx_n$$

be a multiple integral of order n extended over the region D_t which corresponds to a definite region D_0 in the manner just explained. This integral will be an invariant integral if it is independent of t; that is, if we have $I'(t) = 0$. In order to calculate that derivative, we shall give to t an increment h, and we shall calculate the coefficient of h in the development of $I(t+h)$. Let x_i' be the value to which x_i changes when we change t to $t+h$; we have

$$I(t+h) = \iint \cdots \int_{D_t'} M(x_1', x_2', \ldots, x_n') dx_1' dx_2' \cdots dx_n',$$

where the new integral is extended over the region D_t', which corresponds point for point to D_t. Then we may write

$$I(t+h) = \iint \cdots \int_{D_t} M(x_1', \ldots, x_n') \frac{D(x_1', x_2', \ldots, x_n')}{D(x_1, x_2, \ldots, x_n)} dx_1 dx_2 \cdots dx_n.$$

On the other hand, omitting the terms in h of degree higher than the first, we have

$$x_i' = x_i + h X_i + \cdots,$$

$$M(x_1', x_2', \ldots, x_n') = M(x_1, x_2, \ldots, x_n) + h\left(X_1 \frac{\partial M}{\partial x_1} + \cdots + X_n \frac{\partial M}{\partial x_n}\right) + \cdots,$$

$$\frac{D(x_1', x_2', \ldots, x_n')}{D(x_1, x_2, \ldots, x_n)} = \begin{vmatrix} 1+h\frac{\partial X_1}{\partial x_1} & h\frac{\partial X_1}{\partial x_2} & \cdots & h\frac{\partial X_1}{\partial x_n} \\ h\frac{\partial X_2}{\partial x_1} & 1+h\frac{\partial X_2}{\partial x_2} & \cdots & \cdots \\ \cdots & \cdots & \cdots & \cdots \end{vmatrix}$$

$$= 1 + h\left(\frac{\partial X_1}{\partial x_1} + \cdots + \frac{\partial X_n}{\partial x_n}\right) + \cdots,$$

and

$$M(x_1', x_2', \ldots, x_n') \frac{D(x_1', x_2', \ldots, x_n')}{D(x_1, x_2, \ldots, x_n)} = M(x_1, x_2, \ldots, x_n)$$
$$+ h\left[M\left(\frac{\partial X_1}{\partial x_1} + \cdots + \frac{\partial X_n}{\partial x_n}\right) + X_1 \frac{\partial M}{\partial x_1} + \cdots + X_n \frac{\partial M}{\partial x_n}\right] + \cdots.$$

The derivative dI/dt has therefore the value

$$\frac{dI}{dt} = \iint \cdots \int_{D_t} \left[\frac{\partial(MX_1)}{\partial x_1} + \cdots + \frac{\partial(MX_n)}{\partial x_n}\right] dx_1 dx_2 \cdots dx_n.$$

In order that I be an invariant integral, it is necessary and sufficient that dI/dt be identically zero, whatever may be the region D, and therefore that we have

(84) $$\frac{\partial(MX_1)}{\partial x_1} + \cdots + \frac{\partial(MX_n)}{\partial x_n} = 0.$$

This condition is identical with the equation (76), and we obtain Poincaré's theorem: *In order that the multiple integral*

$$\iint \cdots \int M dx_1 \cdots dx_n$$

shall be an invariant integral, it is necessary and sufficient that M be a multiplier.

It follows that if we make any change of variables,

$$x_i = \phi_i(y_1, y_2, \ldots, y_n), \qquad (i = 1, 2, \ldots, n)$$

in the equations (77), we obtain a new system,

(77′) $$\frac{dy_1}{Y_1} = \frac{dy_2}{Y_2} = \cdots = \frac{dy_n}{Y_n} = dt;$$

and if M is a multiplier of the system (77), the n-fold integral

$$\iint \cdots \int M \, dx_1 \, dx_2 \cdots dx_n$$

is an invariant integral of that system, and the n-fold integral which is obtained from it by the same transformation,

$$\iint \cdots \int M(x_1, \cdots, x_n) \frac{D(x_1, x_2, \cdots, x_n)}{D(y_1, y_2, \cdots, y_n)} dy_1 dy_2 \cdots dy_n,$$

is evidently also an invariant integral of the transformed system (77′). Therefore the expression

$$M' = M \frac{D(x_1, x_2, \cdots, x_n)}{D(y_1, y_2, \cdots, y_n)}$$

is a multiplier of the new equations (77′), as we have demonstrated directly.

Example. In order that the volume shall be an invariant integral of the equations (83), $M = 1$ must be a multiplier, which requires that we have

(85) $$\frac{\partial X}{\partial x} + \frac{\partial Y}{\partial y} + \frac{\partial Z}{\partial z} = 0.$$

This is the condition for the *incompressibility* of a fluid for which the equations (83) define a stationary flow.

IV. INFINITESIMAL TRANSFORMATIONS

34. One-parameter groups.[*] Every set of an infinite number of transformations, of any nature whatever, affecting the n variables $x_1, x_2, \cdots, x_n$, form a *group* if the transformation obtained by carrying out any two transformations of this set in succession belongs to the set. For definiteness let us consider two variables x, y, and let T be the transformation defined by the equations

(86) $$x' = f(x, y; a), \qquad y' = \phi(x, y; a),$$

where a denotes an arbitrary parameter. If we regard x and y as the coördinates of a point M in a plane, and x' and y' as the coördinates of another point M', the preceding equations define a point transformation. To each value of the parameter a corresponds thus a definite transformation. Varying this parameter, we obtain an infinite number of different transformations. Let us suppose that we carry out in succession two different transformations of this set, corresponding to any two values a and b of the parameter. The first transformation will carry the pair of values (x, y) over into the pair of values (x', y') given

---

[*] The theory of continuous groups of transformations was developed by Sophus Lie in a great number of papers and in his treatise, *Theorie der Transformationgruppen*.

by the equations (86). The second transformation will then carry the pair of values (x', y') over into a third pair (x'', y'') such that we have

(87) $\qquad x'' = f(x', y'; b), \qquad y'' = \phi(x', y'; b).$

Let us replace x' and y' in these last two equations by their values (86). The resulting equations,

(88) $\qquad x'' = F(x, y; a, b), \qquad y'' = \Phi(x, y; a, b),$

again define a point transformation depending upon the two parameters a and b. We shall say that the set of transformations (86) form a *continuous one-parameter group* if the new transformation (88) belongs to this set. It is necessary and sufficient for this that the equations (88) be of the form

(89) $\qquad x'' = f(x, y; c), \qquad y'' = \phi(x, y; c),$

where c is a value of the parameter depending only upon a and upon b; that is, $c = \psi(a, b)$. The preceding definition evidently applies whatever may be the number of variables, in particular if there is only a single variable.

The relation $x' = x + a$, or, any one of the pairs of relations

$$x' = x + a, \qquad y' = y + 2a;$$
$$x' = x \cos a - y \sin a, \qquad y' = x \sin a + y \cos a;$$
$$x' = ax, \qquad y' = a^2 y$$

represents a one-parameter group. On the contrary, the transformations $x' = x + a$, $y' = y + a^2$ do not form a group, for the transformation resulting from two successive transformations, $x'' = x + a + b$, $y'' = y + a^2 + b^2$, do not belong to the set.

If in the equations (86), which define a group of transformations, we put $a = \Pi(\alpha)$, where α is a new parameter, it is clear that the relations obtained again define a group. The same thing is true also if we make a change of variables, as we easily convince ourselves a priori. In fact, if a set of point transformations in a plane is such that the transformation resulting from two successive transformations belongs to the set, it is clear that this property is independent of the choice of the coördinates by means of which we fix the position of a point in the plane. It is easy to verify this directly. Let us suppose that we put $x = \Pi(u, v)$, $y = \Pi_1(u, v)$, and let the inverse relations be $u = \Pi^{-1}(x, y)$, $v = \Pi_1^{-1}(x, y)$, so that we have identically

$$x = \Pi[\Pi^{-1}(x, y), \Pi_1^{-1}(x, y)], \qquad y = \Pi_1[\Pi^{-1}(x, y), \Pi_1^{-1}(x, y)].$$

By hypothesis, the transformations considered form a group, and the equations (89), where $c = \psi(a, b)$, are a consequence of the equations (86) and (87). Let (u, v), (u', v'), (u'', v'') be the pairs of values of the new variables which correspond respectively to the pairs (x, y), (x', y'), (x'', y''). We have

(90) $\qquad \begin{cases} u' = \Pi^{-1}(x', y') = \Pi^{-1}\{f[\Pi(u, v), \Pi_1(u, v); a], \phi[\Pi(u, v), \Pi_1(u, v); a]\} \\ \qquad = F(u, v; a), \\ v' = \Pi_1^{-1}(x', y') = \Pi_1^{-1}\{f[\Pi(u, v), \Pi_1(u, v); a], \phi[\Pi(u, v), \Pi_1(u, v); a]\} \\ \qquad = \Phi(u, v; a); \end{cases}$

and everything depends on showing that the equations (90) also define a group of transformations. Now we have, for example, $u'' = F(u', v'; b)$, or

$$u'' = \Pi^{-1}\{f[\Pi(u', v'), \Pi_1(u', v'); b], \phi[\Pi(u', v'), \Pi_1(u', v'); b]\}.$$

Since the equations (86) define a group, this value of u'' is equal to
$$\Pi^{-1}[f(x', y'; b), \phi(x', y'; b)] = \Pi^{-1}[f(x, y; c), \phi(x, y; c)];$$
that is, to
$$\Pi^{-1}\{f[\Pi(u, v), \Pi_1(u, v); c], \phi[\Pi(u, v), \Pi_1(u, v); c]\} = F(u, v; c).$$

Similarly, we should find that $v'' = \Phi(u, v; c)$. Two groups of transformations which are carried over one into the other by a change of variables are said to be *similar*. For example, the two groups $x' = ax$, $u' = u + b$ are similar, for we pass from one to the other by putting $u = \log x$, $b = \log a$.

We shall now determine all possible one-parameter groups, supposing that the functions f and ϕ are analytic, and supposing also that the group contains the identical transformation, that is, that for a particular value a_0 of the parameter we have $f(x, y; a_0) = x$, $\phi(x, y; a_0) = y$, whatever x and y may be.

In the equations of condition

(91) $\qquad f(x', y'; b) = f(x, y; c), \qquad \phi(x', y'; b) = \phi(x, y; c)$

we can consider x, y, a, c as independent variables, and b as a function of a and c defined by the relation $c = \psi(a, b)$; x' and y' are functions of x, y, and a defined by the equations (86). Taking derivatives with respect to a, we derive from the relations (91)

(92) $\quad \dfrac{\partial f}{\partial x'}\dfrac{\partial x'}{\partial a} + \dfrac{\partial f}{\partial y'}\dfrac{\partial y'}{\partial a} + \dfrac{\partial f}{\partial b}\dfrac{\partial b}{\partial a} = 0, \qquad \dfrac{\partial \phi}{\partial x'}\dfrac{\partial x'}{\partial a} + \dfrac{\partial \phi}{\partial y'}\dfrac{\partial y'}{\partial a} + \dfrac{\partial \phi}{\partial b}\dfrac{\partial b}{\partial a} = 0.$

But $\partial b/\partial a$ is given by the relation $\partial \psi/\partial a + (\partial \psi/\partial b)(\partial b/\partial a) = 0$, and therefore depends only upon a and b. Solving the preceding equations (92) for $\partial x'/\partial a$, $\partial y'/\partial a$, we obtain, therefore, relations of the form

$$\frac{\partial x'}{\partial a} = \lambda(a, b)\, \xi(x', y', b), \qquad \frac{\partial y'}{\partial a} = \lambda(a, b)\, \eta(x', y', b).$$

Now x' and y' do not depend upon b; the same thing is therefore true of λ, ξ, η if they have been properly chosen. Therefore x' and y' are integrals of the system of differential equations,

(93) $\qquad \dfrac{dx'}{\xi(x', y')} = \dfrac{dy'}{\eta(x', y')} = \lambda(a)\, da,$

which for $a = a_0$ take on respectively the values x and y. Conversely, whatever the functions $\xi(x, y)$, $\eta(x, y)$ may be, the equations $x' = f(x, y, a)$, $y' = \phi(x, y, a)$, which represent the integrals of the preceding system which reduce respectively to x and to y for a particular value a_0 of the parameter, define a continuous group of transformations. In the first place, it will be simpler to introduce a new parameter,

$$t = \int_{a_0}^{a} \lambda(a)\, da,$$

which enables us to write the differential equations (93) in the abridged form

(94) $\qquad \dfrac{dx'}{\xi(x', y')} = \dfrac{dy'}{\eta(x', y')} = dt.$

The general integral of this system can be written, as we have seen above (§ 31), in the form
$$\Omega_1(x', y') = C_1, \qquad \Omega_2(x', y') = t + C_2,$$
where Ω_1 and Ω_2 are definite functions of x', y', and where C_1, C_2 are two arbitrary constants. The solutions which take on the values x and y for $t = 0$ are given by the system of equations

(95) $\qquad \Omega_1(x', y') = \Omega_1(x, y), \qquad \Omega_2(x', y') = \Omega_2(x, y) + t.$

The preceding expressions, indeed, define a continuous group, for if we carry out in succession the two transformations which correspond to the values t_1 and t_2 of the parameter, the resulting transformation corresponds to the value $t_1 + t_2$ of the parameter. The two transformations which correspond to the values t and $-t$ are the inverses of each other. If we have

$$x' = f(x, y; t), \qquad y' = \phi(x, y; t),$$

we may write also

$$x = f(x', y'; -t), \qquad y = \phi(x', y'; -t).$$

If we take for the new variables

$$u = \Omega_1(x, y), \qquad v = \Omega_2(x, y),$$

the equations (95) become

(96) $\qquad u' = u, \qquad v' = v + t;$

and we say that the group is transformed to the reduced form. *Every continuous one-parameter group is therefore similar to a group of translations.*

Let us take, for example, the group $x' = ax$, $y' = a^2 y$. Applying the general method, we have

$$\frac{\partial x'}{\partial a} = x = \frac{x'}{a}, \qquad \frac{\partial y'}{\partial a} = 2ay = 2\frac{y'}{a}.$$

The differential equations (93) are in this case

$$\frac{dx'}{x'} = \frac{dy'}{2y'} = \frac{da}{a} = dt,$$

where $t = \log a$. The finite equations of the group can be written in the form

$$\frac{y'}{x'^2} = \frac{y}{x^2}, \qquad \log x' = \log x + t,$$

and they will be brought to the reduced form by taking for the new variables $\log x$ and y/x^2.

35. Application to differential equations. Let us consider a given differential equation

(97) $\qquad F\left(x, y, \frac{dy}{dx}, \frac{d^2y}{dx^2}, \ldots, \frac{d^n y}{dx^n}\right) = 0,$

and a *known* one-parameter group of transformations of the form (86). Let us suppose that the equation (97) is identical with the equation obtained by carrying out on it the change of its variables x and y defined by the relations (86), whatever may be the numerical value of the parameter a. If this is the case, we shall say that the differential equation (97) *admits* the group of transformations (86). We can make use of this property to simplify the integration. In fact, let us suppose that we carry out a change of variables such that the equations which define the group in question are brought to the simple form $u' = u$, $v' = v + a$. The same change of variables, applied to the proposed differential equation, leads to a new differential equation of the nth order,

$$(98) \qquad \Phi\left(u, v, \frac{dv}{du}, \frac{d^2v}{du^2}, \cdots, \frac{d^n v}{du^n}\right) = 0,$$

which does not change if we replace in it v by $v + a$, whatever may be the numerical value of the constant a. This can happen only if the left-hand side Φ does not contain the variable v. If the equation is of the first order, we obtain the general integral by a quadrature. If $n > 1$, we can lower the order of the equation by unity by taking dv/du for the new unknown dependent variable.

Let us consider, for example, the homogeneous equation of the first order,

$$\frac{dy}{dx} = f\left(\frac{y}{x}\right).$$

This equation does not change if we replace x and y by ax and ay respectively, whatever may be the constant a. Now the formulæ $x' = ax$, $y' = ay$ define a group of transformations, which can also be written in the form

$$\frac{y'}{x'} = \frac{y}{x}, \qquad \log y' = \log y + t.$$

Hence if we put $y/x = u$, $\log y = v$, we are led to an equation that is integrable by a quadrature (see § 3).

Let us now consider linear equations of the first order, and first of all the homogeneous equation $dy/dx + Py = 0$. Since this equation does not change when we replace in it y by ay, whatever may be the constant a, we say that it admits the group of transformations $x' = x$, $y' = ay$. Hence it will be integrable by a quadrature if we take $\log y$ for the dependent variable.

Next, let

$$(99) \qquad \frac{dy}{dx} + Py + Q = 0$$

be the general linear equation of the first order, and let y_1 be a particular integral, not zero, of the equation $dy/dx + Py = 0$. It is easily verified that the equation (99) does not change if we replace y by $y + ay_1$. Hence it admits the group of transformations defined by the equations

$$x' = x, \qquad \frac{y'}{y_1} = \frac{y}{y_1} + a.$$

Taking for the new dependent variable y/y_1, the equation must reduce to an equation integrable by a quadrature. We are led to precisely the calculations of § 4, and it is easy to see in a similar manner that the different cases of reduction of the order of the equation which have been indicated in § 19, for equations of higher order, are essentially only particular cases of the preceding method.

These different methods, which appear at first sight as so many different devices for solution, having no relation one to another, can thus be considered from a common point of view by means of the theory of groups of transformations. To every continuous one-parameter group of transformations on the two variables x and y we can make correspond in this way an infinite number of equations of the first order which can be integrated by a quadrature, and equations of higher order whose order can be depressed by unity.

This fact may be of practical importance in the setting up of the equations in certain problems. Suppose that it is a question of finding the plane curves which possess a certain property, and that we know a priori a one-parameter group (G) of transformations such that, if we apply any transformation of (G) to a curve having the given property, the new curve also has the same property. It is clear that the differential equation of these curves will admit the given group of transformations. If, then, we choose a system of coördinates (u, v) such that the equations of the group (G) shall become $u' = u$, $v' = v + a$, the differential equation of the curves sought in this system of coördinates will contain only u, dv/du, d^2v/du^2, $\cdots$. For example, suppose that we wish to obtain the projections on the xy-plane of the asymptotic lines or the lines of curvature of a helicoid, the axis Oz being the axis of helicoidal movement in the sliding of the surface upon itself. It is clear that if a curve C of the xy-plane is a solution of the problem, then all the curves which we obtain by making C turn through any angle about the origin are also solutions. The differential equation of these curves admits, then, the group formed by the rotations about the origin; the equations of this group in polar coördinates are $\rho' = \rho$, $\omega' = \omega + a$. With the system of variables ρ, ω, the differential equation will contain, therefore, only ρ and $d\omega/d\rho$ (see I, § 243, 2d ed.; § 242, 1st ed.).

So far we have supposed the group G known. We are now led to examine the following problem: *A differential equation being given, to recognize whether or not it admits one or more one-parameter continuous groups of transformations, and to determine these groups.* This is a very important question, which cannot be developed here in detail. We shall limit ourselves to a few particulars.

36. Infinitesimal transformations. Given a system of transformations on n variables, defined by the equations

(100) $$x_i' = f_i(x_1, x_2, \cdots, x_n; a), \qquad (i = 1, 2, \cdots, n),$$

where the functions f_i depend upon an arbitrary parameter a, we say again that these transformations form a *group* if the transformation resulting from any two transformations of the system carried out in succession itself belongs to the system. As above, a group is said to contain the identical transformation if, for some value a_0 of the parameter, we have

$$f_i(x_1, x_2, \cdots, x_n; a_0) = x_i, \qquad (i = 1, 2, \cdots, n),$$

for all values of $x_1, x_2, \cdots, x_n$. It may be shown, as above, that such a group is obtained by integrating a system of differential equations

(101) $$\frac{dx_1'}{\xi_1(x_1', x_2', \cdots, x_n')} = \frac{dx_2'}{\xi_2(x_1', x_2', \cdots, x_n')} = \cdots = \frac{dx_n'}{\xi_n(x_1', \cdots, x_n')} = dt.$$

Let

(102) $$x_i' = \phi_i(x_1, x_2, \cdots, x_n; t), \qquad (i = 1, 2, \cdots, n),$$

be the integrals of this system which reduce to $x_1, x_2, \cdots, x_n$ respectively for $t = 0$. The relations (102) define a continuous one-parameter group, the variable t playing the part of the parameter. Indeed, we have seen (§ 31) that the general integral of this system can be written in the form

$$\Omega_1(x_1', x_2', \cdots, x_n') = C_1, \cdots,$$
$$\Omega_{n-1}(x_1', x_2', \cdots, x_n') = C_{n-1}, \quad \Omega_n(x_1', x_2', \cdots, x_n') = t + C_n,$$

where $\Omega_1, \Omega_2, \cdots, \Omega_n$ are n functions of the variables x_i', which we have defined exactly. The integrals which for $t = 0$ take on the values $x_1, x_2, \cdots, x_n$ are furnished, therefore, by the equations

(103) $\quad \begin{cases} \Omega_i(x_1', \cdots, x_n') = \Omega_i(x_1, \cdots, x_n), & (i = 1, 2, \cdots, n-1), \\ \Omega_n(x_1', \cdots, x_n') = \Omega_n(x_1, \cdots, x_n) + t, \end{cases}$

which are equivalent to the relations (102). In this new form we see immediately that these transformations form a group.

Let $F(x_1, x_2, \cdots, x_n)$ be a function of the n variables x_i; if we replace the variables x_i in it by the functions x_i' given by the relations (102), the result $F(x_1', x_2', \cdots, x_n')$ is a function of $x_1, x_2, \cdots, x_n, t$, which for $t = 0$ reduces to $F(x_1, x_2, \cdots, x_n)$. Let us consider the problem of developing this function according to increasing powers of t. We shall denote by F' the result of replacing x_i by x_i' in $F(x_1, x_2, \cdots, x_n)$, and we shall put

$$X(f) = \xi_1(x_1, x_2, \cdots, x_n)\frac{\partial f}{\partial x_1} + \cdots + \xi_n(x_1, x_2, \cdots, x_n)\frac{\partial f}{\partial x_n},$$

where f is any function of $x_1, x_2, \cdots, x_n$. Similarly, replacing the variables x_i by x_i', we shall put

$$X'(f') = \xi_1(x_1', x_2', \cdots, x_n')\frac{\partial f'}{\partial x_1'} + \cdots + \xi_n(x_1', x_2', \cdots, x_n')\frac{\partial f'}{\partial x_n'}.$$

With these preliminary definitions, we have, by the differential equations (101),

$$\frac{dF'}{dt} = \xi_1(x_1', \cdots, x_n')\frac{\partial F'}{\partial x_1'} + \cdots + \xi_n(x_1', \cdots, x_n')\frac{\partial F'}{\partial x_n'} = X'(F').$$

Likewise we have

$$\frac{d^2 F'}{dt^2} = \frac{d}{dt}[X'(F')] = X'[X'(F')],$$

and in general

$$\frac{d^p F'}{dt^p} = X'^{(p)}(F'),$$

where $X'^{(p)}(F')$ denotes the result of the operation X' carried out p times in succession. Since, for $t = 0$, $x_1', x_2', \cdots, x_n'$ reduce to $x_1, x_2, \cdots, x_n$, it follows that $(d^p F'/dt^p)_{t=0}$ is equal to $X^{(p)}(F)$, and the development of F' is given by the formula

(104) $\quad \begin{cases} F'(x_1', \cdots, x_n') = F(x_1, \cdots, x_n) + tX(F) \\ \qquad + \dfrac{t^2}{2!}X^{(2)}(F) + \cdots + \dfrac{t^p}{p!}X^{(p)}(F) + \cdots. \end{cases}$

If we assume that the function F is regular in the neighborhood of the values $x_1, x_2, \cdots, x_n$, the series on the right is convergent as long as $|t|$ is sufficiently small. We have, in particular,

(105) $\qquad x_i' = x_i + \dfrac{t}{1}\xi_i + \dfrac{t^2}{2!}X(\xi_i) + \dfrac{t^3}{3!}X^{(2)}(\xi_i) + \cdots.$

Let us give to t an infinitesimal value δt. Putting $\delta x_i = x'_i - x_i$, and neglecting infinitesimals of higher order than the first with respect to δt, the preceding formulæ can be written in the form

(106) $\qquad \delta x_1 = \xi_1 \delta t, \qquad \delta x_2 = \xi_2 \delta t, \qquad \cdots, \qquad \delta x_n = \xi_n \delta t.$

We say that these relations define an *infinitesimal transformation* and that $X(f)$, or $\Sigma \xi_i (\partial f / \partial x_i)$, is the symbol of this infinitesimal transformation. To every one-parameter group corresponds an infinitesimal transformation, and conversely. If we choose at pleasure n functions, $\xi_1, \xi_2, \cdots, \xi_n$, of $x_1, x_2, \cdots, x_n$, the resulting expression $X(f)$ is the symbol of an infinitesimal transformation that defines a continuous group whose finite equations would be obtained by integrating the system of differential equations (101).

The introduction of infinitesimal transformations has made it possible to apply the methods of the differential calculus to the theory of groups. Besides, in many questions concerning groups it is the infinitesimal transformation which is concerned, as we shall see from a few examples.

Let us consider $x_1, x_2, \cdots, x_n$ as the coördinates of a point in space of n dimensions, and t as an independent variable which denotes the time. If t varies, the point with the coördinates $x'_1, x'_2, \cdots, x'_n$ describes in a space of n dimensions a curve, or *trajectory*, starting from the point $(x_1, x_2, \cdots, x_n)$. The space of n dimensions, or at least a region of that space, is thus decomposed into an infinite number of one-dimensional manifolds, and each point of the given region belongs to a single one-dimensional manifold. We say that a function $F(x_1, \cdots, x_n)$ is an *invariant* of the group considered if we have

$$F(x'_1, \cdots, x'_n) = F(x_1, \cdots, x_n),$$

whatever may be the value of t. It is easy to obtain all the invariants of a group. In fact, dividing the two sides of the equation (104) by t, and supposing $F' = F$, we obtain the relation

$$X(F) + \frac{t}{2} X^{(2)}(F) + \cdots + \frac{t^{p-1}}{p!} X^{(p)}(F) + \cdots = 0.$$

Since this equality must hold whatever t may be, it is necessary in particular that we have $X(F) = 0$. We say, in this case, that the function F admits the infinitesimal transformation of the group. This condition is, moreover, sufficient; for if we have $X(F) = 0$, we also have $X[X(F)] = 0, \cdots$, and therefore $X^{(p)}(F) = 0$, whatever p may be. *The only invariants of the one-parameter group are therefore the integrals of the equation $X(f) = 0$.*

Let us notice that if two groups have for infinitesimal transformations $X(f)$ and $\Pi . X(f)$ respectively, where $\Pi(x_1, x_2, \cdots, x_n)$ is any function whatever, these two groups have the same invariants, even though they are not identical. If we apply to the same point the transformations of the two groups, this point will indeed describe the same path, but with different velocities. Conversely, if two groups have the same invariants, the two infinitesimal transformations $X(f)$ and $Y(f)$ can differ only by a factor $\Pi(x_1, x_2, \cdots, x_n)$ which depends only upon $x_1, x_2, \cdots, x_n$, for the two equations $X(f) = 0$, $Y(f) = 0$ must have the same integrals.

We shall now introduce another important concept. Let

(107) $\qquad x_1 = f(x, y; a), \qquad y_1 = \phi(x, y; a)$

be a continuous group in two variables. If we apply a transformation of this group to all the points of a plane curve C, we obtain another plane curve C_1. Let y', y'', $\cdots$, $y^{(n)}$ be the successive derivatives of y with respect to x and y_1', y_1'', $\cdots$, $y_1^{(n)}$ the successive derivatives of y_1 with respect to x_1; we have seen (I, § 61, 2d ed.; § 35, 1st ed.) how to calculate these last successive derivatives in terms of x, y, y', $\cdots$, $y^{(n)}$. These calculations lead to formulæ of the form

(108) $$\begin{cases} y_1' = \psi_1(x, y, y'; a), \\ y_1'' = \psi_2(x, y, y', y''; a), \\ \cdots \cdots \cdots \cdots \cdots \cdots \cdots \cdots \\ y_1^{(n)} = \psi_n(x, y, y', \cdots, y^{(n)}; a). \end{cases}$$

The relations (107) and (108) define also a group of transformations in $n+2$ variables, x, y, y', $\cdots$, $y^{(n)}$, which is called the *extended group* of the first. We shall assume this fact, the proof of which presents no other difficulties than the writing of rather long expressions. We shall merely show how the infinitesimal transformation of the extended group can be obtained. Let

$$\xi(x, y)\frac{\partial f}{\partial x} + \eta(x, y)\frac{\partial f}{\partial y}$$

be the infinitesimal transformation of the given group. We can write the equations of this group in the form

(109) $$\begin{cases} x_1 = x + \frac{t}{1}\xi(x, y) + \frac{t^2}{2!}\left(\xi\frac{\partial \xi}{\partial x} + \eta\frac{\partial \xi}{\partial y}\right) + \cdots, \\ y_1 = y + \frac{t}{1}\eta(x, y) + \frac{t^2}{2!}\left(\xi\frac{\partial \eta}{\partial x} + \eta\frac{\partial \eta}{\partial y}\right) + \cdots, \end{cases}$$

and from them we derive

$$y_1' = \frac{dy_1}{dx_1} = \frac{dy + \frac{t}{1}\left(\frac{\partial \eta}{\partial x}dx + \frac{\partial \eta}{\partial y}dy\right) + \cdots}{dx + \frac{t}{1}\left(\frac{\partial \xi}{\partial x}dx + \frac{\partial \xi}{\partial y}dy\right) + \cdots}.$$

The coefficient of t on the right, after expansion in a single power series, is the only thing we need to know. It is obtained by a division and is equal to

$$\frac{\partial \eta}{\partial x} + \left(\frac{\partial \eta}{\partial y} - \frac{\partial \xi}{\partial x}\right)\frac{dy}{dx} - \left(\frac{\partial \xi}{\partial y}\right)\left(\frac{dy}{dx}\right)^2.$$

The symbol of the infinitesimal transformation of the extended group is, therefore, for $n = 1$,

$$\xi(x, y)\frac{\partial f}{\partial x} + \eta(x, y)\frac{\partial f}{\partial y} + \left[\frac{\partial \eta}{\partial x} + \left(\frac{\partial \eta}{\partial y} - \frac{\partial \xi}{\partial x}\right)y' - \frac{\partial \xi}{\partial y}y'^2\right]\frac{\partial f}{\partial y'}.$$

The method is a general one. If the coefficient of t in the development of $y_1^{(n-1)}$ is $\pi(x, y, y', \cdots, y^{(n-1)})$, we have for $y_1^{(n)}$

$$y_1^{(n)} = \frac{dy_1^{(n-1)}}{dx_1} = \frac{dy^{(n-1)} + \frac{t}{1}d\pi + \cdots}{dx + \frac{t}{1}\left(\frac{\partial \xi}{\partial x}dx + \frac{\partial \xi}{\partial y}dy\right) + \cdots},$$

and the coefficient of t on the right-hand side is

$$\frac{d\pi}{dx} - y^{(n)}\left(\frac{\partial \xi}{\partial x} + \frac{\partial \xi}{\partial y}y'\right).$$

Hence we can calculate step by step, to any desired value of n, the infinitesimal transformations of the extended groups which are obtained from the given group.

We say that a system of differential equations

(110) $$\frac{dx_1}{X_1} = \frac{dx_2}{X_2} = \cdots = \frac{dx_n}{X_n}$$

admits the one-parameter group of transformations G defined by the equations (100) if it changes into a system of the same form,

(111) $$\frac{dx_1'}{X_1'} = \frac{dx_2'}{X_2'} = \cdots = \frac{dx_n'}{X_n'},$$

when we take for new variables $x_1', x_2', \cdots, x_n'$ instead of $x_1, x_2, \cdots, x_n$, and if this is true whatever the value of the parameter a may be. Here and below, the symbol X_i' denotes the same function of the variables x_i' that X_i is of the variables x_i. In order that this be true, it follows from the relation which has been established between the system (110) and the partial differential equation

(112) $$X(f) = X_1 \frac{\partial f}{\partial x_1} + X_2 \frac{\partial f}{\partial x_2} + \cdots + X_n \frac{\partial f}{\partial x_n} = 0$$

that it is necessary and sufficient that every transformation of the group G shall carry the equation

$$X'(f') = \sum_{i=1}^{n} X_i'(x_1', x_2', \cdots, x_n') \frac{\partial f'}{\partial x_i'}$$

over into a linear equation equivalent to $X(f) = 0$ for every value of the parameter a. If $f(x_1, x_2, \cdots, x_n)$ is an integral of $X(f) = 0$, $f'(x_1', x_2', \cdots, x_n')$ is also an integral of $X'(f') = 0$; hence, if we replace $x_1', \cdots, x_n'$ by their values given by the expressions (100), $f'(x_1', \cdots, x_n')$ must also be an integral of $X(f) = 0$. It follows that the necessary and sufficient condition that the system of differential equations (110) admit the group of transformations G is that every transformation of that group shall change an integral of the equation $X(f) = 0$ into an integral of the same equation.

Let

(113) $$T(f) = \xi_1 \frac{\partial f}{\partial x_1} + \xi_2 \frac{\partial f}{\partial x_2} + \cdots + \xi_n \frac{\partial f}{\partial x_n}$$

be the infinitesimal transformation of the group G. Replacing the parameter a by the parameter t defined above, we may write

$$f(x_1', x_2', \cdots, x_n') = f(x_1, x_2, \cdots, x_n) + \frac{t}{1} T(f) + \frac{t^2}{2!} T[T(f)] + \cdots.$$

If $f(x_1, \cdots, x_n)$ is any integral of the equation (112), the same thing must be true of $f(x_1', \cdots, x_n')$, and consequently of

$$f(x_1', \cdots, x_n') - f(x_1, \cdots, x_n),$$

or of

$$T(f) + \frac{t}{2} T[T(f)] + \cdots,$$

whatever the value of t may be. In particular, $T(f)$ must be an integral of the equation (112). This condition is sufficient. For, let $f_1, f_2, \cdots, f_{n-1}$ be a system of $n-1$ independent integrals. If $T(f_1), T(f_2), \cdots, T(f_{n-1})$ are also integrals, the same thing must be true of $T(f)$, where f is any other integral. For we have $f = \Pi(f_1, f_2, \cdots, f_{n-1})$, and therefore

$$T(f) = \frac{\partial \Pi}{\partial f_1} T(f_1) + \cdots + \frac{\partial \Pi}{\partial f_{n-1}} T(f_{n-1}).$$

Since $T(f)$ is an integral, the same thing is true of $T[T(f)]$, and so on; the same is therefore true of $f(x_1', x_2', \cdots, x_n')$.

Hence, *in order that the system* (110) *admit the group G of transformations, it is necessary and sufficient that, if f is an integral of $X(f) = 0$, $T(f)$ shall also be an integral.* We say for brevity that the equation $X(f) = 0$ admits the infinitesimal transformation $T(f)$.

Let us now take a differential equation of the first order,

(114) $$\frac{dx}{A} = \frac{dy}{B}.$$

In order that the equation $X(f) = A\, \partial f/\partial x + B\, \partial f/\partial y = 0$ admit the infinitesimal transformation $\xi\, \partial f/\partial x + \eta\, \partial f/\partial y$, it is necessary that we have

$$A \frac{\partial \omega}{\partial x} + B \frac{\partial \omega}{\partial y} = 0, \qquad \xi \frac{\partial \omega}{\partial x} + \eta \frac{\partial \omega}{\partial y} = \Pi(\omega),$$

where $\omega(x, y)$ denotes an integral (other than a constant) of $X(f) = 0$, and where $\Pi(\omega)$ denotes an undetermined function of ω. We derive from these relations

$$\frac{\partial \omega}{\partial x} = -\frac{B\Pi(\omega)}{A\eta - B\xi}, \qquad \frac{\partial \omega}{\partial y} = \frac{A\Pi(\omega)}{A\eta - B\xi},$$

whence,

$$\frac{d\omega}{\Pi(\omega)} = \frac{B\, dx - A\, dy}{B\xi - A\eta}.$$

It follows that $1/(B\xi - A\eta)$ is an integrating factor for $B\, dx - A\, dy$. Conversely, let $\phi(x, y)$ be a function such that its total differential is

$$d\phi = \frac{B\, dx - A\, dy}{B\xi - A\eta}.$$

Then we have, simultaneously,

$$X(\phi) = A \frac{\partial \phi}{\partial x} + B \frac{\partial \phi}{\partial y} = 0, \qquad T(\phi) = \xi \frac{\partial \phi}{\partial x} + \eta \frac{\partial \phi}{\partial y} = 1;$$

hence $T(\phi)$ is also an integral of the equation $X(\phi) = 0$. We can state this result as follows:

In order that the differential equation (114) *admit the group of transformations derived from the infinitesimal transformation $\xi\, \partial f/\partial x + \eta\, \partial f/\partial y$, it is necessary and sufficient that $1/(B\xi - A\eta)$ shall be an integrating factor for $B\, dx - A\, dy$.*

This new method requires only the knowledge of the infinitesimal transformation of the group. As there exist an infinite number of integrating factors, we see that every equation of the first order admits an infinite number of infinitesimal transformations.

Let us return to the general case of the system (110). Let $X(f) = 0$ be the corresponding linear equation and $T(f)$ the symbol of an infinitesimal transformation. Let us consider the equation

(115) $$Z(f) = X[T(f)] - T[X(f)] = 0,$$

where $X[T(f)]$ represents the result of the operation $X(\ )$ applied to $T(f)$, and where $T[X(f)]$ has an analogous meaning; $Z(f)$ is still a linear homogeneous function in the derivatives of the first order $\partial f/\partial x_i$, and it does not contain any derivatives of the second order. To show this, it suffices to prove that the coefficients of a derivative of the second order are the same in $X[T(f)]$ and $T[X(f)]$. Now the coefficient of $\partial^2 f/\partial x_i^2$ is $X_i \xi_i$, and that of $\partial^2 f/\partial x_i \partial x_k$ is $X_i \xi_k + X_k \xi_i$ in $T[X(f)]$, and it is obvious that these coefficients are the same in $X[T(f)]$. The equation $Z(f) = 0$ is therefore an equation of the same type as the equation $X(f) = 0$, which can be written in a form exhibiting the coefficients explicitly:

(116) $$Z(f) = [X(\xi_1) - T(X_1)]\frac{\partial f}{\partial x_1} + \cdots + [X(\xi_i) - T(X_i)]\frac{\partial f}{\partial x_i} + \cdots = 0.$$

If now $T(f)$ is an integral of the equation $X(f) = 0$, whenever f is an integral of the same equation, every integral of $X(f) = 0$ evidently satisfies the linear equation $Z(f) = 0$; we must have, therefore (§ 31),

(117) $$X[T(f)] - T[X(f)] = \rho(x_1, x_2, \cdots, x_n) X(f),$$

where ρ is an undetermined function of $x_1, x_2, \cdots, x_n$. Conversely, if we have an identity of this form, every integral of the equation $X(f) = 0$ satisfies also the equation $X[T(f)] = 0$, and therefore $T(f)$ is also an integral of the equation $X(f) = 0$. *The necessary and sufficient condition that the linear equation $X(f) = 0$ admit the infinitesimal transformation $T(f)$ is expressed by the relation* (117), *where ρ is any function of $x_1, x_2, \cdots, x_n$.* That relation is equivalent to the $(n-1)$ distinct relations

$$\frac{X(\xi_1) - T(X_1)}{X_1} = \frac{X(\xi_2) - T(X_2)}{X_2} = \cdots = \frac{X(\xi_n) - T(X_n)}{X_n}.$$

Given a differential equation of the nth order,

(118) $$\frac{d^n y}{dx^n} = \phi\left(x, y, \frac{dy}{dx}, \frac{d^2 y}{dx^2}, \cdots, \frac{d^{n-1} y}{dx^{n-1}}\right),$$

in order to determine whether it admits the group of transformations deduced from the infinitesimal transformation $\xi(x, y)\partial f/\partial x + \eta(x, y)\partial f/\partial y$, we need only replace the equation (118) by a system of n differential equations of the first order, taking for the auxiliary dependent variables the first $(n-1)$ derivatives $y', y'', \cdots, y^{(n-1)}$, and then determine whether this system admits the infinitesimal transformation of the extended group of G.

Let us consider, for example, the equation of the second order $y'' = \phi(x, y, y')$, which may be replaced by the system

$$\frac{dx}{1} = \frac{dy}{y'} = \frac{dy'}{\phi(x, y, y')}$$

or by the linear equation

$$X(f) = \frac{\partial f}{\partial x} + \frac{\partial f}{\partial y} y' + \frac{\partial f}{\partial y'} \phi(x, y, y').$$

It will be necessary to determine whether this equation admits the infinitesimal transformation

$$\xi(x, y) \frac{\partial f}{\partial x} + \eta(x, y) \frac{\partial f}{\partial y} + \left[\frac{\partial \eta}{\partial x} + \left(\frac{\partial \eta}{\partial y} - \frac{\partial \xi}{\partial x} \right) y' - \frac{\partial \xi}{\partial y} y'^2 \right] \frac{\partial f}{\partial y'}.$$

On carrying through the calculations, we find a condition which contains x, y, and y', and which must be verified for all values of these variables. The equation of the second order being given, if we wish to find the infinitesimal transformations which it admits, we have at our disposal the unknown functions $\xi(x, y)$, $\eta(x, y)$, which do not contain y'. Writing the condition that the preceding relation is independent of y', we may have, according to the given function $\phi(x, y, y')$, a limited or unlimited number of equations which must be satisfied by the functions $\xi(x, y)$ and $\eta(x, y)$. In general these equations will be incompatible, and we see that an equation of the second order, taken arbitrarily, does not admit any infinitesimal transformation. The same thing is true of equations of higher order, and it is seen from this how Sophus Lie was able to classify differential equations according to the number of independent infinitesimal transformations which they admit.

EXERCISES

1*. Let M_0 be the greatest absolute value of $f(x, y_0)$ when x varies from x_0 to $x_0 + a$. If the letters a, b, K, x_0, y_0 have the same meaning as in § 30, the integral of the equation $y' = f(x, y)$ which takes on the value y_0 for $x = x_0$ is continuous in the interval $(x_0, x_0 + \rho)$, where ρ is the smaller of the two numbers a and $\log(1 + Kb/M_0)/K$.

[E. LINDELÖF, *Journal de Mathématiques*, 1894.]

[The inequalities

$$|y_n - y_{n-1}| < M_0 K^{n-1} \frac{(x - x_0)^n}{n!}$$

are established step by step, as in § 27, and y_n will remain between $y_0 - b$ and $y_0 + b$, provided that we have $e^{K(x-x_0)} < 1 + bK/M_0$.]

2. Find two first integrals of the simultaneous systems of differential equations

(α) $\quad \dfrac{dy}{dx} + \phi'(x)y - \psi'(x)z = 0, \quad \dfrac{dz}{dx} + \psi'(x)y + \phi'(x)z = 0,$

(β) $\quad (z - y)^2 \dfrac{dy}{dx} = z, \quad (z - y)^2 \dfrac{dz}{dx} = y,$

(γ) $\quad \dfrac{dy}{y(x + y)} = \dfrac{dz}{(x - y)(2x + 2y + z)} = \dfrac{-dx}{x(x + y)}.$

3. The expression $1/[y - xf(y/x)]$ is an integrating factor for $dy - f(y/x)dx$.

4. The general form of the differential equations of the first order which admit the infinitesimal transformation $y\partial f/\partial x - x\partial f/\partial y$ is

$$\frac{xy' - y}{x + yy'} = \phi(x^2 + y^2).$$

Deduce from this an integrating factor.

5. Find the general form of the differential equations of the first order which admit the infinitesimal transformation $\partial f/\partial x + x(\partial f/\partial y)$; the infinitesimal transformation $x(\partial f/\partial x) + ay(\partial f/\partial y)$.

6. Find a group of transformations for the differential equation

$$\frac{dy}{dx} = \phi(x + ay),$$

where a is constant, and deduce from it an integrating factor.

7*. The differential equations of the elastic space curve,

$$y'z'' - z'y'' = \delta x' - \tfrac{1}{2}\beta y,$$
$$z'x'' - x'z'' = \delta y' - \tfrac{1}{2}\beta x,$$
$$x'y'' - y'x'' = \delta z' - \alpha,$$

where α, β, δ are constants, possess the two first integrals $x'^2 + y'^2 + z'^2 = C$, $\beta(x^2 + y^2) - 4z' = C'$. We then obtain x and y by the integration of a differential equation of the second order.

[HERMITE, *Sur quelques applications des fonctions elliptiques* (p. 93).]

CHAPTER III

LINEAR DIFFERENTIAL EQUATIONS

I. GENERAL PROPERTIES. FUNDAMENTAL SYSTEMS

Linear differential equations have been studied more thoroughly than any other class. They possess a group of characteristic properties which distinguish them sharply and at the same time simplify their study. Moreover, they appear in a great number of important applications of Analysis, and a preliminary study of them is very useful before undertaking the study of differential equations of the most general form. Except when otherwise expressly stated, we shall study here only those equations whose coefficients are analytic functions of the independent variable.

37. Singular points of a linear differential equation. A linear differential equation of the nth order is of the form

$$(1) \quad \frac{d^n y}{dx^n} + a_1 \frac{d^{n-1} y}{dx^{n-1}} + \cdots + a_{n-1} \frac{dy}{dx} + a_n y + a_{n+1} = 0,$$

$a_1, a_2, \cdots, a_{n+1}$ being functions of the single variable x. Its integration is equivalent to that of the system

$$(2) \quad \begin{cases} \dfrac{dy_{n-1}}{dx} + a_1 y_{n-1} + \cdots + a_{n-1} y_1 + a_n y + a_{n+1} = 0, \\ \dfrac{dy}{dx} = y_1, \quad \dfrac{dy_1}{dx} = y_2, \quad \cdots, \quad \dfrac{dy_{n-2}}{dx} = y_{n-1}, \end{cases}$$

obtained by taking for auxiliary dependent variables the first $n-1$ derivatives of y. Let us suppose that the coefficients a_i are analytic in a circle C_0 with the radius R and with its center at the point x_0, and let $y_0, y_0', y_0'', \cdots, y_0^{(n-1)}$ be a system of n arbitrary constants. Applying to the equations (2) a general result established above (§ 23), we see that *the equation* (1) *has an integral analytic in the circle C_0, taking on the value y_0 for $x = x_0$, while its first $n-1$ derivatives take on respectively the values $y_0', y_0'', \cdots, y_0^{(n-1)}$ for $x = x_0$.*

We know also, from the general theory, that it is the only integral of the equation (1) satisfying these initial conditions; we shall say for brevity that it is *defined* by the initial conditions $(x_0, y_0, y_0', y_0'',$

$\cdots, y_0^{(n-1)}$). Let us now suppose at first, for definiteness, that the coefficients a_i are single-valued functions of x, having in the whole plane only isolated singular points. Let L be a path joining two non-singular points x_0 and X, and <u>not passing through any singular point</u>; the integral which is defined by the initial conditions $(x_0, y_0, y_0', \cdots, y_0^{(n-1)})$ is represented by a power series $P(x - x_0)$ convergent in the circle C_0 with the center x_0 and passing through the nearest singular point to x_0. We can follow, by means of this series, the variation of the integral along the path L as long as the path does not go out of the circle C_0. If the path L leaves the circle C_0 at a point a, let us take a point x_1 on the path within the circle C_0 and near enough to a so that the circle C_1 with the center x_1 passing through the nearest singular point does not lie entirely within the circle C_0. From the series $P(x - x_0)$ and from those which we obtain by successive differentiations, we can derive the values of the integral and of its first $n - 1$ derivatives at the point x_1. Let $y_1, y_1', \cdots, y_1^{(n-1)}$ be these values; the integral of the equation (1), which is defined by the initial conditions $(x_1, y_1, y_1', \cdots, y_1^{(n-1)})$, is represented by a power series $P_1(x - x_1)$ convergent in the circle C_1. The values of the two series $P(x - x_0)$ and $P_1(x - x_1)$ are equal in the part common to the two circles C_0 and C_1, since they <u>each</u> represent an integral of the equation (1) satisfying the same initial conditions. It follows that the series $P_1(x - x_1)$ represents the analytic extension in the circle C_1 of the analytic function defined in the circle C_0 by the series $P(x - x_0)$. If all of the portion of L included between x_1 and X does not lie in the circle C_1, we shall take a new point x_2 on the path within C_1, and so on as before.

At the end of a *finite* number of operations we shall certainly arrive at a circle containing the point X. In fact, let S be the length of the path L and δ the lower limit of the distance of any point of L to the singular points. The radii of the successive circles used are at least equal to δ, and we can choose the centers of these circles in such a way that the distance between two successive centers is greater than $\delta/2$. After p operations the length of the broken line obtained by joining these successive centers will be equal to at least $p\delta/2$. If we have $p\delta/2 > S$, the length of the broken line will be greater than the length of L. Hence, after at most $(p - 1)$ operations, we shall have arrived at a circle containing all of the portion of L included between the center of that circle and the point X.

Recapitulating, we see that we can continue the analytic extension of the integral as long as the path described by the variable does not pass through any of the singular points of the coefficients a_i. We know, therefore, a priori, what are the only points which can be singular points for the integrals of a linear equation. It may, however, happen that a point a is a singular point for some of the coefficients a_i without being a singular point for all the integrals. In the particular case where the coefficients are all polynomials or integral functions, all the integrals are analytic functions in the whole plane; that is, they are integral functions and they may reduce to polynomials.

The reasoning may be extended also to the case where the coefficients a_i have any singularities whatever, it being possible for these functions to be multiple-valued. If we start from a point x_0, where these coefficients are analytic, and if we cause the variable x to describe a path L, along the whole length of which we can continue the analytic extension of the coefficients a_i, we can likewise continue the analytic extension of the integrals along this path. The power series which represent the integrals are convergent in the same circles as the series which represent the coefficients.

These results are entirely in accord with those which we have deduced from the method of successive approximations (§ 28).

38. Fundamental systems. Let us consider a linear equation which is also homogeneous, that is, not containing a term independent of y,

$$(3) \quad F(y) = \frac{d^n y}{dx^n} + a_1 \frac{d^{n-1} y}{dx^{n-1}} + \cdots + a_{n-1} \frac{dy}{dx} + a_n y = 0,$$

where $F(y)$ denotes no longer a function of the variable y but the result of an operation carried out on a function y of the variable x. From the definition of this symbol of operation it is clear that, if $y_1, y_2, \cdots, y_p$ are any p functions of x, and $C_1, C_2, \cdots, C_p$ any constants, we have the relation

$$F(C_1 y_1 + C_2 y_2 + \cdots + C_p y_p) = C_1 F(y_1) + C_2 F(y_2) + \cdots + C_p F(y_p).$$

If $y_1, y_2, \cdots, y_p$ are integrals of the equation (3), then $C_1 y_1 + C_2 y_2 + \cdots + C_p y_p$ is also an integral, whatever the numerical values of the constants C_i may be. If we know n particular integrals $y_1, y_2, \cdots, y_n$ of the equation, we can therefore deduce from them an integral

$$(4) \qquad y = C_1 y_1 + C_2 y_2 + \cdots + C_n y_n,$$

in the expression of which appear n arbitrary constants $C_1, C_2, \cdots, C_n$. We cannot conclude from this that the expression (4) really represents the general integral of the equation (3); we must first assure ourselves that we can dispose of the constants $C_1, C_2, \cdots, C_n$ in such a way that, for a particular value x_0 of x, different from a singular point, y and its first $n-1$ derivatives take on any values given in advance. For the sake of brevity, let us indicate by $(y_i^p)_0$ the value which the pth derivative of the particular integral y_i takes on at the point x_0. Setting the values of the integral y, and of its first $n-1$ derivatives at the point x_0, equal to these arbitrary quantities, we obtain a system of n linear equations to determine the constants C_1, C_2, $\cdots$, C_n. The determinant of the coefficients of these unknowns must be different from zero. We shall denote by $\Delta(y_1, y_2, \cdots, y_n)$ the determinant whose elements are the functions $y_1, y_2, \cdots, y_n$, and their derivatives up to those of the $(n-1)$th order:

$$(5) \quad \Delta(y_1, y_2, \cdots, y_n) = \begin{vmatrix} y_1 & y_2 & \cdots & y_n \\ y_1' & y_2' & \cdots & y_n' \\ \cdots & \cdots & \cdots & \cdots \\ y_1^{(n-1)} & y_2^{(n-1)} & \cdots & y_n^{(n-1)} \end{vmatrix}.$$

If this determinant, which is an analytic function of x in the whole region in which the coefficients a_i are analytic, is not identically zero, let us choose for x_0 a point where this determinant is not zero. We can then determine the constants C_i so that y and its first $n-1$ derivatives take on any initial values whatever for x_0. Every integral of the equation (3) is therefore included in the formula (4). We say, for brevity, that this formula represents the *general integral* of the equation (3). The integrals $y_1, y_2, \cdots, y_n$, such that the determinant $\Delta(y_1, y_2, \cdots, y_n)$ is different from zero, form a *fundamental system*.

If this determinant is identically zero, some of the integrals $y_1, y_2, \cdots, y_n$ can be deduced from the others. We shall say, in general, that n functions $y_1, y_2, \cdots, y_n$ of the variable x are not *linearly independent* if there exists between these n functions an identity of the form

$$(6) \quad C_1 y_1 + C_2 y_2 + \cdots + C_n y_n = 0,$$

where $C_1, C_2, \cdots, C_n$ are constants not all of which are zero. *In order that n functions $y_1, y_2, \cdots, y_n$ shall not be linearly independent, it is necessary and sufficient that the determinant $\Delta(y_1, y_2, \cdots, y_n)$ be identically zero.*

The condition is first necessary. For from the relation (6) we obtain the $n-1$ relations of the same form,

$$(7) \quad C_1 y_1^{(p)} + C_2 y_2^{(p)} + \cdots + C_n y_n^{(p)} = 0, \qquad (p = 1, 2, \cdots, n-1)$$

between the derivatives of the first order, of the second order, etc., of the functions y_i. Since the coefficients C_i are not all zero by hypothesis, the equations (6) and (7) cannot be consistent unless the determinant $\Delta(y_1, y_2, \cdots, y_n)$ is identically zero.

Conversely, suppose that $\Delta \equiv 0$, and suppose first that all the first minors of Δ relative to the elements of the last row are not identically zero, for example, that the cofactor of $y_n^{(n-1)}$,

$$\delta = \begin{vmatrix} y_1 & y_2 & \cdots & y_{n-1} \\ y_1' & y_2' & \cdots & y_{n-1}' \\ \cdots & \cdots & \cdots & \cdots \\ y_1^{(n-2)} & y_2^{(n-2)} & \cdots & y_{n-1}^{(n-2)} \end{vmatrix},$$

is different from zero. Let A be a region of the plane of the variable x where the functions y_i are analytic and where this determinant δ does not vanish. Let us put

(8) $\begin{cases} y_n = K_1 y_1 + K_2 y_2 + \cdots + K_{n-1} y_{n-1}, \\ y_n' = K_1 y_1' + K_2 y_2' + \cdots + K_{n-1} y_{n-1}', \\ \cdots \cdots \cdots \cdots \cdots \cdots \cdots \cdots \cdots \cdots \cdots \cdots \\ y_n^{(n-2)} = K_1 y_1^{(n-2)} + K_2 y_2^{(n-2)} + \cdots + K_{n-1} y_{n-1}^{(n-2)}. \end{cases}$

These $n-1$ equations determine $K_1, K_2, \cdots, K_{n-1}$ as analytic functions of x in the region A, since K_i is the quotient of an analytic function divided by the minor δ which is not zero in A. These functions $K_1, \cdots, K_{n-1}$ satisfy also the relation

(9) $\quad y_n^{(n-1)} = K_1 y_1^{(n-1)} + K_2 y_2^{(n-1)} + \cdots + K_{n-1} y_{n-1}^{(n-1)},$

since $\Delta(y_1, y_2, \cdots, y_n)$ is zero at every point of A. Differentiating once each of the equations (8), and taking account of these same relations and of the relation (9), we find

$$K_1' y_1 + \cdots + K_{n-1}' y_{n-1} = 0,$$
$$\cdots \cdots \cdots \cdots \cdots \cdots \cdots,$$
$$K_1' y_1^{(n-2)} + \cdots + K_{n-1}' y_{n-1}^{(n-2)} = 0,$$

and consequently $K_1' = K_2' = \cdots = K_{n-1}' = 0$. The functions $K_1, \cdots, K_{n-1}$ are therefore constants, and we have indeed a relation of the form (6) between the n functions $y_1, y_2, \cdots, y_n$, where all the coefficients are constants and the coefficient C_n is different from zero. Since this relation has been established in the region A, it follows, by analytic extension, that it holds in any region in which the functions $y_1, y_2, \cdots, y_n$ exist and are analytic.

It will be noticed that the minor δ is precisely equal to

$$\Delta(y_1, y_2, \cdots, y_{n-1}).$$

If this minor δ is identically zero without $\Delta(y_1, y_2, \cdots, y_{n-2})$ being also zero, a similar argument would show that the functions $y_1, y_2,$

$\ldots, y_{n-1}$ satisfy a relation of the form (6), where $C_n = 0$, $C_{n-1} \neq 0$. Continuing in this way, we shall therefore surely arrive at a relation of the form (6), in which some of the coefficients may be zero. If, therefore, we know n integrals of the equation (3) such that $\Delta(y_1, y_2, \ldots, y_n) \equiv 0$, one at least of these integrals is a linear combination with constant coefficients of the other integrals. It may also happen that these n integrals reduce in reality to p independent integrals $[p < n-1]$. In order that this may be the case, it is necessary and sufficient that all the determinants analogous to Δ which can be formed with $p+1$ of these integrals shall be zero, one at least of the determinants formed with p integrals being different from zero.

The same lemma enables us also to prove that the general integral of the equation (3) is represented by an expression of the form (4). For, let $(y_1, y_2, \ldots, y_n)$ be a fundamental system of integrals, and y any other integral. From the $(n+1)$ equations

$$F(y) = 0, \qquad F(y_1) = 0, \qquad \ldots, \qquad F(y_n) = 0$$

we derive, by elimination of the coefficients $a_1, a_2, \ldots, a_n$, an equation of condition which is no other than

(10) $\qquad \Delta(y, y_1, y_2, \ldots, y_n) = 0.$

We have, therefore, between these $n+1$ integrals, a relation of the form
$$Cy + C_1 y_1 + \cdots + C_n y_n = 0,$$
where $C, C_1, C_2, \ldots, C_n$ are constants not all of which are zero. Finally, C, the coefficient of y, is certainly different from zero, since the integrals $y_1, y_2, \ldots, y_n$ are linearly independent.

Every linear equation of the nth order has an infinite number of fundamental systems of integrals. In order to obtain such a system, we need only take n integrals such that the determinant formed from the initial values of these n integrals and their first $n-1$ derivatives for a non-singular point x_0 is not zero. If $(y_1, y_2, \ldots, y_n)$ is a first fundamental system, the n integrals $Y_1, Y_2, \ldots, Y_n$, given by the equations

$$Y_i = c_{i1} y_1 + c_{i2} y_2 + \cdots + c_{in} y_n, \qquad (i = 1, 2, \ldots, n)$$

where the coefficients c_{ik} are constants, form a fundamental system, provided that the determinant D formed by the n^2 coefficients c_{ik} is different from zero. We have, in fact, by the rule for the multiplication of determinants,

$$\Delta(Y_1, Y_2, \ldots, Y_n) = D \cdot \Delta(y_1, y_2, \ldots, y_n).$$

It follows from this relation that the quotient $[d\Delta(y_1, \ldots, y_n)/dx]/\Delta$ is the same whatever may be the fundamental system. We shall verify this by calculating this quotient. For this purpose let us observe that the derivative of a function $F(x)$ is equal to the coefficient of h in the development of $F(x+h)$ in powers of h. If we give to x an increment h, and if we replace each element of the determinant Δ by its development, retaining only the terms of the first degree in h, we obtain the determinant

$$\begin{vmatrix} y_1 + hy_1' & y_2 + hy_2' & \cdots & y_n + hy_n' \\ y_1' + hy_1'' & y_2' + hy_2'' & \cdots & y_n' + hy_n'' \\ \cdots & \cdots & \cdots & \cdots \\ y_1^{(n-1)} + hy_1^{(n)} & y_2^{(n-1)} + hy_2^{(n)} & \cdots & y_n^{(n-1)} + hy_n^{(n)} \end{vmatrix}.$$

The coefficient of h is the sum of n determinants which are obtained by taking the coefficients of h in any row and the terms independent of h in the other rows; $n-1$ of these determinants, having two rows identical, are zero, and there remains

$$\frac{d\Delta(y_1, y_2, \cdots, y_n)}{dx} = \begin{vmatrix} y_1 & y_2 & \cdots & y_n \\ \cdots & \cdots & & \cdots \\ y_1^{(n-2)} & y_2^{(n-2)} & \cdots & y_n^{(n-2)} \\ y_1^{(n)} & y_2^{(n)} & \cdots & y_n^{(n)} \end{vmatrix}.$$

This result is true, whatever the functions $y_1, \cdots, y_n$ may be; if these functions are integrals of the equation (3), we can replace $y_1^{(n)}$ in the last row by $-a_1 y_1^{(n-1)} - \cdots - a_n y_1$, and similarly for the others. There remains, after developing with respect to the elements of the last row and taking account of the determinants which have two rows identical,

(11) $$\frac{d\Delta}{dx} = -a_1 \Delta.$$

The quotient which we wish to calculate is therefore equal to $-a_1$, and we derive also from the preceding result the value of the determinant

$$\Delta = \Delta_0 e^{-\int_{x_0}^{x} a_1 dx},$$

where Δ_0 denotes the value of Δ for $x = x_0$. This expression for Δ shows that this determinant is different from zero at every non-singular point, if it is not identically zero — a result which we could also have obtained from the preceding properties.

It should be noticed that every linear equation of which a fundamental system of integrals is $(y_1, y_2, \cdots, y_n)$ can be written in the form (10)

$$\Delta(y, y_1, y_2, \cdots, y_n) = 0,$$

the coefficients containing only the integrals y_i and their derivatives. This shows that any n linearly independent functions $y_1, y_2, \cdots, y_n$ can always be regarded as forming a fundamental system of integrals of a linear equation.

39. The general linear equation. A non-homogeneous linear equation can be written in the form

(12) $$F(y) = \frac{d^n y}{dx^n} + a_1 \frac{d^{n-1} y}{dx^{n-1}} + \cdots + a_{n-1} \frac{dy}{dx} + a_n y = f(x),$$

where the term independent of y has been isolated on the right-hand side. We shall also consider the equation formed by replacing $f(x)$

by zero; the resulting equation, $F(y) = 0$, is called the corresponding homogeneous equation. If we know a particular integral Y of the equation (12), the substitution $y = Y + z$ reduces the integration of that equation to that of the homogeneous equation $F(z) = 0$ by the identity $F(Y + z) = F(Y) + F(z)$. The general integral of the non-homogeneous equation is therefore represented by the expression

$$(13) \qquad y = Y + C_1 y_1 + C_2 y_2 + \cdots + C_n y_n,$$

where $y_1, y_2, \cdots, y_n$ are a fundamental system of n particular integrals of the homogeneous equation, and where $C_1, C_2, \cdots, C_n$ are n arbitrary constants. It often happens in practice that we can easily obtain a particular integral of a linear non-homogeneous equation, and in this case we are led to the integration of the homogeneous equation. The search for a particular integral is facilitated by the following remark, which we need only state: If $f(x)$ is the sum of p functions $f_1(x), f_2(x), \cdots, f_p(x)$, such that we know how to find a particular integral of each of the equations

$$F(y) = f_1(x), \qquad F(y) = f_2(x), \qquad \cdots, \qquad F(y) = f_p(x),$$

the sum $Y_1 + Y_2 + \cdots + Y_p$ of these p particular integrals is an integral of the equation $F(y) = f(x)$.

In general, *if we know the general integral of the homogeneous equation, we can always obtain by quadratures the general integral of the non-homogeneous equation* (supposing, of course, that the left-hand side is the same for the two equations).

The following process, due to Lagrange, is called the *method of the variation of constants*. Let $(y_1, y_2, \cdots, y_n)$ be a fundamental system of integrals of the equation $F(y) = 0$. Imitating as much as possible the process employed for a linear equation of the first order, we shall seek to satisfy the equation (12) by taking for y an expression of the form

$$(14) \qquad y = C_1 y_1 + C_2 y_2 + \cdots + C_n y_n,$$

where $C_1, C_2, \cdots, C_n$ denote n functions of x. We can evidently establish between these n functions $n - 1$ relations chosen at pleasure, provided that they are not inconsistent with the equation (12). If we put

$$(15) \qquad \begin{cases} y_1 C_1' + y_2 C_2' + \cdots + y_n C_n' = 0, \\ y_1' C_1' + y_2' C_2' + \cdots + y_n' C_n' = 0, \\ \cdots \cdots \cdots \cdots \cdots \cdots \cdots \cdots, \\ y_1^{(n-2)} C_1' + y_2^{(n-2)} C_2' + \cdots + y_n^{(n-2)} C_n' = 0, \end{cases}$$

the successive derivatives of y up to the $(n-1)$th derivative have the values

(16)
$$\begin{cases} y' = C_1 y_1' + C_2 y_2' + \cdots + C_n y_n', \\ y'' = C_1 y_1'' + C_2 y_2'' + \cdots + C_n y_n'', \\ \cdots \cdots \cdots \cdots \cdots \cdots \cdots \cdots \cdots, \\ y^{(n-1)} = C_1 y_1^{(n-1)} + C_2 y_2^{(n-1)} + \cdots C_n y_n^{(n-1)}. \end{cases}$$

The first of the relations (15) has been chosen in such a way that the first derivative y' has the same expression as if $C_1, C_2, \cdots, C_n$ were constants, and similarly for those that follow. The derivative of the nth order has a less simple form:

$$\begin{aligned} y^{(n)} = &\; C_1 y_1^{(n)} + C_2 y_2^{(n)} + \cdots + C_n y_n^{(n)} \\ &+ y_1^{(n-1)} C_1' + y_2^{(n-1)} C_2' + \cdots + y_n^{(n-1)} C_n'. \end{aligned}$$

Substituting the preceding values of $y, y', y'', \cdots, y^{(n)}$ in the left-hand side of the equation (12), the coefficients of $C_1, C_2, \cdots, C_n$ are respectively $F(y_1), F(y_2), \cdots, F(y_n)$, and we are led to the new relation

(15′) $\qquad y_1^{(n-1)} C_1' + y_2^{(n-1)} C_2' + \cdots + y_n^{(n-1)} C_n' = f(x),$

which, together with the relations (15), enables us to determine $C_1', \cdots, C_n'$. We can therefore find $C_1, C_2, \cdots, C_n$ by quadratures.

We can also make use of the following method, due to Cauchy.

Let $(y_1, y_2, \cdots, y_n)$ be a fundamental system of integrals of the equation $F(y) = 0$. Let us determine the constants $C_1, C_2, \cdots, C_n$ so that the integral $C_1 y_1 + \cdots + C_n y_n$ and its first $n-2$ derivatives all vanish, while the $(n-1)$th derivative reduces to unity for a value α of x. The integral $\phi(x, \alpha)$ thus obtained depends, of course, upon the variable x and also upon the initial value α, and satisfies the n conditions

(17) $\quad \phi(\alpha, \alpha) = 0, \quad \phi'(\alpha, \alpha) = 0, \quad \phi''(\alpha, \alpha) = 0, \quad \cdots, \quad \phi^{(n-1)}(\alpha, \alpha) = 1,$

where $\phi^{(p)}(\alpha, \alpha)$ denotes the pth derivative of $\phi(x, \alpha)$ with respect to x for the value $x = \alpha$. If we replace α by x in the preceding relations, which amounts to a simple change of notation, they can also be written in the form

(17′) $\quad \phi(x, x) = 0, \quad \phi'(x, x) = 0, \quad \cdots, \quad \phi^{(n-2)}(x, x) = 0, \quad \phi^{(n-1)}(x, x) = 1,$

where $\phi^{(p)}(x, x)$ denotes now the pth derivative of $\phi(x, \alpha)$ with respect to x, in which we have replaced α by x after the differentiation. With this understanding let us consider the function represented by the definite integral

(18) $$Y = \int_{x_0}^{x} \phi(x, \alpha) f(\alpha) \, d\alpha$$

with an arbitrarily fixed lower limit x_0. Applying the general formula for differentiation, and taking account of the conditions (17′), we find successively

$$\frac{dY}{dx} = \int_{x_0}^{x} \phi'(x, \alpha) f(\alpha) \, d\alpha, \quad \cdots, \quad \frac{d^{n-1} Y}{dx^{n-1}} = \int_{x_0}^{x} \phi^{(n-1)}(x, \alpha) f(\alpha) \, d\alpha,$$

$$\frac{d^n Y}{dx^n} = \int_{x_0}^{x} \phi^{(n)}(x, \alpha) f(\alpha) \, d\alpha + f(x).$$

Substituting these values of Y, Y', $\cdots$, $Y^{(n)}$ in $F(Y)$, we find

$$F(Y) = f(x) + \int_{x_0}^{x} [\phi^{(n)}(x, \alpha) + a_1 \phi^{(n-1)}(x, \alpha) + \cdots + a_n \phi(x, \alpha)] f(\alpha) d\alpha.$$

The function under the integral sign on the right is identically zero, since $\phi(x, \alpha)$ is an integral of corresponding homogeneous equation, whatever may be the value of the parameter α. From this it follows that the function Y represented by the definite integral (18) is a particular integral of the non-homogeneous linear equation. It will be noticed that this integral, as well as its first $(n-1)$ derivatives, is zero for the lower limit x_0, which is supposed different from a singular point.*

The application of this method to the equation $d^n y/dx^n = f(x)$ leads to precisely the result obtained above (§ 18).

40. Depression of the order of a linear equation. If we know a certain number of particular integrals of a linear equation, we can make use of them to diminish the order of the equation. Let us consider first a homogeneous equation of the nth order, and let $y_1, y_2, \cdots, y_p$, $(p < n)$ be linearly independent integrals of that equation. The substitution $y = y_1 z$, where z indicates the new dependent variable, reduces the proposed equation $F(y) = 0$ to a new equation of the same type in z, for the expression for any one of the derivatives $d^p y/dx^p$ is itself linear with respect to z and to its derivatives. If y_1 is an integral of the equation $F(y) = 0$, the new equation in z must have $z = 1$ for a solution, which requires that the coefficient of z shall be zero; this fact is verified at once by calculation, for the coefficient of z is precisely $F(y_1)$. The equation in z is therefore of the form

$$(19) \qquad y_1 \frac{d^n z}{dx^n} + b_1 \frac{d^{n-1} z}{dx^{n-1}} + \cdots + b_{n-1} \frac{dz}{dx} = 0,$$

---

*It is easy to verify that the method of the variation of constants and Cauchy's method lead to the same calculations. In fact, the function $\phi(x, \alpha)$ of Cauchy is of the form

$$\phi(x, \alpha) = \phi_1(\alpha) y_1(x) + \phi_2(\alpha) y_2(x) + \cdots + \phi_n(\alpha) y_n(x),$$

where the functions $\phi_i(\alpha)$ are determined by the conditions

$$(A) \quad \begin{cases} \phi_1(\alpha) y_1(\alpha) + \cdots + \phi_n(\alpha) y_n(\alpha) = 0, \\ \cdots \cdots \cdots \cdots \cdots \cdots \cdots \cdots \cdots \cdots \cdots \\ \phi_1(\alpha) y_1^{(n-2)}(\alpha) + \cdots + \phi_n(\alpha) y_n^{(n-2)}(\alpha) = 0, \\ \phi_1(\alpha) y_1^{(n-1)}(\alpha) + \cdots + \phi_n(\alpha) y_n^{(n-1)}(\alpha) = 1, \end{cases}$$

and the particular integral (18) has the value

$$Y = y_1(x) \int_{x_0}^{x} \phi_1(\alpha) f(\alpha) d\alpha + \cdots + y_n(x) \int_{x_0}^{x} \phi_n(\alpha) f(\alpha) d\alpha.$$

But if we compare the conditions (A) with the relations (15) and (15') which determine the C_i' in the method of the variation of constants, we see at once that we have $C_i'(x) = \phi_i(x) f(x)$, and therefore the first method gives us a particular integral by the same quadratures.

where $b_1, b_2, \cdots, b_{n-1}$ are functions of x. This equation reduces to a linear equation of order $n-1$,

$$(20) \qquad y_1 \frac{d^{n-1}u}{dx^{n-1}} + b_1 \frac{d^{n-2}u}{dx^{n-2}} + \cdots + b_{n-1}u = 0,$$

by putting $u = dz/dx$. If $y_1, y_2, \cdots, y_p$ are p integrals of the equation from which we started, the equation (19) has the $p-1$ integrals $y_2/y_1, \cdots, y_p/y_1$, and therefore the equation (20) has the integrals

$$\frac{d}{dx}\left(\frac{y_2}{y_1}\right), \quad \cdots, \quad \frac{d}{dx}\left(\frac{y_p}{y_1}\right).$$

These $p-1$ integrals are linearly independent; otherwise there would exist a relation of the form

$$C_2 \frac{d}{dx}\left(\frac{y_2}{y_1}\right) + \cdots + C_p \frac{d}{dx}\left(\frac{y_p}{y_1}\right) = 0,$$

where $C_2, C_3, \cdots, C_p$ are constants not all of which are zero, and we could conclude from it, by integration, the existence of a relation of the same form, $C_2 y_2 + \cdots + C_p y_p + C_1 y_1 = 0$, where C_1 is a new constant. If $p > 1$, the application of the same process leads from the equation (20) to a new linear equation of order $n-2$, and so on. *The integration of a linear homogeneous equation of which p independent particular integrals are known reduces, therefore, to the integration of a linear homogeneous equation of order $n-p$, followed by quadratures.* When $p = n-1$, the last equation will also be integrable by a quadrature.

Similarly, if we know p integrals, $y_1, y_2, \cdots, y_p$, of a non-homogeneous equation, such that the $p-1$ functions

$$y_2 - y_1, \quad \cdots, \quad y_p - y_1$$

are linearly independent, the substitution $y = y_1 + z$ leads to a homogeneous equation having the $p-1$ integrals $y_2 - y_1, \cdots, y_p - y_1$. It is therefore possible to reduce this equation to a linear homogeneous equation of order $n - p + 1$.

Consider, for example, the linear equation of the second order,

$$(21) \qquad F(y) = \frac{d^2y}{dx^2} + p\frac{dy}{dx} + qy = 0,$$

and let y_1 be a particular integral of this equation. If we put $y = y_1 z$, we find

$$\frac{dy}{dx} = y_1 \frac{dz}{dx} + z\frac{dy_1}{dx}, \qquad \frac{d^2y}{dx^2} = y_1 \frac{d^2z}{dx^2} + 2\frac{dy_1}{dx}\frac{dz}{dx} + z\frac{d^2y_1}{dx^2},$$

and, substituting in the equation (21), we find, since the coefficient of z is zero,

$$(22) \qquad y_1 \frac{d^2 z}{dx^2} + \left(2 \frac{dy_1}{dx} + p y_1\right) \frac{dz}{dx} = 0.$$

Putting $dz/dx = u$, this equation can be written in the form

$$\frac{du}{u} + \left(2 \frac{dy_1}{dx} + p y_1\right) \frac{dx}{y_1} = 0;$$

whence, by integration,

$$\text{Log } u + \int_{x_0}^{x} p\, dx + \text{Log } y_1^2 = \text{Log } C,$$

or

$$u = \frac{C}{y_1^2} e^{-\int_{x_0}^{x} p\, dx}.$$

A second quadrature will give z and consequently y. We see that the equation (21) has the integral y_2 given by the expression

$$(23) \qquad y_2 = y_1 \int_{x_0}^{x} \frac{dx}{y_1^2} e^{-\int_{x_0}^{x} p\, dx},$$

which is independent of y_1. *The general integral of a linear homogeneous equation of the second order is therefore obtained by two quadratures when we know a particular integral.**

This property is a mark of similarity between the linear equation of the second order and Riccati's equation (§ 7). There exists, in fact, a very close relation between these two kinds of differential equations. If we depress the order of the homogeneous equation (21) by the process of § 19, by substituting

$$y = e^{\int z\, dx},$$

we are led to a Riccati equation,

$$(24) \qquad z' + z^2 + pz + q = 0.$$

---

* We can derive from these results a very simple proof of an important theorem of Sturm. Let us suppose that the coefficients p and q are continuous real functions of the real variable x in the interval (a, b), and let x_0, x_1 be two consecutive zeros of a particular integral $y_1(x)$ in the interval (a, b). If $y_2(x)$ is another particular integral independent of $y_1(x)$, the formula which gives u can be written

$$\frac{d}{dx}\left(\frac{y_2}{y_1}\right) = \frac{C}{y_1^2} e^{-\int_{x_0}^{x} p\, dx},$$

which shows that the quotient y_2/y_1 varies always in the same sense when x increases from x_0 to x_1. Now this quotient is infinite for $x = x_0$ and for $x = x_1$; hence it constantly increases from $-\infty$ to $+\infty$ or decreases from $+\infty$ to $-\infty$. *The equation $y_2(x) = 0$ has therefore one and only one root in the interval (x_0, x_1).*

Conversely, any Riccati equation

(25) $$u' + au^2 + bu + c = 0,$$

where a, b, c are functions of $x (a \neq 0)$, may be reduced to the form (24) by putting $u = z/a$, which transforms (25) to an equation of the form (24),

$$z' + z^2 + \left(b - \frac{a'}{a}\right)z + ac = 0.$$

It follows that the general integral of the equation (25) is

(26) $$u = \frac{1}{a}\frac{C_1 y_1' + C_2 y_2'}{C_1 y_1 + C_2 y_2},$$

where y_1 and y_2 are two independent integrals of the linear equation

$$y'' + \left(b - \frac{a'}{a}\right)y' + acy = 0.$$

This expression really contains only a single arbitrary constant, that is, the quotient C_2/C_1, which appears in it to the first degree.*

Example. Legendre's polynomial X_n (I, § 90, 2d ed.; § 88, 1st ed.) satisfies the differential equation

(27) $$(1 - x^2)\frac{d^2 y}{dx^2} - 2x\frac{dy}{dx} + n(n+1)y = 0.$$

---

* It would seem that a quadrature might be necessary to derive the general integral of the linear equation (21) from the general integral of Riccati's equation (24). In reality this is not the case, or, rather, the quadrature reduces to the calculation of $\int p\,dx$. In general, let $z = \phi(x, C)$ be the general integral of a differential equation of the first order, $dz/dx = f(x, z)$. From the relation

$$\frac{\partial \phi}{\partial x} = f(x, \phi)$$

we derive, by differentiating with respect to the constant C,

$$\frac{\partial^2 \phi}{\partial C\, \partial x} = \frac{\partial f}{\partial \phi}\frac{\partial \phi}{\partial C} \quad \text{or} \quad \frac{\partial f}{\partial \phi} = \frac{\partial}{\partial x}\left(\text{Log}\,\frac{\partial \phi}{\partial C}\right).$$

From the last equation we find $\int (\partial f/\partial \phi)\,dx = \text{Log}\,(\partial \phi/\partial C)$, where, of course, the same value of the constant C is to be understood in the two sides of the equation. Applying this to Riccati's equation (24), if $z = \phi(x, C)$ is the general integral of that equation, we conclude that

$$2\int z\,dx + \int p\,dx + \text{Log}\left(\frac{\partial \phi}{\partial C}\right) = 0.$$

If z_1, z_2, z_3 are three integrals of the equation (24), on carrying out the calculation (see § 7) we find that the general integral of the linear equation (21) has the form

$$y = e^{-\frac{1}{2}\int p\,dx}\,\frac{C_1(z_3 - z_1) + C_2(z_3 - z_2)}{\sqrt{(z_1 - z_2)(z_2 - z_3)(z_3 - z_1)}}.$$

To prove this it suffices to notice that, by putting $u = (x^2 - 1)^n$, we have the relation $(x^2 - 1) u' = 2 n x u$, and by taking the $(n + 1)$th derivative of the two sides we have an equation which is identical with the equation (27) when we replace $d^n u/dx^n$ in it by y. In order to obtain a second particular integral of the equation (27), we shall apply the general formula (23) with p equal to $2x/(x^2 - 1) = 1/(x + 1) + 1/(x - 1)$; this gives

$$y_2 = X_n \int \frac{dx}{(x^2 - 1) X_n^2}.$$

It might seem necessary to know the n roots, $\alpha_1, \alpha_2, \cdots, \alpha_n$, of the polynomial X_n in order to calculate this integral, but this is not the case. For, let us write the integrand in a form which exhibits the simple fractions which come from the roots $+1$ and -1 of the denominator:

$$\frac{1}{(x^2 - 1) X_n^2} = \frac{1}{2}\left(\frac{1}{x-1} - \frac{1}{x+1}\right) + \frac{P_n}{X_n^2},$$

where P_n is a polynomial of degree $2n - 2$, the quotient obtained by dividing $1 - X_n^2$ by $x^2 - 1$. It follows that

$$y_2 = \frac{1}{2} X_n \operatorname{Log}\left(\frac{x-1}{x+1}\right) + X_n \int \frac{P_n}{X_n^2} dx.$$

This last integral is a rational function, for if it contained a logarithmic term such as $\operatorname{Log}(x - \alpha_i)$, the point α_i would be a singular point for y_2, and the integrals of the equation (27) can have no other singular points than $x = \pm 1$ (§ 37). We can therefore calculate this integral by rational operations (I, § 104, 2d ed.; § 109, 1st ed.). Since the integral must be of the form Q_{n-1}/X_n, where Q_{n-1} may be taken as a polynomial of degree not greater than $n-1$, we can determine the coefficients of this polynomial, for example, by the condition

$$Q'_{n-1} X_n - Q_{n-1} X'_n = P_n.$$

Having once obtained the polynomial Q_{n-1}, we may write the general integral of the equation (27) in the form

$$y = C_1 X_n + C_2 \left[Q_{n-1} + \frac{1}{2} X_n \operatorname{Log}\left(\frac{x-1}{x+1}\right)\right].$$

41. Analogies with algebraic equations. The preceding properties establish an evident analogy between the theory of linear differential equations and the theory of algebraic equations. This analogy persists in a large number of questions. As an example of this we shall show how we can extend to linear equations the theory of the greatest common divisor. In general, let

$$F(y) = a_0 \frac{d^n y}{dx^n} + a_1 \frac{d^{n-1} y}{dx^{n-1}} + \cdots + a_{n-1} \frac{dy}{dx} + a_n y$$

be a symbolic polynomial where $a_0, a_1, \cdots, a_n$ are given functions of x. If a_0 is not zero, we shall say for brevity that $F(y)$ is of the nth order. If $G(y)$ is a symbolic polynomial of the same nature and of the pth order, it is clear that $G[F(y)]$ is again a symbolic polynomial of the same kind and of the $(n + p)$th order. Let now

$$F_1(y) = b_0 \frac{d^m y}{dx^m} + b_1 \frac{d^{m-1} y}{dx^{m-1}} + \cdots + b_{m-1} \frac{dy}{dx} + b_m y$$

be another polynomial of order $m\ (m \leqq n)$. We can find a third polynomial $G(y)$ of order $n - m$ such that $F(y) - G[F_1(y)]$ is at most of order $m - 1$ (a polynomial of order zero is of the form ay, where a is a function of x). Let us put

$$G(y) = \lambda_0 \frac{d^{n-m}y}{dx^{n-m}} + \lambda_1 \frac{d^{n-m-1}y}{dx^{n-m-1}} + \cdots + \lambda_{n-m} y.$$

The coefficient of $d^n y/dx^n$ in $G[F_1(y)]$ is $\lambda_0 b_0$, and if we take $\lambda_0 = a_0/b_0$, the difference $F(y) - \lambda_0 d^{n-m}[F_1(y)]/dx^{n-m}$ will be at most of order $n - 1$. Let a_1' be the coefficient of $d^{n-1}y/dx^{n-1}$ in this difference. If we take $\lambda_1 = a_1'/b_0$, the difference

$$F(y) - \lambda_0 \frac{d^{n-m}}{dx^{n-m}}[F_1(y)] - \lambda_1 \frac{d^{n-m-1}}{dx^{n-m-1}}[F_1(y)]$$

will be at most of order $n - 2$. Continuing in this way, we see that we can determine, step by step, the coefficients $\lambda_0, \lambda_1, \cdots, \lambda_{n-m}$ in such a way as to obtain an identity of the form

(28) $$F(y) - G[F_1(y)] = F_2(y),$$

where $F_2(y)$ is at most of order $m - 1$. This operation is entirely analogous to the division of one algebraic polynomial by another.

Now suppose that we wish to obtain the integrals common to two linear equations

(29) $$F(y) = 0, \qquad F_1(y) = 0.$$

The identity (28) shows that these integrals are the same as the integrals common to the two equations $F_1(y) = 0$, $F_2(y) = 0$. If $F_2(y)$ is not identically zero, the same operations can be repeated on $F_1(y)$ and $F_2(y)$, and so on until we arrive at two consecutive polynomials, $F_{k-1}(y)$ and $F_k(y)$, such that $F_{k-1}(y) = G_{k-1}[F_k(y)]$. This last symbolic polynomial $F_k(y)$ is the analogue of the algebraic greatest common divisor: *all the integrals common to the two equations* (29) *satisfy the linear equation* $F_k(y) = 0$, *and conversely*. If $F_k(y)$ is of the degree zero, the two equations have no other common integral than the trivial solution $y = 0$.

If in the relation (28) $F_2(y)$ is identically zero, the equation $F(y) = 0$ has all the integrals of $F_1(y) = 0$. Conversely, in order that $F(y) = 0$ shall have all the integrals of $F_1(y) = 0$, it is necessary that $F_2(y)$ be identically zero, for a linear equation of order not greater than $m - 1$ cannot have all the integrals of a linear equation of the mth order. Hence in this case we have identically

$$F(y) = G[F_1(y)],$$

and if we put $F_1(y) = z$, the integration of $F(y) = 0$ is reduced to the successive integration of the two linear equations

$$G(z) = 0, \qquad F_1(y) = z,$$

of orders $n - m$ and m, of which only the second is non-homogeneous.

We can deduce from this observation another solution of a problem already treated. Suppose that we know p independent integrals $y_1, y_2, \cdots, y_p\ (p < n)$ of $F(y) = 0$. We can form a linear equation of the pth order having these p integrals (§ 38). Let $F_1(y) = 0$ be this equation of the pth order; then we have identically $F(y) = G[F_1(y)]$, and if the equation $G(z) = 0$ of order $n - p$ has been integrated, we can integrate $F_1(y) = z$ by quadratures alone, since we know the general integral of $F_1(y) = 0$. The reduction is the same as by the first method, but the new process is more symmetric.

Appel, Laguerre, Halphen, E. Picard, and many others after them have extended to linear equations the theory of symmetric functions of the roots, the theory of invariants, and the very fundamental work of Galois relative to the group of an algebraic equation. The theory of invariants is founded on the easily verified fact that a linear homogeneous equation is changed into a new equation of the same kind by every transformation of the type

$$x = f(t), \quad y = z\phi(t),$$

where t is the new independent variable and z the new dependent variable, whatever the functions $f(t)$ and $\phi(t)$ may be.

We can sometimes make use of this transformation to simplify a linear equation. For example, if we wish to make the coefficient of the derivative of order $n-1$ disappear, we find that it suffices to put

$$y = z e^{-\frac{1}{n}\int a_1 dx},$$

retaining the variable x. Since we have two arbitrary functions f and ϕ at our disposal, it would seem that we could take advantage of them to make two coefficients disappear; but this reduction, although theoretically possible, is illusory in most cases. For example, we can always choose the functions f and ϕ so as to reduce any linear equation of the second order to the simple form $z'' = 0$, but the actual determination of these functions presents the same difficulties as the problem of integrating the original equation.

42. The adjoint equation. Lagrange extended the theory of integrating factors to linear equations in the following way. Let $F(y)$ be a linear function of y and of its first n derivatives,

$$F(y) = a_0 y^{(n)} + a_1 y^{(n-1)} + \cdots + a_{n-1} y' + a_n y,$$

where $a_0, a_1, \cdots, a_n$ are any functions of x, and where $y', y'', \cdots, y^{(n)}$ denote the successive derivatives of y. Let us try to find a function z of x such that the product $zF(y)$ shall be the derivative with respect to x of another function linear in y and in its derivatives up to those of order $n-1$. The general formula for integration by parts (I, § 87, 2d ed.; § 85, 1st ed.), applied to each of the terms of the product $zF(y)$, gives us

$$(30) \quad \begin{cases} zF(y) = \dfrac{d}{dx}\left[a_0 z y^{(n-1)} - \dfrac{d}{dx}(a_0 z) y^{(n-2)} + \cdots \pm y \dfrac{d^{n-1}(a_0 z)}{dx^{n-1}} \right] \\[4pt] \quad + \dfrac{d}{dx}\left[a_1 z y^{(n-2)} - \dfrac{d}{dx}(a_1 z) y^{(n-3)} + \cdots \mp y \dfrac{d^{n-2}(a_1 z)}{dx^{n-2}} \right] \\[4pt] \quad + \cdots\cdots\cdots\cdots\cdots\cdots\cdots\cdots\cdots\cdots\cdots\cdots\cdots\cdots\cdots \\[4pt] \quad + \dfrac{d}{dx}[a_{n-1} z y] + y G(z), \end{cases}$$

where we have put

$$(31) \quad G(z) = (-1)^n \frac{d^n(a_0 z)}{dx^n} + (-1)^{n-1} \frac{d^{n-1}(a_1 z)}{dx^{n-1}} + \cdots - \frac{d(a_{n-1} z)}{dx} + a_n z.$$

If we denote by $\Psi(y, z)$ the expression which appears on the right-hand side of the equation (30) which is bilinear with respect to y and z and to their derivatives, we can write that equation in the abridged form

$$(32) \qquad zF(y) - yG(z) = \frac{d}{dx}[\Psi(y, z)],$$

so that for all the possible forms of the functions y and z the binomial $zF(y) - yG(z)$ is the derivative of $\Psi(y, z)$. If we now take for z an integral of the equation $G(z) = 0$, the product $zF(y)$ is the derivative of an expression of the same form, linear with respect to $y, y', \cdots, y^{(n-1)}$, and the equation $F(y) = 0$ is equivalent to a linear equation of order $n - 1$,

$$(33) \qquad \Psi(y, z) = C,$$

which we obtain by replacing in Ψ the function z by the integral in question. Now the equation $G(z) = 0$ is likewise a linear equation of the nth order; it is called the *adjoint equation* of $F(y) = 0$, and the symbolic polynomial $G(z)$ is called the *adjoint polynomial* of $F(y)$.

We see, then, that if we know an integral of the adjoint equation, the integration of the given equation is reduced to the integration of a linear equation of order $n - 1$ whose right-hand side is an arbitrary constant. If we know p independent integrals, $z_1, z_2, \cdots, z_p$, of the adjoint equation, every integral of the given equation satisfies p relations of the form

$$(34) \quad \Psi(y, z_1) = C_1, \qquad \Psi(y, z_2) = C_2, \quad \cdots, \quad \Psi(y, z_p) = C_p,$$

where $C_1, C_2, \cdots, C_p$ denote p constants. Eliminating the derivatives $y^{(n-1)}$, $y^{(n-2)}, \cdots, y^{(n-p+1)}$ from these p equations, we obtain a linear equation of order $n - p$ whose right-hand side depends upon the p arbitrary constants C_1, $C_2, \cdots, C_p$. In particular, if $p = n$ (that is, if we know the general integral of the adjoint equation), we can solve the n equations (34) for $y, y', \cdots, y^{(n-1)}$, and we can obtain the general integral of the given equation without any quadrature.

There exist between the integrals of the two equations $F(y) = 0$, $G(z) = 0$ some remarkable relations, which we cannot develop here.* We shall only show that there exists a *reciprocal relation* between these two equations. More precisely, if $G(z)$ is the adjoint polynomial of $F(y)$, then, conversely, $F(y)$ is the adjoint polynomial of $G(z)$. For if $F_1(y)$ denotes the adjoint polynomial of $G(z)$, we have a relation between $F_1(y)$ and $G(z)$ of the same form as the relation (32),

$$(32') \qquad yG(z) - zF_1(y) = \frac{d}{dx}[\Psi_1(y, z)].$$

Adding the relations (32) and (32'), we find

$$z[F(y) - F_1(y)] = \frac{d}{dx}[\Psi(y, z) + \Psi_1(y, z)].$$

If $F(y) - F_1(y)$ were not zero, the product $z[F(y) - F_1(y)]$ would be the derivative of a function containing z and some of its derivatives. Now the derivative of a function containing $z, z', \cdots, z^{(p)}$ contains at least one derivative of z, namely, $z^{(p+1)}$. The preceding relation is therefore possible only if $F_1(y)$ is identical with $F(y)$.

---

* See DARBOUX, *Théorie des surfaces*, Vol. II, Bk. IV, chap. v. See also Exercise 17, p. 171, at the end of this chapter.

II. THE STUDY OF SOME PARTICULAR EQUATIONS

43. Equations with constant coefficients. Linear differential equations with constant coefficients were integrated by Euler. Consider first a homogeneous equation

$$(35) \quad F(y) = y^{(n)} + a_1 y^{(n-1)} + \cdots + a_{n-1} y' + a_n y = 0,$$

where $a_1, a_2, \cdots, a_n$ are any constants. By the general theory (§ 37) none of the integrals of this equation have a singular point in the finite plane; that is, they are *integral* functions of x. Let

$$(36) \quad y = c_0 + c_1 \frac{x}{1} + c_2 \frac{x^2}{2!} + \cdots + c_m \frac{x^m}{m!} + \cdots$$

be the development in series of an integral. The series which represent the successive derivatives have an analogous form. Replacing y and its successive derivatives by their developments in series in the left-hand side of the equation (35), and equating to zero the coefficient of any power of x, say x^p, we obtain the following relation between $n + 1$ consecutive coefficients:

$$(37) \quad c_{n+p} + a_1 c_{n+p-1} + a_2 c_{n+p-2} + \cdots + a_{n-1} c_{p+1} + a_n c_p = 0.$$

If we substitute in it successively $p = 0, 1, 2, \cdots$, we can calculate, step by step, all the coefficients $c_n, c_{n+1}, \cdots$, in terms of the n first coefficients $c_0, c_1, \cdots, c_{n-1}$, which may be taken arbitrarily. The series (36) thus obtained is convergent in the whole plane and represents the integral which for $x = 0$ is equal to c_0, while the first $n - 1$ derivatives take on respectively the values $c_1, c_2, \cdots, c_{n-1}$ for $x = 0$. We shall show that this integral can be expressed in terms of exponential functions when it does not reduce to a polynomial.

The equation (37) is a recurrent formula with constant coefficients which connects the $n + 1$ consecutive coefficients. Now it is easy to find particular solutions of that equation. For this purpose, let us consider the algebraic equation

$$(38) \quad f(r) = r^n + a_1 r^{n-1} + a_2 r^{n-2} + \cdots + a_{n-1} r + a_n = 0,$$

which, for the sake of brevity, we shall call the *auxiliary equation*, the left-hand side $f(r)$ being the *auxiliary polynomial*. If r is a root of this equation, it is clear that the relation (37) is satisfied, whatever may be the value of the integer p, by putting $c_m = r^m$. The particular integral thus obtained is equal to e^{rx}, and we see that e^{rx} *is a particular integral of the equation* (35) *if r is a root of the auxiliary equation* $f(r) = 0$. The verification is immediate, for if we replace y by e^{rx} in the left-hand side of the equation (35), the result of the substitution is $e^{rx}f(r)$.

If the equation (38) has n distinct roots $r_1, r_2, \cdots, r_n$, we know n particular integrals $e^{r_1 x}, e^{r_2 x}, \cdots, e^{r_n x}$, and therefore an integral

(39) $$y = C_1 e^{r_1 x} + C_2 e^{r_2 x} + \cdots + C_n e^{r_n x},$$

the expression for which contains n arbitrary constants $C_1, C_2, \cdots, C_n$. This expression represents the general integral, for the determinant $\Delta(e^{r_1 x}, e^{r_2 x}, \cdots, e^{r_n x})$ can be written in the form

$$\Delta = e^{(r_1 + r_2 + \cdots + r_n)x} \begin{vmatrix} 1 & r_1 & r_1^2 & \cdots & r_1^{n-1} \\ 1 & r_2 & r_2^2 & \cdots & r_2^{n-1} \\ \cdots & \cdots & \cdots & \cdots & \cdots \\ 1 & r_n & r_n^2 & \cdots & r_n^{n-1} \end{vmatrix},$$

and the determinant on the right is, except for sign, the product of the differences $r_i - r_k$.

Before studying the case in which the auxiliary equation has multiple roots, we shall prove a lemma. Let us make the substitution $y = e^{\alpha x} z$ in the equation (35), where α is any constant and z the new dependent variable; by Leibnitz's formula we have

(40) $$\begin{cases} y' = e^{\alpha x}(\alpha z + z'), \\ \cdots \cdots \cdots \cdots \cdots, \\ y^{(p)} = e^{\alpha x}\left(\alpha^p z + \frac{p}{1}\alpha^{p-1}z' + \frac{p(p-1)}{2!}\alpha^{p-2}z'' + \cdots + z^{(p)}\right), \\ \cdots \cdots \cdots \cdots \cdots \cdots \cdots \cdots \cdots \cdots \cdots \cdots \end{cases}$$

Substituting these values of $y, y', y'', \cdots$ in the left-hand side of the equation (35), $e^{\alpha x}$ appears as a factor, and we have

$$F(e^{\alpha x} z) = e^{\alpha x} G(z),$$

where $G(z)$ is a linear expression in $z, z', \cdots, z^{(n)}$ with constant coefficients. In order to calculate the coefficients of $G(z)$, let us observe that if we replace in $F(y)$ the indices which indicate differentiation by exponents, and y itself by $y^0 = 1$, the result obtained is identical with $f(y)$. If we carry out the same transformation with the function z, the formulæ (40) may be written symbolically

$$y^p = e^{\alpha x}(\alpha + z)^p;$$

hence $G(z)$ can also be written, in the same symbolic notation, $f(\alpha + z)$, and, replacing the exponents of z by the indices which indicate differentiation, we see that the new equation in z is

(41) $$F(e^{\alpha x} z) = e^{\alpha x}\left[f(\alpha) z + f'(\alpha) z' + \frac{f''(\alpha)}{2!} z'' + \cdots + \frac{f^{(n)}(\alpha)}{n!} z^{(n)}\right] = 0.$$

Now let r be a p-fold root of the auxiliary equation; if we replace a by that root r in the equation (41), the coefficients of $z, z', z'', \cdots, z^{(p-1)}$ in this equation are zero, and we obtain an integral by taking for z a function whose pth derivative is zero, that is, an arbitrary polynomial of degree $p-1$. Consequently, *if r is a p-fold root of the auxiliary equation, to that root corresponds p particular independent integrals of the linear equation* (35), $e^{rx}, xe^{rx}, \cdots, x^{p-1}e^{rx}$.

Let the k distinct roots of $f(r) = 0$ be $r_1, r_2, \cdots, r_k$, and let their respective orders of multiplicity be denoted by $\mu_1, \mu_2, \cdots, \mu_k$ ($\Sigma\mu_i = n$). From these roots we can form n particular integrals of the linear equation. These n integrals are independent, for any linear relation with constant coefficients between these n integrals would lead to an identity of the form

$$e^{r_1 x}\phi_1(x) + e^{r_2 x}\phi_2(x) + \cdots + e^{r_k x}\phi_k(x) = 0,$$

where $\phi_1, \phi_2, \cdots, \phi_n$ denote polynomials not all of which vanish identically. Such a relation is impossible if the k numbers $r_1, r_2, \cdots, r_k$ are distinct. For, let $n_1, n_2, \cdots, n_k$ be the respective degrees of these polynomials. It is understood that any term in the identity is simply omitted if the corresponding polynomial is zero. Dividing by $e^{r_1 x}$, we can again write this relation in the form

$$\phi_1(x) + e^{(r_2 - r_1)x}\phi_2(x) + \cdots + e^{(r_k - r_1)x}\phi_k(x) = 0.$$

Differentiating both sides of this equation, we have

$$\phi_1'(x) + e^{(r_2 - r_1)x}[\phi_2'(x) + (r_2 - r_1)\phi_2(x)] + \cdots = 0.$$

The degree of the polynomial which multiplies $e^{(r_2 - r_1)x}$ is again equal to n_2, and the polynomial does not vanish; and similarly for the others. After having differentiated $(n_1 + 1)$ times, we shall have, therefore, a relation of the same form as the relation from which we started, but with one term less,

$$e^{s_2 x}\psi_2(x) + e^{s_3 x}\psi_3(x) + \cdots + e^{s_k x}\psi_k(x) = 0,$$

where the $k - 1$ numbers $s_2, \cdots, s_k$ are different, and where $\psi_2, \psi_3, \cdots, \psi_k$ are polynomials of degrees $n_2, n_3, \cdots, n_k$ respectively. Continuing in this way, we arrive finally at a relation of the form $e^{tx}\pi(x) = 0$, where $\pi(x)$ is a polynomial not identically zero. But this is evidently absurd. The general integral of the linear equation (35) is therefore represented by the expression

(42) $$y = e^{r_1 x}P_{\mu_1 - 1} + e^{r_2 x}P_{\mu_2 - 1} + \cdots + e^{r_k x}P_{\mu_k - 1},$$

where $P_{\mu_1 - 1}, \cdots, P_{\mu_k - 1}$ are polynomials with arbitrary coefficients, of degrees equal to their subscripts.

If the auxiliary equation has imaginary roots, the general integral (42) contains imaginary symbols, but we can make these imaginaries disappear if the coefficients $a_1, a_2, \cdots, a_n$ are real. For in this case, if the equation $f(r) = 0$ has the root $\alpha + \beta i$ of multiplicity p, $\alpha - \beta i$ is also a root of the same degree of multiplicity. The sum of the two terms of the formula (42) coming from these two roots can be written

$$e^{\alpha x}[(\cos \beta x + i \sin \beta x)\Phi(x) + (\cos \beta x - i \sin \beta x)\Psi(x)],$$

where $\Phi(x)$ and $\Psi(x)$ are two polynomials of degree $p - 1$ with arbitrary coefficients, or in the equivalent form

$$e^{\alpha x}[\cos \beta x \Phi_1(x) + \sin \beta x \Psi_1(x)],$$

where Φ_1 and Ψ_1 are also two arbitrary polynomials of degree $p - 1$.

Note. In order to express the general integral of the equation (35) in terms of exponential functions, we observe that it is first necessary to solve the equation $f(r) = 0$. If this equation is not solved, the recurrent relations (37) enable us always to calculate, step by step, as many as we wish of the coefficients of the power series which represents the integral corresponding to the given initial conditions.

We can determine in advance the number of coefficients which it suffices to calculate in order to obtain the value of the integral with a certain degree of approximation. Let A be the largest of the numbers $1, |a_1|, \cdots, |a_n|$, and B the largest of the numbers $|c_0|, |c_1|, \cdots, |c_{n-1}|$. It is easy to prove, step by step, that we have $|c_{n+p}| < B(An)^{p+1}$. The absolute value of the remainder of the series which represents the integral, commencing with the term in x^{n+p}, is therefore less than the value of the series

$$B\left[\frac{(An)^{p+1}\rho^{n+p}}{(n+p)!} + \frac{(An)^{p+2}\rho^{n+p+1}}{(n+p+1)!} + \cdots\right],$$

where $\rho = |x|$, and consequently less than

$$\frac{B(An)^{p+1}\rho^{n+p}}{(n+p)!} e^{An\rho}.$$

Consider now a non-homogeneous linear equation with constant coefficients. We can avoid the use of the general method and find a particular integral directly if the right-hand side, $\phi(x)$, is a polynomial. For if the coefficient a_n of y in the equation

$$\frac{d^n y}{dx^n} + a_1 \frac{d^{n-1} y}{dx^{n-1}} + \cdots + a_{n-1}\frac{dy}{dx} + a_n y = b_0 x^m + b_1 x^{m-1} + \cdots$$

is not zero, we can find another polynomial of degree m,

$$y = \psi(x) = c_0 x^m + c_1 x^{m-1} + \cdots,$$

which, substituted for y in the left-hand side of the preceding equation, gives a result identical with $\phi(x)$. The $m + 1$ coefficients

$c_0, c_1, c_2, \cdots, c_m$ are determined, step by step, by the relations

$$a_n c_0 = b_0, \qquad a_n c_1 + m a_{n-1} c_0 = b_1,$$
$$a_n c_2 + (m-1) a_{n-1} c_1 + m(m-1) a_{n-2} c_0 = b_2, \qquad \cdots,$$

where a_n is different from zero by hypothesis. This computation is not applicable when $a_n = 0$. More generally, suppose that the derivative of the lowest order which appears in the left-hand side is the derivative of the pth order. Taking for the dependent variable $z = d^p y/dx^p$, the given equation is transformed into a linear equation of order $n - p$, where the coefficient of z is not zero. According to the case which has just been treated, this equation in z has a polynomial of the mth degree for a particular integral. Hence one particular integral of the equation in y itself is a polynomial of degree $m + p$. The coefficients of this polynomial can again be determined by a direct substitution. It should be noticed that the coefficients of $x^{p-1}, x^{p-2}, \cdots, x$, and the constant term are arbitrary.

If the right-hand side $\phi(x)$ is of the form $e^{\alpha x} P(x)$, where α is constant and $P(x)$ denotes a polynomial, we reduce this case to the preceding by putting $y = e^{\alpha x} z$, which leads to the equation

$$(43) \quad \frac{f^{(n)}(\alpha)}{n!} \frac{d^n z}{dx^n} + \frac{f^{(n-1)}(\alpha)}{(n-1)!} \frac{d^{n-1} z}{dx^{n-1}} + \cdots + f'(\alpha) \frac{dz}{dx} + f(\alpha) z = P(x).$$

This equation has for a particular integral, as we have just seen, a polynomial whose degree we can determine a priori; the equation in y has therefore a particular integral of the form $e^{\alpha x} Q(x)$, where $Q(x)$ also is a polynomial. Suppose in particular that $P(x)$ reduces to a constant factor C. If α is not a root of the auxiliary equation, the equation (43) has the particular integral $z = C/f(\alpha)$, and the equation in y has the particular integral $C e^{\alpha x}/f(\alpha)$. If α is a multiple root of multiplicity p of the auxiliary equation, the equation (43) is satisfied by putting

$$f^{(p)}(\alpha) \frac{d^p z}{dx^p} = p!\, C,$$

or

$$z = \frac{C x^p}{f^{(p)}(\alpha)},$$

and consequently the equation in y has the particular integral $C x^p e^{\alpha x}/f^{(p)}(\alpha)$. By virtue of a general remark (§ 38) we can therefore find a particular integral directly whenever the right-hand side is the sum of products of exponentials and polynomials. This is the case in particular if the right-hand side is of the form $P(x) \cos \alpha x$ or $P(x) \sin \alpha x$, for we need only express $\cos \alpha x$ and $\sin \alpha x$ in terms of $e^{\alpha x i}$ and of $e^{-\alpha x i}$. Having once recognized by the preceding considerations the form of a particular integral, it is not necessary to pass through all the indicated transformations in order to calculate the coefficients upon which it depends; it is often preferable to substitute directly in the left-hand side of the given equation.

Example. Let it be required to find the general integral of the equation

(44) $$F(y) = \frac{d^4 y}{dx^4} - y = ae^x + be^{2x} + c \sin x + g \cos 2x,$$

where a, b, c, g are constants. The auxiliary equation $r^4 - 1 = 0$ has the simple roots $1, -1, +i, -i$; the general integral of the homogeneous equation is therefore

(45) $$y = C_1 e^x + C_2 e^{-x} + C_3 \cos x + C_4 \sin x.$$

We must next find a particular integral of each of the four equations obtained by taking successively for right-hand sides $ae^x, be^{2x}, c \sin x, g \cos 2x$. Since unity is a simple root of $f(r) = r^4 - 1 = 0$, the first of these equations has the particular integral $axe^x/f'(1) = axe^x/4$. Since 2 is not a root of the equation $f(r) = 0$, the second equation has the particular integral $be^{2x}/f(2) = be^{2x}/15$.

In the third equation, $F(y) = c \sin x$, we can replace $\sin x$ by $(e^{xi} - e^{-xi})/2i$, and we have to seek a particular integral of each of the two equations

$$F(y) = \frac{c}{2i} e^{xi}, \qquad F(y) = -\frac{c}{2i} e^{-xi}.$$

Now, since $+i$ and $-i$ are simple roots of $f(r) = 0$, we know, a priori, that they have respectively two particular integrals of the form Mxe^{xi}, Nxe^{-xi}. The sum of these two integrals is of the form $x(m \cos x + n \sin x)$, and we can determine the coefficients m and n by substituting in $F(y)$ and equating the result identically to $c \sin x$. This method avoids the use of the symbol i. It turns out that it is necessary to take $m = c/4, n = 0$. We find similarly that the last equation $F(y) = g \cos 2x$ has the particular integral $g \cos 2x/15$. Adding all these particular integrals to the right-hand side of the equation (45), we obtain the general integral of the given equation (44).

44. D'Alembert's method. A large number of methods have been devised for the integration of linear equations with constant coefficients, particularly in the case where the auxiliary equation has multiple roots. One of the most interesting, which is applicable to many questions of the same kind, consists in considering a linear equation, in which $f(r) = 0$ has multiple roots, as the limit of a linear equation in which all the roots of $f(r) = 0$ are distinct. In general, let

(46) $$F(y) = \frac{d^n y}{dx^n} + a_1 \frac{d^{n-1} y}{dx^{n-1}} + \cdots + a_{n-1} \frac{dy}{dx} + a_n y = 0$$

be a linear equation, where the coefficients $a_1, a_2, \cdots, a_n$ are functions of x which depend also upon certain variable parameters $\alpha_1, \alpha_2, \cdots, \alpha_p$. Suppose that there exists a function $f(x, r)$ having the following property: for q values of r, depending upon the parameters $\alpha_1, \alpha_2, \cdots, \alpha_p$,

and in general distinct, the function $f(x, r)$ of x is an integral of the equation (46). Let $r_1, r_2, \cdots, r_q$ be these q values of r such that the functions
$$f(x, r_1), \quad f(x, r_2), \quad \cdots, \quad f(x, r_q)$$
form q independent particular integrals of the equation (46), whatever the values of the parameters $\alpha_1, \alpha_2, \cdots, \alpha_p$ may be. If for certain particular values of these parameters the q values $r_1, r_2, \cdots, r_q$ are not distinct, the number of the known integrals is diminished. Suppose, for example, that r_2 becomes equal to r_1. If r_2 is different from r_1, the equation has the two integrals $f(x, r_1)$, $f(x, r_2)$, and consequently
$$\frac{f(x, r_2) - f(x, r_1)}{r_2 - r_1}$$
is also an integral. Now, if r_2 approaches r_1, the preceding function has for its limit the derivative $[f'_r(x, r)]_{r_1}$. If a third root r_3 becomes equal to r_1, we take, similarly, supposing first that r_3 differs a little from r_1, the integral
$$\frac{f(x, r_3) - f(x, r_1) - (r_3 - r_1)[f'_r(x, r)]_{r_1}}{(r_3 - r_1)^2},$$
and this integral has for its limit $[f''_{r^2}(x, r)]_{r_1}/2$ when r_3 approaches r_1. This reasoning is perfectly general: if, for certain values of the parameters $\alpha_1, \cdots, \alpha_p$, k of the roots are equal to r_1, the corresponding equation (46) has the k particular integrals
$$f(x, r_1), \quad \left(\frac{\partial f}{\partial r}\right)_{r_1}, \quad \left(\frac{\partial^2 f}{\partial r^2}\right)_{r_1}, \quad \cdots, \quad \left(\frac{\partial^{k-1} f}{\partial r^{k-1}}\right)_{r_1}.$$

In the case of a linear equation with constant coefficients the parameters $\alpha_1, \alpha_2, \cdots, \alpha_p$ are the coefficients themselves, and the function $f(x, r)$ is e^{rx}. This leads again to the results which we obtained before directly.

45. Euler's linear equation. The linear equation

(47) $\quad x^n \dfrac{d^n y}{dx^n} + A_1 x^{n-1} \dfrac{d^{n-1} y}{dx^{n-1}} + \cdots + A_{n-1} x \dfrac{dy}{dx} + A_n y = 0,$

where $A_1, A_2, \cdots, A_n$ are constants, reduces to the preceding by the change of variable* $x = e^t$. Since $dt/dx = 1/x$, we have
$$\frac{dy}{dx} = \frac{dy}{dt}\frac{dt}{dx} = \frac{1}{x}\frac{dy}{dt}, \qquad \frac{d^2 y}{dx^2} = \frac{1}{x^2}\left(\frac{d^2 y}{dt^2} - \frac{dy}{dt}\right),$$

---

*The general theory (§ 37) tells us that the integrals of the equation (47) can have no other singular point than $x = 0$. Now e^t cannot be zero for any value of t. The integrals obtained by the change of variable $x = e^t$ must therefore be integral functions.

and we easily verify, step by step, that the product $x^p[d^p y/dx^p]$ is a linear expression with constant coefficients in dy/dt, $d^2 y/dt^2$, $\cdots$, $d^p y/dt^p$. The given linear equation is therefore transformed by this change of variable into an equation with constant coefficients.

To obtain the general integral of the equation (47), it is not necessary to carry out the calculations of this change of variable, for we know that the transformed equation has integrals of the form e^{rt}. The given equation has therefore a certain number of integrals of the form $(e^t)^r = x^r$. Replacing y by x^r in the left-hand side of the equation (47), the result of the substitution is $x^r f(r)$, where

$$f(r) = r(r-1)\cdots(r-n+1) + A_1 r(r-1)\cdots(r-n+2) + \cdots + A_{n-1} r + A_n.$$

If the equation $f(r) = 0$, which here plays the same rôle as the auxiliary equation, has n distinct roots $r_1, r_2, \cdots, r_n$, the general integral is

$$y = C_1 x^{r_1} + C_2 x^{r_2} + \cdots + C_n x^{r_n}.$$

If r is a multiple root of multiplicity μ of $f(r) = 0$, to that root corresponds, by D'Alembert's method, the μ particular integrals

$$x^r, \quad \frac{\partial}{\partial r}(x^r) = x^r \operatorname{Log} x, \quad \cdots, \quad \frac{\partial^{\mu-1} x^r}{\partial r^{\mu-1}} = x^r (\operatorname{Log} x)^{\mu-1}.$$

The general integral of the equation (47) is therefore in all cases

$$(48) \qquad y = x^{r_1} P_{\mu_1 - 1}(\operatorname{Log} x) + \cdots + x^{r_k} P_{\mu_k - 1}(\operatorname{Log} x),$$

where $r_1, r_2, \cdots, r_k$ are the k distinct roots of $f(r) = 0$, where $\mu_1, \mu_2, \cdots, \mu_k$ are their orders of multiplicity, and where $P_{\mu_i - 1}(\operatorname{Log} x)$ is a polynomial in $\operatorname{Log} x$ with arbitrary coefficients of degree $\mu_i - 1$.

If, in the equation (47), we replace the right-hand side by an expression of the form $x^m Q(\operatorname{Log} x)$, where Q denotes a polynomial, it can be shown, as in the case of the equations with constant coefficients, that the new equation thus obtained has as a particular integral an expression of the same form, whose unknown coefficients can be calculated by a substitution.

46. Laplace's equation. We can sometimes represent the integrals of a linear equation by definite integrals in which the independent variable appears as a parameter under the integral sign. One of the most important applications of this method is due to Laplace and affects the equation

$$(49) \quad F(y) = (a_0 + b_0 x)\frac{d^n y}{dx^n} + (a_1 + b_1 x)\frac{d^{n-1} y}{dx^{n-1}} + \cdots + (a_n + b_n x)y = 0,$$

whose coefficients are at most of the first degree. Let us try to find a solution of this equation by taking for y an expression of the form

$$\tag{50} y = \int_{(L)} Z e^{zx} dz,$$

where Z is a function of the variable z and where L is a definite path of integration independent of x. We have, in general,

$$\frac{d^p y}{dx^p} = \int_{(L)} Z z^p e^{zx} dz,$$

and, replacing y and its derivatives in the left-hand side of the equation (49) by the preceding expressions, we find

$$\tag{51} F(y) = \int_{(L)} Z e^{zx} (P + Qx) dz,$$

where we have set, for brevity,

$$P = a_0 z^n + a_1 z^{n-1} + \cdots + a_{n-1} z + a_n,$$
$$Q = b_0 z^n + b_1 z^{n-1} + \cdots + b_{n-1} z + b_n.$$

The function under the integral sign in the expression (51) is the derivative with respect to z of $Z e^{zx} Q$, provided that we have

$$\tag{52} \frac{d(ZQ)}{dz} = ZP, \quad \text{or} \quad \frac{d}{dz}[\text{Log}(ZQ)] = \frac{P}{Q}.$$

We derive from this condition

$$Z = \frac{1}{Q} e^{\int_{z_0}^{z} \frac{P}{Q} dz},$$

where the lower limit z_0 does not cause $Q(z)$ to vanish. The function Z having thus been determined, the definite integral (51) is equal to the variation of the auxiliary function

$$V = e^{zx} ZQ = e^{zx + \int_{z_0}^{z} \frac{P}{Q} dz}$$

along the path L. It will suffice, therefore, in order to obtain an integral of the given equation (49), to choose the path of integration L in such a way that the function V takes on the same value at the end as at the beginning, and so that the integral (50) has a finite value different from zero.

Let $a, b, c, \cdots, l$ be the roots of the equation $Q(z) = 0$. The auxiliary function V is of the form

$$\tag{53} V = e^{zx + R(z)} (z-a)^\alpha (z-b)^\beta \cdots (z-l)^\lambda,$$

where $R(z)$ is a rational function whose denominator has no other roots than the roots $a, b, c, \cdots, l$ of $Q(z)$, and of a multiplicity one unit less. Let $\mathcal{A}, \mathcal{B}, C, \cdots$ denote loops described about $a, b, c, \cdots$, in the positive sense, starting from an arbitrary initial point, and let $\mathcal{A}_{-1}, \mathcal{B}_{-1}, C_{-1}, \cdots$ denote the same loops described in the opposite sense. The function V is multiplied by $e^{2\pi i \alpha}$ when z describes the loop $\mathcal{A}$, and by $e^{-2\pi i \alpha}$ when z describes the loop $\mathcal{A}_{-1}$, and similarly for the others. It follows that if we make the variable z describe the loops $\mathcal{A}, \mathcal{B}, \mathcal{A}_{-1}, \mathcal{B}_{-1}$ in succession, the function V takes on again its initial value. The definite integral (50), taken over this path $\mathcal{A}\mathcal{B}\mathcal{A}_{-1}\mathcal{B}_{-1}$, is not, in general, zero. It gives a particular integral of the given equation. By associating the p points $a, b, c, \cdots, l$ in pairs in all possible ways, we obtain $p(p-1)/2$ integrals, which in reality reduce to $p-1$ independent integrals.

We do not find n particular integrals in this way. In order to obtain others, we may look for the paths L having their extremities at certain of the singular points $a, b, c, \cdots, l$ and such that the function V vanishes at the two extremities. If a is a simple root of $Q(z) = 0$, the function Z contains the factor $(z-a)^{\alpha-1}$, and it will be possible for the integral (50) to have a finite value when one of the extremities of the path L is at the point a only if the real part of α is positive, and in this case V does approach zero at the same time as $|z-a|$. If a is an m-fold root of $Q(z) = 0$, the rational function $R(z)$ contains a term of the form $A_{m-1}/(z-a)^{m-1}$. In order to determine the behavior of the absolute value of V in the neighborhood of the point $z = a$, we need only study the absolute value of the following important factor:

Setting
$$(z-a)^\alpha e^{\frac{A_{m-1}}{(z-a)^{m-1}}}.$$

$$z - a = \rho(\cos\phi + i\sin\phi), \quad A_{m-1} = \Lambda(\cos\psi + i\sin\psi), \quad \alpha = \alpha' + \alpha''i,$$

we may write the absolute value of this factor in the form

$$e^{-\alpha''\phi}\rho^{\alpha'}e^{\Lambda\rho^{1-m}\cos[\psi-(m-1)\phi]}.$$

In order that V shall approach zero with $|z-a|$, it will suffice to make z describe a curve such that the angle ϕ which the tangent makes with the real axis satisfies the condition $\cos[\psi - (m-1)\phi] < 0$. For example, we may take $\phi = [\psi + (2k+1)\pi]/(m-1)$. If the angle ϕ has been taken in this way, the product Ze^{zx} also approaches zero with $|z-a|$. Proceeding in the same way with the other points $b, c, \cdots, l$, we see that we can determine new paths L, closed or not, giving other particular integrals.

Finally, we can also take, for the paths of integration, curves going off to infinity. We are again led to determine a path L having an infinite branch such that the function V approaches zero when the point z goes off indefinitely on this branch. If, for example, the rational function $R(z)$ is zero, and if the angle of x lies between 0 and $\pi/2$, it will suffice to make z describe an infinite branch asymptotic to a line that makes an angle of $3\pi/4$ with the real axis.

Leaving these general considerations,[*] let us consider in particular Bessel's equation,

(54) $$x\frac{d^2y}{dx^2} + (2n+1)\frac{dy}{dx} + xy = 0,$$

where n is a given constant. We have here

$$P = (2n+1)z, \quad Q = 1 + z^2,$$

and consequently

$$Z = (1+z^2)^{n-\frac{1}{2}}, \quad V = e^{zx}(1+z^2)^{n+\frac{1}{2}}.$$

The definite integral

(55) $$y = \int_{(L)} e^{zx}(1+z^2)^{n-\frac{1}{2}}dz$$

---

[*] See an important paper by Poincaré in the *American Journal of Mathematics*, Vol. VII.

is therefore a particular integral of the equation (54) if the function

$$e^{zx}(1+z^2)^{n+\frac{1}{2}}$$

takes on the same values at the extremities of the path of integration. We can first take a succession of two loops described, the first in the positive sense about the point $z = + i$, the second in the reverse sense about the point $z = -i$. For the second path of integration we can take next a curve surrounding one of the singular points $\pm i$ and having two infinite branches with an asymptotic direction such that the real part of zx approaches $-\infty$.

The real part of the constant n may be supposed positive or zero, for if we put $y = x^{-2n}z$, the equation in z does not differ from the equation (54) except in the change from n to $-n$. When this is the case, we can also take for the path of integration the straight line joining the two points $+i$ and $-i$. Moreover, the integral thus obtained is identical with the first except for a constant factor. In order to reduce this integral to the usual form, let us put $z = it$. It then takes the form

$$y = \int_{-1}^{+1} e^{ixt}(1-t^2)^{n-\frac{1}{2}}dt,$$

or

(56) $$y = \int_{-1}^{+1} \cos xt\,(1-t^2)^{n-\frac{1}{2}}dt.$$

The remarkable particular case in which n is half an odd number deserves mention. If n is positive, the integral (56) always exists and can even be calculated explicitly, since $n - 1/2$ is a positive integer. But if the path L is a closed curve, the definite integral (55) is always zero. It seems, then, that in this case the application of the general method gives only one particular integral. However, in this apparently unfavorable case we can express the general integral in terms of elementary functions. For, let us make the inverse transformation to the preceding, so that n shall be half of a negative odd number. Then $n - 1/2$ is a negative integer, and the definite integral (55), taken along any closed curve, is a particular integral of the linear equation (54). Taking for the path L a circle having one of the points $\pm i$ for center, we see that the residue of the function

$$e^{zx}(1+z^2)^{n-\frac{1}{2}}$$

with respect to each of these poles is an integral of the linear equation. Now, it is clear that the residue with respect to the pole $z = + i$ is the product of e^{ix} and a polynomial, and, similarly, that the residue with respect to the pole $z = - i$ is the product of e^{-ix} and a polynomial. These two particular integrals are independent, for their quotient is equal to the product of e^{2ix} and a rational function. It is clear that their sum is a real integral, as is also the product of their difference and i.

Note. The linear equation with constant coefficients is a particular case of Laplace's equation, which is obtained by supposing all the coefficients b_i zero. If we suppose also $a_0 = 1$, we have $Q(z) = 0$, while $P(z)$ reduces to the auxiliary polynomial $f(z)$. The general method appears to fail, since the expression for Z becomes illusory. But it requires only a little care to recognize how the method must be modified. In fact, the reasoning proves that the definite integral

$\int_{(L)} Z e^{zx} dz$ is a particular integral of the linear equation, provided that the definite integral $\int_{(L)} Z f(z) e^{zx} dz$, taken along the same path L, is zero. Now, if we take for L a closed curve, it is sufficient that the product $Z f(z)$ be an analytic function of z in the interior of this curve. If, therefore, $\Pi(z)$ denotes any analytic function in a region R of the plane, the definite integral

$$ y = \int_{(L)} \frac{\Pi(z)}{f(z)} e^{zx} dz, $$

taken along any closed curve L lying in this region, is a particular integral of the linear equation with constant coefficients. We see how this result, due to Cauchy, is thus easily brought into close relation with Laplace's method.

As a verification, it is easy to find the known particular integrals. Let $z = a$ be a p-fold root of the auxiliary equation $f(z) = 0$. Let us take for the path of integration a circle about a as center not containing any other roots of $f(z) = 0$, and let $\Pi(z)$ be an analytic function in this circle. The residue of the function $\Pi(z) e^{zx}/f(z)$ or $\Pi(z) e^{zx}/[(z-a)^p f_1(z)]$ is equal to the coefficient of h^{p-1} in the development of the product

$$ \Pi(a + h) e^{ax} \frac{e^{xh}}{f_1(a+h)} $$

according to powers of h. If we have

$$ \frac{\Pi(a+h)}{f_1(a+h)} = A_0 + A_1 h + \cdots + A_{p-1} h^{p-1} + \cdots, $$

the coefficients $A_0, A_1, \cdots, A_{p-1}$ are arbitrary, since the function $\Pi(z)$ is any function analytic in the neighborhood of the point a. The residue sought is therefore equal to

$$ e^{ax} \left[A_0 \frac{x^{p-1}}{(p-1)!} + A_1 \frac{x^{p-2}}{(p-2)!} + \cdots + A_{p-1} \right], $$

that is, to the product of the exponential e^{ax} and an arbitrary polynomial of degree $p - 1$. This agrees with the result already known.

III. REGULAR INTEGRALS. EQUATIONS WITH PERIODIC COEFFICIENTS

Aside from the very elementary cases which we have just treated, it is, in general, impossible to determine, simply from the form of a linear equation, whether the general integral is algebraic or whether it can be expressed in terms of the classic transcendentals. Mathematicians have therefore been led to study the properties of these integrals directly from the equation itself, instead of trying to express them (somewhat at random) as combinations of a finite number of known functions. We have already seen (Chap. III, Part I) that the nature of the singular points of an analytic function is an essential element enabling us in certain cases to characterize these functions completely. We know a priori (§ 37) the singular points of the integrals of a linear equation. We shall now show how we can make a complete study of the integrals in the neighborhood of a singular point in a special case, which is nevertheless rather general and very important.

47. Permutation of the integrals around a critical point.
Let a be an isolated singular point of some of the coefficients $p_1, p_2, \cdots, p_n$ of the linear equation

$$(57) \qquad F(y) = \frac{d^n y}{dx^n} + p_1 \frac{d^{n-1} y}{dx^{n-1}} + \cdots + p_{n-1} \frac{dy}{dx} + p_n y = 0.$$

We shall suppose also that the coefficients are single-valued in the neighborhood of a. Let C be a circle with the center a in the interior of which $p_1, p_2, \cdots, p_n$ have no other singular point than a and are otherwise analytic. Let x_0 be a point within C near a. All the integrals are analytic in the neighborhood of the point x_0. Let $y_1, y_2, \cdots, y_n$ be n particular integrals of a fundamental system. If the variable x describes in the positive sense a circle passing through the point x_0 about a as center, we can follow the analytic extension of the integrals $y_1, y_2, \cdots, y_n$ along the whole of this path, and we return to the point x_0 with n functions $Y_1, Y_2, \cdots, Y_n$ which are again integrals of the equation (57), where Y_i indicates the function into which y_i passes after a circuit around the point a in the positive sense. We have, therefore, since $Y_1, Y_2, \cdots, Y_n$ are integrals of the equation (57), n relations of the form

$$(58) \qquad \begin{cases} Y_1 = a_{11} y_1 + a_{12} y_2 + \cdots + a_{1n} y_n, \\ Y_2 = a_{21} y_1 + a_{22} y_2 + \cdots + a_{2n} y_n, \\ \cdots \cdots \cdots \cdots \cdots \cdots \cdots \cdots \cdots \\ Y_n = a_{n1} y_1 + a_{n2} y_2 + \cdots + a_{nn} y_n, \end{cases}$$

where the coefficients a_{ik} are constants which of course depend upon the fundamental system chosen. It is easy to obtain the value of the determinant D formed by these n^2 coefficients. For we have, by § 38,

$$\Delta(y_1, y_2, \cdots, y_n) = C e^{-\int_{x_0}^{x} p_1 \, dx}.$$

If x describes the circle γ with the center a in the positive sense, y_i changes into Y_i; hence we have

$$\Delta(Y_1, Y_2, \cdots, Y_n) = \Delta(y_1, y_2, \cdots, y_n) e^{-\int_{\gamma} p_1 \, dx}.$$

But the quotient of the two Wronskians is equal to D (§ 38), so that $D = e^{-2\pi i R}$, where R indicates the residue of p_1 with respect to the point a. This determinant is therefore never zero.

Since the coefficients in the equations (58) depend upon the fundamental system chosen, it is natural to seek a particular system of integrals such that these expressions are as simple as possible. Let us seek first to determine a particular integral $u = \lambda_1 y_1 + \lambda_2 y_2 + \cdots + \lambda_n y_n$, such that a circuit around the point a reproduces that integral multiplied by a constant factor. It is necessary for this that we have $U = su$, where U is the value of u after the circuit, and where s is a constant factor, that is,

$$\lambda_1 (a_{11} y_1 + a_{12} y_2 + \cdots + a_{1n} y_n) + \cdots \\ + \lambda_n (a_{n1} y_1 + a_{n2} y_2 + \cdots + a_{nn} y_n) - s(\lambda_1 y_1 + \cdots + \lambda_n y_n) = 0.$$

Such a relation cannot exist between the n integrals unless the coefficients of $y_1, y_2, \cdots, y_n$ all vanish separately. The $n+1$ unknown coefficients

$\lambda_1, \lambda_2, \cdots, \lambda_n$, s must satisfy the n conditions

(59) $\begin{cases} \lambda_1(a_{11}-s) + \lambda_2 a_{21} + \cdots + \lambda_n a_{n1} = 0, \\ \lambda_1 a_{12} + \lambda_2(a_{22}-s) + \cdots + \lambda_n a_{n2} = 0, \\ \cdots \cdots \cdots \cdots \cdots \cdots \cdots \cdots \cdots \cdots \cdots, \\ \lambda_1 a_{1n} + \lambda_2 a_{2n} + \cdots + \lambda_n(a_{nn}-s) = 0. \end{cases}$

Since the quantities $\lambda_1, \lambda_2, \cdots, \lambda_n$ cannot all be zero at the same time without having $u = 0$, we see that s must be a root of the equation of the nth degree,

(60) $F(s) = \begin{vmatrix} a_{11}-s & a_{21} & \cdots & a_{n1} \\ a_{12} & a_{22}-s & \cdots & a_{n2} \\ \cdots & \cdots & \cdots & \cdots \\ a_{1n} & a_{2n} & \cdots & a_{nn}-s \end{vmatrix} = 0,$

which we shall call the *characteristic equation*; according to a remark made a moment ago, this equation cannot have the root $s = 0$, for the determinant D of the n^2 coefficients a_{ik} would be zero.

Conversely, let s be a root of this equation; the relations (59) determine values for the coefficients λ_i not all zero, and the integral $u = \lambda_1 y_1 + \cdots + \lambda_n y_n$ is multiplied by s after the circuit around the point a. This being the case, let us suppose first that the characteristic equation has n distinct roots $s_1, s_2, \cdots, s_n$. We shall have n particular integrals $u_1, u_2, \cdots, u_n$ such that, after the circuit in the direct sense around the point a, we have

(61) $\qquad U_1 = s_1 u_1, \quad U_2 = s_2 u_2, \quad \cdots, \quad U_n = s_n u_n,$

where U_i denotes the final value of u_i after the circuit. *These n integrals $u_1, u_2, \cdots, u_n$ form a fundamental system.* For, suppose that we have a relation of the form

(62) $\qquad C_1 u_1 + C_2 u_2 + \cdots + C_n u_n = 0,$

where the constant coefficients $C_1, C_2, \cdots, C_n$ are not all zero. After one, two, $\cdots$, $(n-1)$ circuits, we should have the relations of the same form,

(63) $\begin{cases} C_1 s_1 u_1 + C_2 s_2 u_2 + \cdots + C_n s_n u_n = 0, \\ C_1 s_1^2 u_1 + C_2 s_2^2 u_2 + \cdots + C_n s_n^2 u_n = 0, \\ \cdots \cdots \cdots \cdots \cdots \cdots \cdots \cdots \cdots \cdots \cdots, \\ C_1 s_1^{n-1} u_1 + C_2 s_2^{n-1} u_2 + \cdots + C_n s_n^{n-1} u_n = 0. \end{cases}$

The linear relations (62) and (63) can be satisfied only if we have at the same time $C_1 u_1 = 0, \cdots, C_n u_n = 0$, since the corresponding determinant is different from zero.

It is easy to form an analytic function which is multiplied by a constant factor s *different from zero* after a circuit around the point a. In fact, the function $(x-a)^r$ or $e^{r \text{Log}(x-a)}$ is multiplied by $e^{2\pi i r}$ after such a circuit, and if we determine r by the condition $r = \text{Log}(s)/2\pi i$, this function $(x-a)^r$ is indeed multiplied by s after a circuit around a. Every other function u having the same property is of the form $(x-a)^r \phi(x-a)$, where the function $\phi(x-a)$ is single-valued in the neighborhood of the point a, since the product $u(x-a)^{-r}$ comes back to its initial value after a circuit around the point a. The integral u_k is therefore of the form

$$u_k = (x-a)^{r_k} \phi_k(x-a),$$

where $r_k = \text{Log}(s_k)/2\pi i$ and where the functions ϕ_k are single-valued in the neighborhood of the point a. In a circle C with the radius R about the point a as center and in which the coefficients $p_1, \cdots, p_n$ are analytic except at the point a, the integral u_k cannot have any other singular point than a. The same thing is therefore true of the function $\phi_k(x-a)$, and the point a is an ordinary point or an isolated singular point for that function. We can dismiss the possibility that a is a pole. In fact, if the point a were a pole of order m, since the exponent r_k is determined except for an integer, we can write

$$u_k = (x-a)^{r_k - m}[(x-a)^m \phi_k(x-a)],$$

and the product $(x-a)^m \phi_k(x-a)$ is analytic for $x = a$. If the point a is not an essentially singular point for $\phi_k(x-a)$, we say that *the integral is regular for* $x = a$. We can then suppose that the function $\phi_k(x-a)$ has a finite value, different from zero, for $x = a$.

48. Examination of the general case. It remains to examine the case where the characteristic equation has multiple roots. We shall show that we can always find n integrals forming a fundamental system and breaking up into a certain number of groups such that if $y_1, y_2, \cdots, y_p$ denote the p integrals of the same group, we have, after a circuit in the positive sense around the point a,

(64) $\quad Y_1 = sy_1, \qquad Y_2 = s(y_1 + y_2), \qquad \cdots, \qquad Y_p = s(y_{p-1} + y_p).$

The different values of s are the roots of the characteristic equation, and to the same root may correspond several different groups. If the n roots are distinct, which is the case we have just examined, each group is composed of a single integral.

The problem reduces in reality to showing that we can reduce the linear substitutions defined by the equations (58) to a canonical form such as we have just indicated by replacing $y_1, y_2, \cdots, y_n$ by suitably chosen linear combinations of these variables. Assuming that the theorem has been proved for the case of $n-1$ variables, we shall show that it is also true for n variables.

From what has been shown in the preceding paragraph, we can always find a particular integral u such that we have $U = \mu u$. Replacing one of the integrals, y_1 for example, by this integral u, the expressions (58) take the form

(65) $\quad \begin{cases} U = \mu u, \\ Y_2 = b_2 u + b_{22} y_2 + \cdots + b_{2n} y_n, \\ \cdots\cdots\cdots\cdots\cdots\cdots\cdots\cdots, \\ Y_n = b_n u + b_{n2} y_2 + \cdots + b_{nn} y_n. \end{cases}$

If in the last $n-1$ expressions we neglect the terms $b_2 u, \cdots, b_n u$, these equations define a linear substitution carried out on the $n-1$ variables $y_2, y_3, \cdots, y_n$. The determinant D' of this substitution in $n-1$ variables is not zero, for the determinant D of the linear substitution in n variables is equal to $\mu D'$ and cannot be zero. Since the theorem is assumed to hold for $n-1$ variables, we may suppose this auxiliary substitution reduced to the canonical form. This amounts to replacing $y_2, y_3, \cdots, y_n$ by $n-1$ linearly independent combinations $z_1, z_2, \cdots, z_{n-1}$ such that the equations which define the linear substitution

$$Y_i = b_{i2} y_2 + \cdots + b_{in} y_n \qquad (i = 2, 3, \cdots, n)$$

are replaced by a certain number of groups of equations such as

$$Z_1 = sz_1, \quad Z_2 = s(z_1 + z_2), \quad \cdots, \quad Z_p = s(z_{p-1} + z_p).$$

If we carry out the same transformations on the equations (65), it will be necessary to add to the right-hand side of the preceding relations terms containing u as a factor. In other words, we can find $n-1$ integrals that form with u a fundamental system, and that separate into a certain number of groups such that we have for the integrals $z_1, z_2, \cdots, z_p$ of a single group

(66) $\quad Z_1 = sz_1 + K_1 u, \quad Z_2 = s(z_1 + z_2) + K_2 u, \quad \cdots, \quad Z_p = s(z_{p-1} + z_p) + K_p u,$

where $K_1, K_2, \cdots, K_p$ are constants. We shall first try to make as many as possible of these coefficients disappear. For this purpose let us put

$$u_1 = z_1 + \lambda_1 u, \quad u_2 = z_2 + \lambda_2 u, \quad \cdots, \quad u_p = z_p + \lambda_p u,$$

where $\lambda_1, \lambda_2, \cdots, \lambda_p$ are p constant coefficients. An easy calculation shows that we have for these new integrals

(67) $\quad \begin{cases} U_1 = su_1 + [K_1 + (\mu - s)\lambda_1] u, \\ U_i = s(u_{i-1} + u_i) + [K_i + (\mu - s)\lambda_i - s\lambda_{i-1}] u. \end{cases} \quad (i>1)$

If $\mu - s$ is not zero, we can choose $\lambda_1, \lambda_2, \cdots, \lambda_p$ in such a way that the coefficients of u on the right are zero, and we have for the new integrals u_i

$$U_1 = su_1, \quad U_2 = s(u_1 + u_2), \quad \cdots, \quad U_p = s(u_{p-1} + u_p).$$

The substitution to which this group of integrals is subjected after a circuit around a is of the canonical form. If $\mu = s$, since s cannot be zero, we can choose $\lambda_1, \lambda_2, \cdots, \lambda_{p-1}$ in such a way as to make the coefficients of u in the expressions for $U_2, U_3, \cdots, U_p$ disappear. But we may have several groups of variables z_i subjected to a transformation of the canonical form for which the value of s is equal to μ. Suppose, for definiteness, that there are two such groups, containing respectively p and q variables. After the preceding change of variables the substitutions which these two groups undergo are of the form

(I) $\quad U_1 = su_1 + K_1 u, \quad U_2 = s(u_2 + u_1), \quad \cdots, \quad U_p = s(u_p + u_{p-1}),$
(II) $\quad U_1' = su_1' + K_1' u, \quad U_2' = s(u_2' + u_1'), \quad \cdots, \quad U_q' = s(u_q' + u_{q-1}').$

If $K_1' = K_1 = 0$, we have three groups of integrals, u, $(u_1, u_2, \cdots, u_p)$, $(u_1', u_2', \cdots, u_q')$, subjected to a substitution of the canonical form. If we suppose that $p \geqq q$, and if K_1 is not zero, by putting $v_i = u_i' - K_1' u_i / K_1$ the second group of integrals is replaced by a group of q integrals v_i which undergo a substitution of the canonical form. Next, putting $u_0 = K_1 u/s$, the $(p+1)$ integrals $u_0, u_1, \cdots, u_p$ form a single group which undergoes a transformation of the canonical form. If $K_1 = 0$, while K_1' is not zero, putting $u_0' = K_1' u/s$, we have two groups of integrals, $(u_1, u_2, \cdots, u_p)$, $(u_0', u_1', \cdots, u_q')$, which undergo a substitution of the canonical form. The theorem stated is therefore true in general.*

---

*For a full treatment of the application of Weierstrass's theory of *elementary divisors* to linear differential equations the paper by L. Sauvage (*Annales de l'École Normale supérieure*, 1891, p. 285) may be consulted.

49. Formal expressions for the integrals.

It remains for us to find a formal expression which will show clearly the law of permutation of the integrals of the same group after a circuit around the point a. Let $y_1, y_2, \cdots, y_p$ be a group of integrals which undergo the permutations (64). Let us put $y_k = (x-a)^r z_k$, where r is equal to $\operatorname{Log} s/2\pi i$. The p functions $z_1, z_2, \cdots, z_p$ must be such that we have

$$Z_1 = z_1, \qquad Z_2 = z_1 + z_2, \qquad \cdots, \qquad Z_p = z_{p-1} + z_p.$$

Hence the function z_1 must be a single-valued function $\phi_1(x-a)$ in the neighborhood of the point a. As to the function z_2, we derive from the preceding equalities $Z_2/Z_1 = z_2/z_1 + 1$; hence the difference $z_2/z_1 - \operatorname{Log}(x-a)/2\pi i$ is a single-valued function $\psi_1(x-a)$, and we have also

$$z_2 = \frac{1}{2\pi i} \operatorname{Log}(x-a) \phi_1(x-a) + \phi_2(x-a),$$

where $\phi_2(x-a)$ is another single-valued function. Let us put $t = \operatorname{Log}(x-a)/2\pi i$ and consider the general case. When x describes a loop in the positive sense around the point a, t increases by unity, and $z_1, z_2, \cdots, z_p$, considered as functions of t, must satisfy the relations

(68) $\quad \begin{cases} z_1(t+1) = z_1(t), \quad z_2(t+1) = z_2(t) + z_1(t), \quad \cdots, \\ z_p(t+1) = z_p(t) + z_{p-1}(t). \end{cases}$

In order to find the most general solution of the equations (68), we may remark that these relations can be satisfied by taking $z_1 = 1$, $z_2 = t$, and by choosing for $z_i(t)$ a polynomial of degree $i-1$ in t whose coefficients are determined step by step. The calculation is facilitated by observing that the relation

$$z_i(t+1) - z_i(t) = z_{i-1}(t) \qquad (i \geqq 3)$$

is satisfied for $t = 0, 1, 2, \cdots, i-3$ if we take for $z_i(t)$ a polynomial of the form $K_i t(t-1) \cdots (t-i+2)$. In order that it may be satisfied identically, it will suffice if it is satisfied by another value of t, for example, by $t = i-2$, since the two sides are polynomials of degree $i-2$ in t. We thus find the condition $(i-1) K_i = K_{i-1}$, whence we derive $K_i = 1/(i-1)!$. We therefore obtain a particular solution of the equations (68) by putting

$$\theta_1 = 1, \qquad \theta_i(t) = \frac{t(t-1)\cdots(t-i+2)}{(i-1)!}. \qquad (i = 2, 3, \cdots, p)$$

In order to obtain the general solution, let us indicate by $\phi_k(t)$ functions such that $\phi_k(t+1) = \phi_k(t)$. The first of the equations (68) shows that $z_1(t)$ is a function of this kind, say $\phi_1(t)$. The second shows, similarly, that the difference $z_2(t) - \theta_2(t) z_1(t)$ does not change when we change t to $t+1$; hence $z_2(t)$ is of the form $z_2(t) = \phi_2(t) + \theta_2 \phi_1(t)$. We can continue the reasoning step by step. Suppose that we have shown that $z_{k-1}(t)$ is of the form

$$z_{k-1}(t) = \phi_{k-1}(t) + \theta_2 \phi_{k-2}(t) + \cdots + \theta_{k-1} \phi_1(t). \qquad (k = 3, 4, \cdots, i)$$

The general relation $z_i(t+1) - z_i(t) = z_{i-1}(t)$ shows that the difference

$$z_i(t) - \theta_2 \phi_{i-1}(t) - \theta_3 \phi_{i-2}(t) - \cdots - \theta_i \phi_1(t)$$

does not change when t changes to $t+1$; hence the function $z_i(t)$ is of the form

$$z_i(t) = \phi_i(t) + \theta_2 \phi_{i-1}(t) + \cdots + \theta_i \phi_1(t).$$

Combining these results, the general solution of the equations (68) is given by the relations

(69) $\begin{cases} z_1(t) = \phi_1(t), \\ z_2(t) = \theta_2 \phi_1(t) + \phi_2(t), \\ z_3(t) = \theta_3 \phi_1(t) + \theta_2 \phi_2(t) + \phi_3(t), \\ \cdots \cdots \cdots \cdots \cdots \cdots \cdots \cdots \cdots \cdots, \\ z_p(t) = \theta_p \phi_1(t) + \theta_{p-1} \phi_2(t) + \cdots + \theta_2 \phi_{p-1}(t) + \phi_p(t), \end{cases}$

where the functions $\phi_1, \phi_2, \cdots, \phi_p$ do not change when t is changed to $t + 1$.

Let us return now to the variable x, and let us indicate by $\Theta_i[\text{Log}(x - a)]$ the polynomial in $\text{Log}(x - a)$ obtained by replacing t by $\text{Log}(x - a)/2\pi i$ in $\theta_i(t)$. We see that the p integrals $y_1, y_2, \cdots, y_p$ of the group under consideration, which undergo the substitution (64) after a circuit in the positive sense around the point a, are represented by formal expressions of the following type:

(70) $\begin{cases} y_1 = (x - a)^r \Phi_1(x - a), \\ y_2 = (x - a)^r [\Theta_2 \{\text{Log}(x - a)\} \Phi_1(x - a) + \Phi_2(x - a)], \\ \cdots \cdots \cdots \cdots \cdots \cdots \cdots \cdots \cdots \cdots, \\ y_p = (x-a)^r [\Theta_p\{\text{Log}(x-a)\}\Phi_1(x-a) + \Theta_{p-1}\{\text{Log}(x-a)\}\Phi_2(x-a) + \cdots], \end{cases}$

where $\Phi_1(x - a), \Phi_2(x - a), \cdots, \Phi_p(x - a)$ are single-valued functions in the neighborhood of the point a.

It will be observed that all the integrals of this group can be deduced from the last of them, y_p, which is of the form

$$y_p = (x - a)^r [\psi_0(x - a) + \psi_1(x - a) \text{Log}(x - a) + \cdots \\ + \psi_{p-1}(x - a) \{\text{Log}(x - a)\}^{p-1}],$$

where $\psi_0, \psi_1, \cdots, \psi_{p-1}$ are single-valued functions in the neighborhood of the point a, the last of which, ψ_{p-1}, is *different from zero*. From the relations (64) we have

$$y_{p-1} = \frac{Y_p}{s} - y_p,$$

and consequently y_{p-1} is the product of $(x - a)^r$ and a polynomial of degree $p - 2$ in $\text{Log}(x - a)$, the coefficients of which are single-valued functions in the neighborhood of the point a. In the same way we derive y_{p-2} from y_{p-1}, and so on.

If the point a is not an essentially singular point for any of the functions $\Phi_1, \Phi_2, \cdots, \Phi_p$, all the integrals of the group considered (70) are said to be *regular* for $x = a$. By the remark made on page 131, we can then suppose that all the functions $\Phi_i(x - a)$ are analytic for $x = a$, replacing r, if necessary, by another exponent which differs from it only by an integer.

50. Fuchs' theorem. The determination of the numbers $s_1, s_2, \cdots, s_n$, or, what amounts to the same thing, the corresponding exponents $r_1, r_2, \cdots, r_n$, is in general a very difficult problem. We can obtain these exponents r_i by algebraic calculations whenever all the integrals of the equation considered are regular in the neighborhood of the point a. This results from an important theorem due to Fuchs: *In order that the equation* (57) *shall have n independent integrals, regular in the neighborhood of the point a, it is necessary and sufficient that the coefficient p_i of $d^{n-i}y/dx^{n-i}$ in this equation be of the form $(x - a)^{-i} P_i(x)$, where the function $P_i(x)$ is analytic in the neighborhood of the point a.*

If $P_i(a)$ is not zero, the point a is a pole of order i for p_i; but if $P_i(a) = 0$, the point a is a pole of order less than i. It may even happen that the point a is an ordinary point for some of the coefficients p_i. The preceding conditions may be restated as follows: *The linear equation must be of the form*

$$(71) \quad \begin{cases} (x-a)^n \dfrac{d^n y}{dx^n} + (x-a)^{n-1} P_1(x) \dfrac{d^{n-1} y}{dx^{n-1}} + \cdots \\ \qquad + (x-a) P_{n-1}(x) \dfrac{dy}{dx} + P_n(x) y = 0, \end{cases}$$

where $P_1, P_2, \cdots, P_n$ are analytic functions in the neighborhood of the point a.

We shall develop the proof only for the case of an equation of the second order, and we shall suppose, for simplicity, that $a = 0$. In this particular case the first part of Fuchs' theorem may be stated as follows: *Every equation of the second order, which has two independent and regular integrals in the neighborhood of the origin, is of the form*

$$(72) \quad x^2 y'' + x P(x) y' + Q(x) y = 0,$$

where $P(x)$ and $Q(x)$ are analytic in this neighborhood.

If the corresponding equation in s (60) has two distinct roots s_1, s_2, the equation (72) has two regular integrals of the form

$$(\text{I}) \qquad y_1 = x^{r_1} \phi_1(x), \qquad y_2 = x^{r_2} \phi_2(x),$$

where the exponents r_1, r_2 are different and where $\phi_1(x)$, $\phi_2(x)$ are two analytic functions *which are not zero for $x = 0$*. If the equation in s has a double root, without causing the appearance of logarithmic terms in the expression for the general integral, we have again two particular integrals of the preceding form, where the difference $r_2 - r_1$ is an integer. We can always suppose that that difference is not zero; for if we had $r_2 = r_1$, we could replace y_2 by the combination $\phi_1(0) y_2 - \phi_2(0) y_1$, which is divisible by x^{r_1+1}. Finally, if the expression for the integral contains a logarithmic term in the neighborhood of the origin, we can take a fundamental system of the form

$$(\text{II}) \qquad y_1 = x^{r_1} \phi_1(x), \qquad y_2 = x^{r_1} [\phi_1(x) \operatorname{Log}(x) + \psi_1(x)],$$

where $\phi_1(x)$ is an analytic function which is not zero for $x = 0$, and where $\psi_1(x)$ is a single-valued function in the neighborhood of the origin, which may have the point $x = 0$ for a pole. We have to show that every equation which has two independent integrals of the form (I) or of the form (II) in the neighborhood of the origin belongs to the Fuchs type. The direct verification does not offer any difficulty, but we can abridge the work as follows: If we put $y = x^{r_1} \phi_1(x) u$, the linear equation in u obtained by this transformation has a general integral of one of the forms

$$u = C_1 + C_2 x^\rho \pi(x), \qquad u = C_1 + C_2 [\operatorname{Log}(x) + \pi(x)],$$

where $\pi(x)$ is analytic for $x = 0$ or has this point for a pole. This equation is of the form (72), for the derivative u' is of the form

$$u' = C_2 x^\mu \zeta(x),$$

where $\zeta(x)$ is an analytic function which is not zero for $x = 0$. The linear equation in u is therefore

$$\frac{u''}{u'} = \frac{\mu}{x} + \frac{\zeta'(x)}{\zeta(x)},$$

which is of the Fuchs type. Now it is easy to see that this type is preserved after a transformation such as $y = x^{r_1}\phi_1(x)u$. The first part of the proposition is therefore established.

In order to prove the converse, let us substitute for y on the left-hand side of the equation (72) a development of the form

(73) $$y = c_0 x^r + c_1 x^{r+1} + \cdots + c_n x^{r+n} + \cdots, \qquad (c_0 \neq 0)$$

and let
$$P(x) = a_0 + a_1 x + \cdots, \qquad Q(x) = b_0 + b_1 x + \cdots$$

be the developments of the functions P and Q. The coefficient of x^r in the resulting equation is

$$[r(r-1) + a_0 r + b_0] c_0.$$

Since, by hypothesis, the first coefficient c_0 is not zero, we must take for r one of the roots of the equation of the second degree

(74) $$D(r) = r(r-1) + a_0 r + b_0 = 0.$$

Having taken a root of this equation for r, we can choose c_0 arbitrarily. Let us take, for example, $c_0 = 1$. Similarly, the coefficient of x^{r+p} after the substitution is

$$c_p[(r+p)(r+p-1) + a_0(r+p) + b_0] + F = c_p D(r+p) + F,$$

where F is a polynomial with integral coefficients in $c_0, c_1, \cdots, c_{p-1}, a_1, a_2, \cdots, a_p, b_1, b_2, \cdots, b_p$. Putting successively $p = 1, 2, 3, \cdots$, we shall be able to calculate, step by step, the successive coefficients $c_1, c_2, \cdots, c_n$, unless $D(r+p)$ is zero for a positive value of the integer p, that is, unless the equation (74) has a second root r' equal to the first r increased by a positive integer. Discarding this case for the moment, we shall obtain a particular integral represented by a series of the form (73), the convergence of which will be demonstrated later. If the equation $D(r) = 0$ has two distinct roots r, r', whose difference is not an integer, the preceding method enables us to obtain two independent integrals, and the general integral is represented in the neighborhood of the origin by the expression

(75) $$y = C_1 x^r \phi(x) + C_2 x^{r'} \psi(x),$$

where $\phi(x)$ and $\psi(x)$ are two analytic functions which do not vanish for $x = 0$.

This is no longer the case if the two roots of the equation (74) are equal or if their difference is an integer. Let r and $r - p$ be these two roots, where p is a positive integer or zero. We can always obtain a first integral of the form $y_1 = x^r \phi(x)$. A second integral y_2 is given by the general formula (23), which becomes here

$$y_2 = x^r \phi(x) \int \frac{dx}{x^{2r}[\phi(x)]^2} e^{-\int \left(\frac{a_0}{x} + a_1 + a_2 x + \cdots\right) dx}.$$

The sum of the roots of the equation (74), or $1 - a_0$, is equal in this case to $2r - p$; hence $a_0 = p + 1 - 2r$, and accordingly

$$e^{-\int \left(\frac{a_0}{x} + a_1 + a_2 x + \cdots\right) dx} = x^{2r-(p+1)} S(x),$$

where $S(x)$ is a regular function in the neighborhood of the origin, which is not zero for $x = 0$. The second integral y_2 can therefore be written in the form

$$y_2 = x^r \phi(x) \int \frac{T(x)\,dx}{x^{p+1}},$$

where $T(x)$ is an analytic function which is not zero for $x = 0$.

If A is the coefficient of x^p in $T(x)$, we see that the integral y_2 is of the form

$$y_2 = x^r \phi(x)\left[A \operatorname{Log} x + \frac{\phi_1(x)}{x^p}\right] = x^{r-p}\psi(x) + A x^r \phi(x) \operatorname{Log} x,$$

where $\psi(x)$ denotes a new analytic function in the neighborhood of the origin. This result agrees precisely with the general theory. As a particular case, it may happen that we have $A = 0$; the general integral does not then contain logarithms in the neighborhood of the origin. But since $T(0)$ is not zero, it is to be noticed that this case never arises when $p = 0$, that is, when the equation (74) has a double root.*

To complete the demonstration, it remains only to prove the convergence of the series (73) obtained by taking for r a root of the equation (74) such that the second root r' is not equal to r increased by a positive integer. To simplify the proof, we may suppose that $r = 0$ and that the second root r' is not equal to a positive integer; for if we put $y = x^\mu z$, the equation analogous to $D(r) = 0$ for the linear equation in z has the roots of the equation (74) reduced by μ. We shall suppose, therefore, that such a transformation has already been made, so that the equation (74) has the root $r = 0$ and that the second root is not a positive integer. For this it is necessary that b_0 be zero. Modifying the notation somewhat, and dividing all the terms by x, we shall write the equation (72) in the form

(76) $\qquad xy'' + a_0 y' = xy'(a_1 + a_2 x + \cdots) + y(b_1 + b_2 x + \cdots),$

where the coefficients $a_1, b_1, a_2, \cdots$ are not the same as before. We are to prove that this equation (76) has an analytic integral in the neighborhood of the origin, which does not vanish for $x = 0$, provided that $1 - a_0$ is not a positive integer. Now, if we try to satisfy this equation *formally* by a series of the form

(77) $\qquad\qquad y = 1 + c_1 x + \cdots + c_n x^n + \cdots,$

we obtain successively relations between the coefficients of the form

(78) $\qquad \begin{cases} nc_n\{n - 1 + a_0\} = P_n\{a_1, a_2, \cdots, b_1, b_2, \cdots, b_n, c_1, c_2, \cdots, c_{n-1}\}, \\ (n = 1, 2, \cdots) \end{cases}$

where P_n is a polynomial whose coefficients are all real positive numbers. By hypothesis, the coefficient $n - 1 + a_0$ does not vanish for any positive integral

---

*Let us suppose that the functions $P(x)$ and $Q(x)$ in the equation (72) are *even* functions of x, and that the difference between the roots of $D(r) = 0$ is an *odd integer* $2n+1$. In this case the logarithmic term always disappears in the integral y_2. In fact, if we take for the independent variable $t = x^2$, the equation (72) is replaced by an equation of the same form,

(72′) $\qquad 4t^2\dfrac{d^2y}{dt^2} + 2t\left[1 + P(\sqrt{t})\right]\dfrac{dy}{dt} + Q(\sqrt{t})\,y = 0,$

and the roots of the equation analogous to $D(r) = 0$ are, as is easily verified, half of the roots of $D(r) = 0$. Since their difference is not an integer, it follows that the general integral of the equation (72′) does not contain any logarithmic term in the neighborhood of the origin. The same thing is therefore true of the equation (72).

value of n. We can therefore determine a positive number μ such that we have, for every positive integral value of n, $|n - 1 + a_0| > \mu(n + 1)$, since the quotient $(n - 1 + a_0)/(n + 1)$ approaches unity as n becomes infinite. Let us replace, on the other hand, the coefficients of xy' and y on the right-hand side of the equation (76) by dominant functions, and let us consider the auxiliary equation

(79) $\quad \mu(xY'' + 2Y') = xY'(A_1 + A_2 x + \cdots) + Y(B_1 + B_2 x + \cdots).$

If we attempt to satisfy this new equation by a series of the form

(80) $\quad\quad\quad Y = 1 + C_1 x + \cdots + C_n x^n + \cdots,$

we are led to the relations analogous to the relations (78),

(81) $\quad\quad n\mu C_n(n + 1) = P_n(A_1, A_2, \cdots, B_1, B_2, \cdots, C_1, \cdots, C_{n-1}).$

If we compare the expressions which give the values of the coefficients c_n and C_n,

$$c_n = \frac{P_n(a_1, a_2, \cdots, b_1, b_2, \cdots, c_1, \cdots, c_{n-1})}{n(n - 1 + a_0)}, \quad C_n = \frac{P_n(A_1, A_2, \cdots, C_{n-1})}{n\mu(n + 1)},$$

the conditions $A_i \geqq |a_i|$, $B_i \geqq |b_i|$, $|n - 1 + a_0| \geqq \mu(n + 1)$ show successively that

$$|c_1| < C_1, \quad |c_2| < C_2, \quad \cdots, \quad |c_n| < C_n;$$

hence it will suffice to show the convergence of the auxiliary series or to show that the equation (79) has an analytic integral, in the neighborhood of the origin, not vanishing for $x = 0$. If we take for the dominant functions an expression of the form $M/(1 - x/r)$, the auxiliary equation (79) can be written

$$\frac{xY'' + 2Y'}{xY' + Y} = \frac{M}{\mu} \frac{1}{1 - \dfrac{x}{r}};$$

whence we derive, by a first integration,

$$xY' + Y = C\left(1 - \frac{x}{r}\right)^{-\frac{Mr}{\mu}}$$

and then

$$xY = C \int_0^x \left(1 - \frac{x}{r}\right)^{-\frac{Mr}{\mu}} dx + C'.$$

We have only to take $C' = 0$, $C = 1$ in order to have an analytic integral, in the neighborhood of the origin, not vanishing for $x = 0$.

Extension to the general case. The proof of Fuchs' theorem for the general case can be based on the same principles by showing that if it is true for an equation of order $(n - 1)$, it is also true for an equation of order n.

If the equation (57) has n particular integrals separating into a certain number of groups of the form (70), it has at least one particular integral of the form $(x - a)^r \phi(x - a)$, where $\phi(x - a)$ is an analytic function in the neighborhood of the point a, which does not vanish for $x = a$. The substitution

$$y = (x - a)^r \phi(x - a) u$$

will lead to a linear equation in u which has the particular integral $u = 1$; hence the derivative u' satisfies a linear homogeneous equation of order $n - 1$. The theorem being supposed true for a linear equation of order $n - 1$, this equation in u' is of the Fuchs form; the same thing is evidently true of the equation in u and therefore of the equation in y.

Conversely, let us consider an equation of the form (71), in which $a = 0$. This equation is formally satisfied by a series of the form

$$y = c_0 x^r + c_1 x^{r+1} + \cdots, \qquad (c_0 \neq 0)$$

where r denotes a root of the *fundamental characteristic equation*

(82) $\quad \begin{cases} D(r) = r(r-1)\cdots(r-n+1) \\ \quad + P_1(0)\, r(r-1)\cdots(r-n+2) + \cdots + P_n(0) = 0 \end{cases}$

such that no other root of this same equation is equal to r increased by a positive integer. In order to establish the convergence of this series, it is easy to show, by an artifice analogous to the one employed for $n = 2$, that it suffices to prove that a linear equation of the form

$$\frac{d^n}{dx^n}(x^{n-1} Y) = \frac{M}{1-\dfrac{x}{r}}\frac{d^{n-1}}{dx^{n-1}}(x^{n-1} Y)$$

has an analytic integral in the neighborhood of the origin not vanishing for $x = 0$. Now this equation has the particular integral (§§ 18 and 39)

$$Y = \frac{1}{(n-2)!\, x^{n-1}} \int_0^x (x-t)^{n-2} \left(1 - \frac{t}{r}\right)^{-Mr} dt,$$

which actually satisfies the preceding condition. If the equation (82) has n distinct roots, $r_1, r_2, \cdots, r_n$, such that none of the differences $r_i - r_k$ is equal to an integer, the general integral of the linear equation is of the form

$$y = C_1 x^{r_1} \phi_1(x) + C_2 x^{r_2} \phi_2(x) + \cdots + C_n x^{r_n} \phi_n(x),$$

where $\phi_1, \phi_2, \cdots, \phi_n$ are analytic in the neighborhood of the origin. If the equation (82) has equal roots or, more generally, roots such that some of the differences $r_i - r_k$ are integers, these roots separate into a certain number of groups, the difference between two roots of the same group being an integer, while the difference between two roots of different groups is never an integer. Let r be the largest root of one of these groups. We have just seen that the equation (71) has a particular integral of the form $x^r \phi(x)$, where $\phi(x)$ is an analytic function in the neighborhood of the origin and such that $\phi(0)$ is not zero. By putting $y = x^r \phi(x) u$, then $du/dx = v$, we are led to a linear differential equation of order $n - 1$ in v, which is again of the Fuchs form. The theorem being supposed true for an equation of order $n - 1$, that equation in v has $n - 1$ particular independent integrals of the form

$$v = x^{\alpha}[\psi_0(x) + \psi_1(x) \operatorname{Log} x + \cdots + \psi_q(x)(\operatorname{Log} x)^q],$$

where $\psi_0, \psi_1, \cdots, \psi_q$ are analytic functions for $x = 0$. If α is not an integer, we easily see, by a succession of integrations by parts, that $\int v\, dx$ is an expression of the same kind as v. If α is an integer, $\int v\, dx$ contains also a logarithmic term

$$C (\operatorname{Log} x)^{q+1},$$

where C is a constant coefficient. Fuchs' theorem is therefore true for an equation of the nth order.[*]

---

[*] For greater detail see the paper by Fuchs in *Crelle's Journal* or the thesis of Jules Tannery (*Annales de l'Ecole Normale*, 2d series, Vol. IV, 1875).

51. Gauss's equation. Let us apply the general method to the equation

(83) $\quad x(1-x)y'' + [\gamma - (\alpha + \beta + 1)x]y' - \alpha\beta y = 0$,

where α, β, γ are constants. The singular points in the finite plane are $x = 0$ and $x = 1$. The characteristic equation for the point $x = 0$ is $r(r + \gamma - 1) = 0$, and its roots are $r = 0$, $r = 1 - \gamma$. If γ is not zero nor equal to a negative integer, it follows from the preceding theory that the equation has an analytic integral in the neighborhood of the origin corresponding to the root $r = 0$. In order to determine this integral, let us substitute in the equation the series

$$y = c_0 + c_1 x + \cdots + c_n x^n + \cdots$$

and equate to zero the coefficient of x^{n-1}. This gives a recurrent relation between any two consecutive coefficients

$$n(\gamma + n - 1)c_n = (\alpha + n - 1)(\beta + n - 1)c_{n-1};$$

hence the analytic integral is the series

$$F(\alpha, \beta, \gamma, x) = 1 + \frac{\alpha \cdot \beta}{1 \cdot \gamma} x + \frac{\alpha(\alpha+1)\beta(\beta+1)}{1 \cdot 2 \cdot \gamma(\gamma+1)} x^2 + \cdots,$$

which is called the *hypergeometric series*. This series is convergent in the circle Γ_0 with unit radius about the origin as center. In order to obtain a second integral, let us make the transformation $y = x^{1-\gamma} z$. This leads to an equation of the same form,

(84) $\quad \begin{cases} x(1-x)z'' + [2 - \gamma - (\alpha + \beta + 3 - 2\gamma)x]z' \\ \quad - (\alpha + 1 - \gamma)(\beta + 1 - \gamma)z = 0, \end{cases}$

which differs from the first only in the substitution of $\alpha + 1 - \gamma$, $\beta + 1 - \gamma$, $2 - \gamma$ for α, β, γ respectively. If $2 - \gamma$ is not zero nor equal to a negative integer, the equation (83) has therefore the second integral $x^{1-\gamma} F(\alpha + 1 - \gamma, \beta + 1 - \gamma, 2 - \gamma, x)$; and if γ is not an integer, the general integral is represented in the circle Γ_0 by the expression

(85) $\quad y = C_1 F(\alpha, \beta, \gamma, x) + C_2 x^{1-\gamma} F(\alpha + 1 - \gamma, \beta + 1 - \gamma, 2 - \gamma, x)$.

If γ is an integer, the difference between two roots of the characteristic equation is zero or equal to an integer, and the integral contains in general a logarithmic term in the neighborhood of the origin. We shall study only the case where $\gamma = 1$. The two integrals

$$F(\alpha, \beta, \gamma, x), \qquad x^{1-\gamma} F(\alpha + 1 - \gamma, \beta + 1 - \gamma, 2 - \gamma, x)$$

reduce in this case to the single integral $F(\alpha, \beta, 1, x)$.

In order to find a second integral, let us first suppose that γ differs but little from unity, say $\gamma = 1 - h$, where h is very small; then the equation (83) has the two integrals

$$F(\alpha, \beta, 1 - h, x), \qquad x^h F(\alpha + h, \beta + h, 1 + h, x),$$

and consequently the quotient

$$\frac{x^h F(\alpha + h, \beta + h, 1 + h, x) - F(\alpha, \beta, 1 - h, x)}{h}$$

is also an integral. As h approaches zero, this quotient approaches as a limit the derivative of the numerator with respect to h at the point $h = 0$. The derivative of the factor x^h gives us a logarithmic term which, for $h = 0$, reduces to $F(\alpha, \beta, 1, x) \log x$. To find the derivative with respect to h of any coefficient in the two series, such as the coefficient

$$\frac{(\alpha+h)(\alpha+h+1)\cdots(\alpha+h+n-1)(\beta+h)(\beta+h+1)\cdots(\beta+h+n-1)}{n!\,(1+h)(2+h)\ldots(n+h)},$$

it is convenient to calculate first the logarithmic derivative. We find thus a new integral which has the form

(86) $\begin{cases} \psi_1(x) = F(\alpha, \beta, 1, x)\operatorname{Log} x \\ \qquad + \Sigma A_n \dfrac{\alpha(\alpha+1)\cdots(\alpha+n-1)\beta(\beta+1)\cdots(\beta+n-1)}{(n!)^2} x^n, \end{cases}$

where we put

$$A_n = \frac{1}{\alpha} + \frac{1}{\alpha+1} + \cdots + \frac{1}{\alpha+n-1} + \frac{1}{\beta} + \cdots + \frac{1}{\beta+n-1}$$
$$\qquad - 2\left(1 + \frac{1}{2} + \cdots + \frac{1}{n}\right).$$

We might study in the same way the integrals of Gauss's equation in the neighborhood of the point $x=1$, but it suffices simply to notice that if we replace x by $1-x$, the equation does not change in form, but γ is replaced by $\alpha+\beta+1-\gamma$. The general integral is therefore represented in the circle Γ_1 with unit radius about the point $x=1$ as center by the expression

$$y = C_1 F(\alpha, \beta, \alpha+\beta+1-\gamma, 1-x)$$
$$\qquad + C_2(1-x)^{\gamma-\alpha-\beta}F(\gamma-\alpha, \gamma-\beta, \gamma+1-\alpha-\beta, 1-x),$$

provided that $\gamma - \alpha - \beta$ is not an integer.

In order to study the integrals for values of x of very large absolute value, we put $x=1/t$, and we are then led to study the integrals of a new linear equation in the neighborhood of the origin. The integrals of this equation are likewise regular in the neighborhood of the origin, and the roots of the characteristic equation are precisely α and β. If we substitute simultaneously $x=1/t$, $y=t^\alpha z$, the equation obtained is again of the form (83), but β is replaced by $\alpha+1-\gamma$, and γ by $\alpha+1-\beta$. Gauss's equation has therefore the integral

$$x^{-\alpha}F\left(\alpha, \alpha+1-\gamma, \alpha+1-\beta, \frac{1}{x}\right).$$

By symmetry it has also the integral obtained from this one by interchanging α and β, and therefore the general integral is represented in the region exterior to the circle Γ_0 by the expression

$$y = C_1 x^{-\alpha}F\left(\alpha, \alpha+1-\gamma, \alpha+1-\beta, \frac{1}{x}\right) + C_2 x^{-\beta}F\left(\beta+1-\gamma, \beta, \beta+1-\alpha, \frac{1}{x}\right),$$

provided that $\alpha - \beta$ is not an integer.

Note. Every linear equation of the form

(87) $\qquad (x-a)(x-b)y'' + (lx+m)y' + ny = 0,$

where a, b, l, m, n are any constants ($a \neq b$), reduces to Gauss's equation by the change of variable $x = a + (b-a)t$. For, to identify the resulting equation

(88) $\qquad t(1-t)\dfrac{d^2y}{dt^2} - \left(\dfrac{la+m}{b-a} + lt\right)\dfrac{dy}{dt} - ny = 0$

with the equation (83), we need only put $\gamma = -(la+m)/(b-a)$, and then determine α and β by the two conditions $\alpha+\beta+1 = l$, $\alpha\beta = n$.

52. Bessel's equation.

Let us consider in particular the equation

(89) $$x(1-kx)y'' + (c-bx)y' - ay = 0,$$

which has the two singular points $x=0$, $x=1/k$, and which can be reduced to Gauss's equation by the change of variable $kx = t$. If we make the parameter k approach zero while a, b, c approach finite limits A, B, C, the singular point $x=1/k$ goes off to infinity, and we obtain at the limit the linear equation

(90) $$xy'' + (C - Bx)y' - Ay = 0,$$

whose only singular point at a finite distance is the point $x=0$. If B is not zero, replacing Bx by x we are led to an equation of the same form, where $B = 1$. Likewise, if $B = 0$ and A is different from zero, we can suppose $A = 1$. Finally, disregarding the trivial case $A = B = 0$, the equation (90) can be replaced by one of the two forms

(91) $$xy'' + (\gamma - x)y' - \alpha y = 0,$$

(92) $$xy'' + \gamma y' - y = 0.$$

Studying the integrals of these two equations in the neighborhood of the origin, as we have done for Gauss's equation, we are led to introduce the two series

$$G(\alpha, \gamma, x) = 1 + \frac{\alpha}{1 \cdot \gamma}x + \frac{\alpha(\alpha+1)}{1 \cdot 2 \cdot \gamma(\gamma+1)}x^2 + \cdots,$$

$$J(\gamma, x) = 1 + \frac{1}{\gamma}x + \frac{1}{1 \cdot 2 \cdot \gamma(\gamma+1)}x^2 + \cdots,$$

which may be considered as degenerate cases of the hypergeometric series. If we replace in $F(\alpha, \beta, \gamma, x)$ the variable x by kx and β by $1/k$, the coefficient of x^n in $F(\alpha, 1/k, \gamma, kx)$ approaches the coefficient of x^n in $G(\alpha, \gamma, x)$ as k approaches zero. Similarly, the coefficient of x^n in $F(1/k, 1/k, \gamma, k^2 x)$ approaches the coefficient of x^n in $J(\gamma, x)$ as k approaches zero.

If γ is not an integer, the general integral of the equation (91) is given by the expression

(93) $$y = C_1 G(\alpha, \gamma, x) + C_2 x^{1-\gamma} G(\alpha + 1 - \gamma, 2 - \gamma, x).$$

Likewise, the general integral of the equation (92) is

(94) $$y = C_1 J(\gamma, x) + C_2 x^{1-\gamma} J(2 - \gamma, x).$$

These formulæ are valid in the whole plane.

If γ is an integer, the general integral of the equation (92) always contains a logarithmic term. For example, if $\gamma = 1$, we obtain an integral different from $J(1, x)$ by finding the limit for $h = 0$ of the quotient

$$\frac{x^h J(1+h, x) - J(1-h, x)}{h},$$

which gives for the general integral

$$y = C_1 J(1, x) + C_2 \left[J(1, x) \log x - 2 \sum_1^{+\infty} \left(1 + \frac{1}{2} + \cdots + \frac{1}{n}\right) \frac{x^n}{(n!)^2} \right].$$

We can reduce to the form (92) a certain linear equation which appears in a large number of questions of mathematical physics. Let us put in the equation (92) $x = -t^2/4$; replacing γ by $n+1$, the equation obtained is identical with the equation already studied (§ 46),

(95) $$t\frac{d^2y}{dt^2} + (2n+1)\frac{dy}{dt} + ty = 0.$$

If, in this last equation, we put $y = t^{-n}z$, we obtain a new form of Bessel's equation,

(96) $$t^2\frac{d^2z}{dt^2} + t\frac{dz}{dt} + (t^2 - n^2)z = 0.$$

The three equations (92), (95), (96), where $\gamma = n+1$, are therefore absolutely equivalent to one another. If n is not an integer, the preceding development shows that the general integral of Bessel's equation (96) is

$$z = C_1 t^n J\left(n+1, -\frac{t^2}{4}\right) + C_2 t^{-n} J\left(1-n, -\frac{t^2}{4}\right).$$

We have shown above (§ 46) that if n is half an *odd* integer, the general integral of the equation (95) can be expressed in terms of elementary transcendental functions. Hence the transcendental function $J(\gamma, x)$ is expressible in terms of exponential functions if γ is half of an odd integer.

Note. The equation studied by Riccati,

(97) $$\frac{du}{dx} + Au^2 - Bx^m = 0,$$

where A, B, m are given constants, can also be reduced to any one of the equivalent equations (92), (95), (96). Indeed, we have seen (§ 40) that the general integral of the equation (97) is z'/Az, where z is the general integral of the linear equation

(98) $$\frac{d^2z}{dx^2} - ABx^m z = 0.$$

If we make the change of variable $x = \lambda t^\mu$, where λ and μ are two undetermined quantities, the last equation becomes

(99) $$t\frac{d^2z}{dt^2} - (\mu - 1)\frac{dz}{dt} - AB\lambda^{m+2}\mu^2 t^{(m+2)\mu - 1} z = 0.$$

In order to identify this equation with the equation (95), we need only take $\mu = 2/(m+2)$, and determine λ by the condition $AB\lambda^{m+2}\mu^2 = -1$. The corresponding value of n is $-\mu/2$ or $-1/(m+2)$. We can therefore express the general integral of Riccati's equation (97) in finite terms whenever $1/(m+2)$ is half of a positive or negative odd integer $2i+1$, that is, whenever m is equal to $-4i/(1+2i)$, where i denotes a positive or negative integer.

53. Picard's equations. Given a linear differential equation with coefficients analytic except for poles, we can determine by Fuchs' method whether the general integral is itself an analytic function except for poles. For this it is necessary and sufficient: (1) that the integrals shall be regular in the neighborhood of each of the singular points; (2) that all the roots of the characteristic equation, relative to each of these singular points, shall be integers; finally, (3) that all the logarithmic terms shall disappear from the expression for the general integral in the neighborhood of a singular point.

Suppose that all these conditions are satisfied. The general integral is then a single-valued analytic function except for poles in the whole plane. If the coefficients of the equation are rational functions, there are only a finite number of singular points $a_1, a_2, \cdots, a_n$. In order for the general integral to be a rational function, it is sufficient that the equation obtained by putting $x = 1/t$ shall itself have all its integrals regular in the neighborhood of the point $t = 0$, since the general integral is single-valued and therefore cannot contain logarithmic terms nor fractional powers of t. If this last condition is satisfied, we can obtain the general integral by equating coefficients according to the method of undetermined coefficients. In fact, let $-m_i$ be the smallest root of the characteristic equation relative to the point $x = a_i$, and N the smallest root of the characteristic equation relative to the point $t = 0$ for the transformed equation. It is clear that the product of any integral y and the expression

$$(x - a_1)^{m_1}(x - a_2)^{m_2}\cdots(x - a_n)^{m_n}$$

is a rational function having no poles in the finite portion of the plane. This product is therefore a polynomial $P(x)$, whose degree is at most equal to

$$m_1 + m_2 + \cdots + m_n - N.$$

Since we know an upper bound for the degree of this polynomial, the coefficients can be determined by replacing y by an expression of the form $P(x)\prod(x - a_i)^{-m_i}$, where $P(x)$ is the most general polynomial of this degree, in the left-hand side of the given equation, and then equating the result identically to zero.

Picard has given another very important case where the general integral can be expressed in terms of the classic transcendental functions. *Given a linear homogeneous differential equation, whose coefficients are elliptic functions of the independent variable with identical periods, if its general integral is an analytic function except for poles, that integral can be expressed in terms of the standard transcendental functions of the theory of elliptic functions.*

For simplicity in writing, let us develop the proof for an equation of the second order only. Let $f_1(x), f_2(x)$ be two independent integrals of a linear homogeneous equation $y'' + p(x)y' + q(x)y = 0$, where $p(x)$ and $q(x)$ are elliptic functions with the periods 2ω and $2\omega'$. By hypothesis, $f_1(x)$ and $f_2(x)$ are single-valued functions analytic except for poles. Since the given equation does not change when we replace x by $x + 2\omega$, $f_1(x + 2\omega)$ and $f_2(x + 2\omega)$ are also integrals, and we have the relations

(100) $\quad f_1(x + 2\omega) = af_1(x) + bf_2(x), \quad f_2(x + 2\omega) = cf_1(x) + df_2(x),$

where a, b, c, d are constant coefficients whose determinant $ad - bc$ is not zero. For if we had $ad - bc = 0$, we could derive from (100) a relation between $f_1(x + 2\omega)$ and $f_2(x + 2\omega)$ of the form $C_1 f_1(x + 2\omega) + C_2 f_2(x + 2\omega) = 0$, where C_1 and C_2 are constants not both equal to zero. This is impossible, since f_1 and f_2 are two independent integrals. For the same reason, we have another system of relations

(101) $\quad f_1(x + 2\omega') = a'f_1(x) + b'f_2(x), \quad f_2(x + 2\omega') = c'f_1(x) + d'f_2(x),$

where a', b', c', d' are constant coefficients, and $a'd' - b'c'$ is not zero. Let us try to find, as in § 47, an integral $\phi(x) = \lambda f_1(x) + \mu f_2(x)$ such that $\phi(x + 2\omega) = s\phi(x)$. We have for the determination of λ, μ, s the two equations

$$\lambda(a - s) + \mu c = 0, \qquad \lambda b + \mu(d - s) = 0;$$

whence we derive the equation of the second degree for s,

$$F(s) = s^2 - (a+d)s + ad - bc = 0.$$

If this equation has two distinct roots s_1, s_2, there exist two independent integrals $\phi_1(x)$, $\phi_2(x)$ such that we have

(102) $\qquad \phi_1(x + 2\omega) = s_1 \phi_1(x), \qquad \phi_2(x + 2\omega) = s_2 \phi_2(x),$

and the relations (101) can be replaced by the two relations of the same form

(103) $\quad \phi_1(x + 2\omega') = k\phi_1(x) + l\phi_2(x), \qquad \phi_2(x + 2\omega') = m\phi_1(x) + n\phi_2(x).$

By means of the relations (102) and (103), we can now obtain two different expressions for $\phi_1(x + 2\omega + 2\omega')$ and $\phi_2(x + 2\omega + 2\omega')$. We have, on the one hand,

$$\phi_1(x + 2\omega + 2\omega') = s_1 \phi_1(x + 2\omega') = s_1 k \phi_1(x) + s_1 l \phi_2(x).$$

On the other hand, proceeding in the inverse order, we may also write

$$\phi_1(x + 2\omega + 2\omega') = k\phi_1(x + 2\omega) + l\phi_2(x + 2\omega) = ks_1 \phi_1(x) + ls_2 \phi_2(x).$$

Since these two expressions must be identical, we have $l = 0$, for $s_1 - s_2$ is not zero. Similarly, by considering the two expressions for $\phi_2(x + 2\omega + 2\omega')$, we find $m = 0$. The integrals $\phi_1(x)$, $\phi_2(x)$ are therefore analytic functions except for poles, which reproduce themselves multiplied by a constant factor when the variable x increases by a period; these are called *doubly periodic functions of the second kind*. Every function $\phi(x)$ analytic except for poles which possesses this property can be expressed in terms of the transcendental functions p, ζ, σ, since the logarithmic derivative $\phi'(x)/\phi(x)$ is an elliptic function, and we have seen that the integration does not introduce any new transcendental (II, Part I, § 75). Moreover, we can prove this without any integration. Let $\phi(x)$ be an analytic function except for poles such that

$$\phi(x + 2\omega) = \mu \phi(x), \qquad \phi(x + 2\omega') = \mu' \phi(x).$$

Consider the auxiliary function $\psi(x) = e^{\rho x} \sigma(x - a)/\sigma(x)$, where a and ρ are any two constants. From the properties of the function σ (see Vol. II, Part I, § 72) we have

$$\psi(x + 2\omega) = e^{2\omega\rho - 2\eta a} \psi(x), \qquad \psi(x + 2\omega') = e^{2\omega'\rho - 2\eta' a} \psi(x).$$

In order for the quotient $\phi(x)/\psi(x)$ to be an elliptic function, it is sufficient that

$$2\omega\rho - 2a\eta = \text{Log } \mu, \qquad 2\omega'\rho - 2a\eta' = \text{Log } \mu'.$$

These relations determine ρ and a (II, Part I, p. 161). It should be noticed that we can take $a = 0$ if Log μ and Log μ' are proportional to the corresponding periods 2ω, $2\omega'$.

Let us now turn to the case where the equation $F(s) = 0$ has a double root s. We can find (§ 48) two independent integrals $\phi_1(x)$, $\phi_2(x)$ such that

(104) $\qquad \phi_1(x + 2\omega) = s\phi_1(x), \qquad \phi_2(x + 2\omega) = s\phi_2(x) + C\phi_1(x).$

If $C = 0$, all the integrals of the equation, and in particular $f_1(x)$ and $f_2(x)$, are multiplied by s when x is increased by 2ω. Assuming $C = 0$, let us try to find a linear combination $\lambda f_1(x) + \mu f_2(x)$ which reproduces itself multiplied by s' when

x increases by $2\omega'$. Starting from the equations (103), we find two independent integrals $\phi_1(x)$, $\phi_2(x)$ such that either

$$\phi_1(x+2\omega') = s_1'\phi_1(x), \qquad \phi_2(x+2\omega') = s_2'\phi_2(x)$$

or

$$\phi_1(x+2\omega') = s'\phi_1(x), \qquad \phi_2(x+2\omega') = s'\phi_2(x) + C'\phi_1(x),$$

where C' is not zero. In the first case the integrals $\phi_1(x)$, $\phi_2(x)$ are again doubly periodic functions of the second kind. In the second case the integral $\phi_1(x)$ alone is a doubly periodic function of the second kind. As for the integral $\phi_2(x)$, the quotient $\phi_2(x)/\phi_1(x)$ increases by a constant C' when x increases by $2\omega'$, and it does not change when x increases by 2ω. Now the function $A\zeta(x) + Bx$, where A and B are two constant coefficients, possesses the same property, provided that we have

$$2A\eta + 2B\omega = 0, \qquad 2A\eta' + 2B\omega' = C'.$$

The difference $\phi_2/\phi_1 - A\zeta(x) - Bx$ is therefore an elliptic function.

If the coefficient C is not zero in the equations (104), we have relations between the integrals $\phi_1(x)$, $\phi_2(x)$, $\phi_1(x+2\omega')$, $\phi_2(x+2\omega')$ of the form (103), and we can again deduce from them two different expressions for $\phi_1(x+2\omega+2\omega')$ and $\phi_2(x+2\omega+2\omega')$. By writing that they are identical, we obtain the conditions $l = 0$, $k = n$. The integral $\phi_1(x)$ is again a doubly periodic function of the second kind, while the integral $\phi_2(x)$ satisfies the two relations

$$\frac{\phi_2(x+2\omega)}{\phi_1(x+2\omega)} = \frac{\phi_2(x)}{\phi_1(x)} + \frac{C}{s}, \qquad \frac{\phi_2(x+2\omega')}{\phi_1(x+2\omega')} = \frac{\phi_2(x)}{\phi_1(x)} + \frac{m}{k}.$$

Let us determine just as before the two coefficients A and B in such a way that $2A\eta + 2B\omega = C/s$, $2A\eta' + 2B\omega' = m/k$. Then the difference

$$\frac{\phi_2(x)}{\phi_1(x)} - A\zeta(x) - Bx$$

is again an elliptic function. We see, therefore, that the general integral is in all cases expressible in terms of the single transcendentals e^x, $p(x)$, $\zeta(x)$, $\sigma(x)$.

Let us consider, for example, *Lamé's equation*

(105) $$\frac{d^2y}{dx^2} - [n(n+1)p(x) + h]y = 0,$$

where n is an integer and h is an arbitrary constant. The integration of this equation by Hermite was the starting point for the preceding theory. The general integral of this equation is a function analytic except for poles. In fact, the only singular points are the origin and the points $2m\omega + 2m'\omega'$. In the neighborhood of the origin the integrals are regular, and the roots of the characteristic equation are $r' = -n$, $r'' = n+1$. Their difference is an odd integer, and the coefficient of y is an even function; therefore the expression for the general integral does not contain any logarithmic term (see ftn., p. 137).

54. Equations with periodic coefficients. In many important questions of mechanics, linear equations with periodic coefficients occur. We shall indicate rapidly their more important properties. Let

(106) $$\frac{d^n y}{dt^n} + p_1 \frac{d^{n-1}y}{dt^{n-1}} + \cdots + p_n y = 0$$

be a linear equation whose coefficients are continuous functions of the real variable t, having a period ω, which we may always suppose *positive*. If the integrals $y_1(t), y_2(t), \cdots, y_n(t)$ form a fundamental system, it is clear that $y_1(t+\omega), y_2(t+\omega), \cdots, y_n(t+\omega)$ are also integrals of the equation (106), since that equation remains unchanged when we replace t by $t+\omega$. Hence we have n relations of the form

(107) $\qquad y_i(t+\omega) = a_{i1}y_1(t) + a_{i2}y_2(t) + \cdots + a_{in}y_n(t). \qquad (i=1, 2, \cdots, n)$

The determinant H of the coefficients a_{ik} is different from zero. For, by repeating the reasoning of page 129, we find that this determinant has the value

(108) $$H = e^{-\int_0^\omega p_1 \, dt}.$$

The equations (107) define a linear substitution with constant coefficients, whose determinant is not zero. We are therefore led to a study entirely similar to the one which has already been made in detail in §§ 48, 49. Instead of making the complex variable x describe a circuit in the positive sense around a singular point a, the variable t describes a segment of the real axis of length ω. It follows from that study that we can always choose a fundamental system of integrals such that the relations (107) reduce to a simple canonical form. The actual formation of this system depends first of all on the solution of the *characteristic equation*

(109) $$F(s) = \begin{vmatrix} a_{11}-s & a_{12} & \cdots & a_{1n} \\ a_{21} & a_{22}-s & \cdots & a_{2n} \\ \cdots & \cdots & \cdots & \cdots \\ a_{n1} & a_{n2} & \cdots & a_{nn}-s \end{vmatrix} = 0.$$

All the roots of this equation are different from zero, since their product is equal to the determinant H, whose value we have just written down. If the n roots of that equation are distinct, there exists a fundamental system of integrals such that the equations (107) take the form

(110) $\qquad y_1(t+\omega) = s_1 y_1(t), \quad \cdots, \quad y_n(t+\omega) = s_n y_n(t).$

If the equation (109) has multiple roots, we can always find a fundamental system of integrals which separate into a certain number of groups such that the p integrals $y_1, y_2, \cdots, y_p$ of the same group satisfy relations of the form

(111) $$\begin{cases} y_1(t+\omega) = s y_1(t), \\ y_2(t+\omega) = s[y_2(t)+y_1(t)], \\ \cdots \cdots \cdots \cdots \cdots \cdots \cdots, \\ y_p(t+\omega) = s[y_p(t)+y_{p-1}(t)]. \end{cases}$$

In order to find expressions for these integrals, let us seek first the general form of a single-valued continuous function $f(t)$ such that $f(t+\omega) = sf(t)$, where the factor s is not zero. Let α be a determination of $(1/\omega)\operatorname{Log} s$. It is clear that the product $f(t)e^{-\alpha t}$ has the period ω; hence $f(t)$ is of the form $f(t) = e^{\alpha t}\phi(t)$, where $\phi(t)$ is a continuous function with the period ω. Accordingly, if s_i is a root of the characteristic equation, we shall put $\alpha_i = (1/\omega)\operatorname{Log} s_i$. The constants α_i, which are determined except for multiples of $2\pi\sqrt{-1}/\omega$, are called the *characteristic exponents*. The real parts of these exponents, which are determined without ambiguity, are called the *characteristic numbers*. If the equation (109) has n distinct roots $s_1, s_2, \cdots, s_n$, the equation (106) has then n independent particular integrals of the form

(112) $\quad y_1 = e^{\alpha_1 t}\phi_1(t), \quad y_2 = e^{\alpha_2 t}\phi_2(t), \quad \cdots, \quad y_n = e^{\alpha_n t}\phi_n(t),$

where $\alpha_1, \alpha_2, \cdots, \alpha_n$ are the characteristic exponents, and where $\phi_1, \phi_2, \cdots, \phi_n$ are continuous functions with the period ω.

In the general case it is evidently sufficient to find expressions for the integrals of a group which satisfy the relations (111). Now if we substitute in these relations $y_i = e^{\alpha t} z_i$, where α is equal to $(1/\omega) \log s$, they become

(111') $\quad \begin{cases} z_1(t + \omega) = z_1(t), \\ z_2(t + \omega) = z_2(t) + z_1(t), \\ \cdots \cdots \cdots \cdots \cdots \cdots \cdots, \\ z_p(t + \omega) = z_p(t) + z_{p-1}(t). \end{cases}$

When t increases by ω, the variable $\tau = t/\omega$ increases by unity. Taking τ for a new variable, the problem is reduced to one solved above (§ 49). If we set

$$P_i(t) = \frac{t(t - \omega) \cdots (t - i\omega + \omega)}{\omega^i i!}, \quad (i = 1, 2, \cdots, p)$$

the general expressions for the functions $y_1, y_2, \cdots, y_p$ are

(113) $\quad \begin{cases} y_1 = e^{\alpha t}\phi_1(t), \quad y_2 = e^{\alpha t}[P_1(t)\phi_1(t) + \phi_2(t)], \quad \cdots, \\ y_i(t) = e^{\alpha t}[P_{i-1}(t)\phi_1(t) + P_{i-2}(t)\phi_2(t) + \cdots + P_1(t)\phi_{i-1}(t) + \phi_i(t)], \\ (i = 1, 2, \cdots, p) \end{cases}$

where $\phi_1, \phi_2, \cdots, \phi_p$ are continuous functions with the period ω, the first of which, $\phi_1(t)$, is not zero. We see again here, as in § 49, that all these integrals can be deduced from the last one of the group. For $z_{p-1}(t)$ is equal to the difference $z_p(t + \omega) - z_p(t)$, and, similarly, $z_{p-2}(t) = z_{p-1}(t + \omega) - z_{p-1}(t)$, and so on. We can therefore write the equations (113) in the form

(114) $\quad \begin{cases} y_p(t) = e^{\alpha t} z_p(t), \\ y_{p-1}(t) = e^{\alpha t} \Delta_1(z_p), \\ y_{p-2}(t) = e^{\alpha t} \Delta_2(z_p), \\ \cdots \cdots \cdots \cdots \cdots, \\ y_1(t) = e^{\alpha t} \Delta_{p-1}(z_p), \end{cases}$

where $\Delta_1(z_p), \Delta_2(z_p), \cdots$ indicate the successive differences of $z_p(t)$ when we change t to $t + \omega$. Let us observe that $z_p(t)$ is a polynomial in t of degree $p-1$, whose coefficients are periodic functions of t. The successive differences $\Delta_1(z_p), \Delta_2(z_p), \cdots$ are therefore polynomials of the same kind with decreasing degrees, the pth difference being zero. Let us indicate by $Dz_p, D^2 z_p, \cdots, D^i z_p$ the successive derivatives of z_p taken with respect to t, considering the coefficients of this polynomial as constants. From the theory of finite differences, we know that the successive differences $\Delta_1(z_p), \Delta_2(z_p), \cdots$ are linear combinations *with numerical coefficients* of the derivatives $Dz_p, D^2 z_p, \cdots, D^i z_p$, and conversely.*

---

*Without resorting to this theory, we may observe that Taylor's formula gives us, step by step,
$$\Delta_i(z_p) = \omega^i D^i z_p + \cdots,$$
where the terms not written contain only the derivatives $D^{i+1}, \cdots$. We can therefore express, conversely, the derivatives $D^i z_p$ as linear functions of the differences $\Delta_i, \Delta_{i+1}, \cdots$.

We can therefore replace the system of integrals (114) by the equivalent system

(115)
$$\begin{cases} Y_p(t) = e^{\alpha t} z_p(t), \\ Y_{p-1}(t) = e^{\alpha t} D z_p, \\ Y_{p-2}(t) = e^{\alpha t} D^2 z_p, \\ \cdots\cdots\cdots\cdots, \\ Y_1(t) = e^{\alpha t} D^{p-1} z_p. \end{cases}$$

Note 1. The integrals of the group (113), corresponding to the characteristic exponent α, approach zero when t becomes infinite passing through positive values, if and only if the real part of α is *negative*. In order that all the integrals of the equation (106) shall approach zero as t becomes infinite, it is therefore necessary and sufficient that all the *characteristic numbers shall be negative*, or, what amounts to the same thing, that the absolute value of each of the roots of the equation (109) is *less than unity*.

Note 2. If s is a real positive root of the equation (109), it is natural to take for α the real determination of $(1/\omega) \log s$. If the coefficients of the equation (106) are real, the same thing will evidently be true in this case of the integrals $y_1, y_2, \cdots, y_p$ of the group (113) and consequently of the periodic functions $\phi_1(t), \phi_2(t), \cdots$.

Let $s = \lambda + \mu \sqrt{-1}$ be a p-fold root of the equation (109), where $\mu \neq 0$, and let $\alpha = \alpha' + \alpha'' \sqrt{-1}$ be a corresponding determination of the exponent α. To the group of integrals (113) we can adjoin a conjugate group obtained by replacing α by $\alpha' - \alpha'' \sqrt{-1}$ and the functions $\phi_i(t)$ by the conjugate functions. It is clear that by combining these $2p$ integrals linearly in pairs we can derive from them a system of $2p$ real integrals.

Finally, suppose that s is a real *negative* root. Then we can write the value of $\alpha = \alpha' + (\pi/\omega) \sqrt{-1}$, and to that root corresponds a particular integral of the form

$$y = e^{\alpha' t} \left(\cos \frac{\pi t}{\omega} + \sqrt{-1} \sin \frac{\pi t}{\omega} \right) (\psi_1(t) + \sqrt{-1} \psi_2(t)),$$

where the functions ψ_1 and ψ_2 are real and periodic. If the coefficients of (106) are real, it is clear that the real part and the coefficient of $\sqrt{-1}$ must each satisfy separately the linear equation. We would proceed similarly with the other integrals of the group (113) if p is greater than unity.

Moreover, the case where s is real and negative reduces to the case where s is real and positive by considering the period 2ω instead of the period ω. It is clear, in fact, that if an integral is multiplied by s when we change t to $t + \omega$, it will be multiplied by s^2 when we change t to $t + 2\omega$.

Note 3. When the coefficients p_i are analytic functions of the complex variable $t = t' + t'' \sqrt{-1}$, analytic in the strip R included between the two parallels to the real axis $t'' = \pm h$, the integrals of the equation (106) are analytic functions in the same strip. The reasoning used under the supposition that the variable t moves along the real axis applies without modification to the case in which that variable moves in the strip R. It follows that the functions $\phi_i(t)$, which appear in the expressions of (113), are periodic analytic functions in the strip R. They can therefore be developed in series of sines and cosines of multiples of the angle $2\pi t/\omega$ (see Vol. II, Part I, § 65).

55. Characteristic exponents. The investigation of the characteristic exponents is in general very difficult.* The solution of this problem evidently reduces to the determination of the coefficients a_{ik} which appear in the equations (107), which, in turn, is equivalent to the following: knowing the initial values, for $t = t_0$, of the n integrals $y_1, y_2, \cdots, y_n$ and their first $n-1$ derivatives, to find the values of these integrals and of their derivatives for $t = t_0 + \omega$. The coefficients a_{ik} are then obtained by the solution of the n systems of linear equations

(116) $\begin{cases} y_i(t_0 + \omega) = a_{i1} y_1(t_0) + a_{i2} y_2(t_0) + \cdots + a_{in} y_n(t_0) \\ y_i^{(p)}(t_0 + \omega) = a_{i1} y_1^{(p)}(t_0) + \cdots + a_{in} y_n^{(p)}(t_0). \\ \qquad\qquad (p = 1, 2, \ldots, (n-1)) \quad (i = 1, 2, \ldots, n) \end{cases}$

We cannot in general solve this last problem except by the use of general methods, for example, by successive approximations. Let us replace p_i by λp_i in the equation (106), where λ denotes a variable parameter, and then develop in powers of λ the integral of that equation which together with its first $(n-1)$ derivatives takes on preassigned values independent of λ for $t = t_0$,

(117) $\qquad y = f_0(t) + \lambda f_1(t) + \cdots + \lambda^n f_n(t) + \cdots,$

where $f_0(t)$ is a polynomial in t, of degree $n-1$ at most, which can be written down immediately from the initial conditions. Substituting this value of y in (106), we see that the other coefficients $f_1(t), f_2(t), \cdots$ are determined, step by step, by relations of the form

$$\frac{d^n f_i}{dt^n} = \Phi[t, f_1(t), f_1'(t), \cdots],$$

in which the right-hand sides depend only upon the functions $f_1, f_2, \cdots, f_{i-1}$, and upon their derivatives. Moreover, these coefficients, together with their first $n-1$ derivatives, must vanish for $t = t_0$. Hence these coefficients can be found by quadratures. We have already noticed (§ 28) that the series obtained is convergent for any value λ. If we put $\lambda = 1$ in the relation (117) and in all those which we obtain from it by differentiation, we shall have the developments of the integral under consideration and of its derivatives in series which are convergent for all real values of t. Hence we can obtain in this way the quantities $y_i(t_0 + \omega)$, $y_i^{(p)}(t_0 + \omega)$ which appear in the equations (116), and consequently we can determine the coefficients a_{ik}.†

Example. Let us consider, for example, the equation

(118) $\qquad\qquad \dfrac{d^2 y}{dt^2} = p(t) y,$

where $p(t)$ is a continuous function of t with the period ω. The product of the roots of the characteristic equation is here equal to one, by formula (108).

---

* When the coefficients p_i are analytic integral functions of the complex variable t, the change of variable $e^{2\pi t i/\omega} = x$ replaces the given equation by a linear equation whose coefficients are single-valued in the neighborhood of the origin, and we are led to study the law of the permutation of the integrals when the variable x describes a loop around the origin. But the equation thus obtained is not in general of the Fuchs form.

† If we allow the parameter λ to have any value, it follows, from the process used above, that the coefficients a_{ik}, and consequently the coefficients of the characteristic equation, are *integral functions* of this parameter.

That equation is therefore of the form

(119) $$s^2 - As + 1 = 0.$$

In order to determine the coefficient A, let us denote by $f(t)$ and $\phi(t)$ the integrals of the equation (118) which satisfy the initial conditions $f(0) = 1$, $f'(0) = 0$, $\phi(0) = 0$, $\phi'(0) = 1$. From the relations

$$f(t + \omega) = a_{11} f(t) + a_{12} \phi(t),$$
$$\phi(t + \omega) = a_{21} f(t) + a_{22} \phi(t),$$
$$\phi'(t + \omega) = a_{21} f'(t) + a_{22} \phi'(t),$$

we derive, by putting $t = 0$, $a_{11} = f(\omega)$, $a_{22} = \phi'(\omega)$. The characteristic equation in this special case is

$$(a_{11} - s)(a_{22} - s) - a_{12} a_{21} = 0,$$

whence $A = a_{11} + a_{22} = f(\omega) + \phi'(\omega)$.

If we now replace $p(t)$ by $\lambda p(t)$, we obtain the developments of the integrals $f(t)$, $\phi(t)$ in the form

$$f(t) = 1 + \lambda f_1(t) + \cdots + \lambda^n f_n(t) + \cdots,$$
$$\phi(t) = t + \lambda \phi_1(t) + \cdots + \lambda^n \phi_n(t) + \cdots,$$

where the functions f_n and ϕ_n, together with f_n' and ϕ_n', vanish for $t = 0$. Substituting these developments in the two sides of the equation (118), after having replaced p by λp, we find

$$\frac{d^2 f_n}{dt^2} = p(t) f_{n-1}(t), \qquad \frac{d^2 \phi_n}{dt^2} = p(t) \phi_{n-1}(t),$$

whence we derive the recurrent relations

$$f_n(t) = \int_0^t dt \int_0^t p(t) f_{n-1}(t) \, dt, \qquad \phi_n(t) = \int_0^t dt \int_0^t p(t) \phi_{n-1}(t) \, dt,$$

which enable us to calculate step by step all these functions by starting with $f_0(t) = 1$, $\phi_0(t) = t$. It follows that we may write

(120) $$A = 2 + \sum_{n=1}^{+\infty} [f_n(\omega) + \phi_n'(\omega)].$$

If the function $p(t)$ is never negative, we see at once that all the functions $f_n(t)$, $\phi_n(t)$, $\phi_n'(t)$ are positive for $t > 0$. It follows that $A > 2$, and the equation (119) has two real and positive roots, one greater and the other smaller than unity. The conclusion is much less evident in the other cases. If $p(t)$ never takes on a positive value, it follows from a thorough study made by Liapunof * that the absolute value of A is less than 2, if the absolute value of

$$\omega \int_0^\omega p \, dt$$

is less than or equal to 4. The equation (119) has in this case two conjugate imaginary roots, the absolute value of each of which is unity.

---

*LIAPUNOF, *Problème général de la stabilité du mouvement* (*Annales de la Faculté des Sciences de Toulouse*, 2d series, Vol. IX, p. 403). On the general theory of linear equations with periodic coefficients, in addition to the preceding paper, see also Floquet's *Annales de l'École Normale supérieure*, 1883, and Poincaré's *Les Méthodes nouvelles de la Méchanique céleste* (Vol. I, chap. iv).

IV. SYSTEMS OF LINEAR EQUATIONS

56. General properties. Most of the theorems established for a linear equation can be extended without difficulty to systems of linear equations in several dependent variables. We shall assume in what follows, as we may without loss of generality, that these equations are of the first order (§ 22). Let $y_1, y_2, \cdots, y_n$ be the n dependent functions, and x the independent variable. It follows from a general theorem (§ 37) that the integrals have no other singular points than those of the coefficients. If we assign the initial values $y_1^0, y_2^0, \cdots, y_n^0$ for a point $x = x_0$ which is not a singular point, we can follow the analytic extension of these integrals along the whole of any path starting from x_0 and not passing through any of these singular points, which are known in advance.

We shall suppose, only for simplification in writing, that we have a system of three equations with three dependent variables. Let us consider first the system of three homogeneous equations,

(121)
$$\begin{cases} \dfrac{dy}{dx} + ay + bz + cu = 0, \\ \dfrac{dz}{dx} + a_1 y + b_1 z + c_1 u = 0, \\ \dfrac{du}{dx} + a_2 y + b_2 z + c_2 u = 0, \end{cases}$$

where $a, b, c, \cdots$ are functions of the single variable x. If we know a particular system of integrals (y_1, z_1, u_1), the functions (Cy_1, Cz_1, Cu_1) also form a system of integrals for any value of the constant C. Similarly, if we know two particular systems of integrals, (y_1, z_1, u_1) and (y_2, z_2, u_2), we can derive from them a new system of integrals depending upon two arbitrary constants,

$$C_1 y_1 + C_2 y_2, \qquad C_1 z_1 + C_2 z_2, \qquad C_1 u_1 + C_2 u_2.$$

Finally, if we know three particular systems of integrals,

$$(y_1, z_1, u_1), \qquad (y_2, z_2, u_2), \qquad (y_3, z_3, u_3),$$

the equations

(122)
$$\begin{cases} y = C_1 y_1 + C_2 y_2 + C_3 y_3, \\ z = C_1 z_1 + C_2 z_2 + C_3 z_3, \\ u = C_1 u_1 + C_2 u_2 + C_3 u_3 \end{cases}$$

represent also a system of integrals, where C_1, C_2, C_3 are arbitrary constants. In order to assert that the expressions (122) represent the general integral of the system (121), we must make sure that we can choose the constants C_1, C_2, C_3 in such a way that, for a given point $x = x_0$ not a singular point, y, z, u take on any preassigned values y_0, z_0, u_0 whatever. In order for this to be true, it is necessary and sufficient that the determinant of the nine functions y_i, z_i, u_i ($i = 1, 2, 3$),

$$\Delta = \begin{vmatrix} y_1 & z_1 & u_1 \\ y_2 & z_2 & u_2 \\ y_3 & z_3 & u_3 \end{vmatrix},$$

shall not vanish identically. If this is true, we shall say that the set of three particular systems of integrals form a *fundamental system*.

If Δ vanishes identically, the three particular systems of integrals reduce to two, or even to a single system. Suppose, first, that not all the first minors of Δ vanish simultaneously, for example, that the minor $\delta = y_1 z_2 - y_2 z_1$ is not identically zero. Let A be a region of the plane where δ does not vanish. We shall determine two auxiliary functions K_1 and K_2, analytic in the region A, such that we have

(123) $\qquad y_3 = K_1 y_1 + K_2 y_2, \qquad z_3 = K_1 z_1 + K_2 z_2,$

and since the determinant Δ is zero, these functions K_1 and K_2 also satisfy the relation

(124) $\qquad\qquad\qquad u_3 = K_1 u_1 + K_2 u_2.$

If we replace y, z, and u in the first two equations of the system (121) by the preceding expressions for y_3, z_3, u_3, observing that (y_1, z_1, u_1) and (y_2, z_2, u_2) form two particular systems of integrals, we obtain, after simplification, the equations

$$y_1 K_1' + y_2 K_2' = 0, \qquad z_1 K_1' + z_2 K_2' = 0,$$

from which we derive $K_1' = K_2' = 0$. The functions K_1 and K_2 are therefore constants, and the relations (123) and (124) remain true in the whole region of existence of the functions y_i, z_i, u_i. It follows that the system of integrals (y_3, z_3, u_3) is a combination of the other two.

If all the first minors of Δ vanish identically, the three systems of integrals reduce to a single system. Since the elements of Δ cannot all vanish simultaneously, let us suppose that y_1 is different from zero, and let us put $y_2 = K y_1$. From the relations $y_1 z_2 - z_1 y_2 = 0$, $y_1 u_2 - u_1 y_2 = 0$ we derive also $z_2 = K z_1$, $u_2 = K u_1$. Replacing y, z, u in the first of the equations (121) by $K y_1$, $K z_1$, $K u_1$ respectively, there remains $y_1 K' = 0$. Hence K is constant, and the system (y_2, z_2, u_2) differs from the system (y_1, z_1, u_1) only by a constant factor. Similarly, the third system of integrals is identical with the first. It should be observed that y_1, y_2, y_3 are not necessarily linearly independent; for example, one or two of these functions may be zero, but not all three may be zero.

The value of the determinant Δ may be calculated as follows. The derivative Δ' is the sum of the three determinants

$$\Delta' = \begin{vmatrix} y_1' & z_1 & u_1 \\ y_2' & z_2 & u_2 \\ y_3' & z_3 & u_3 \end{vmatrix} + \begin{vmatrix} y_1 & z_1' & u_1 \\ y_2 & z_2' & u_2 \\ y_3 & z_3' & u_3 \end{vmatrix} + \begin{vmatrix} y_1 & z_1 & u_1' \\ y_2 & z_2 & u_2' \\ y_3 & z_3 & u_3' \end{vmatrix}.$$

Replacing the derivatives y_i', z_i', u_i' by their values obtained from the equations (121), these three determinants reduce, respectively, to $-a\Delta$, $-b_1\Delta$, $-c_2\Delta$. We have, therefore, the relation $\Delta' = -(a + b_1 + c_2)\Delta$, and consequently

$$(125) \qquad \Delta(x) = \Delta(x_0) e^{-\int_{x_0}^{x}(a+b_1+c_2)dx}.$$

When we know the general integral of the homogeneous system (121), we can deduce from it by quadratures the general solution of the non-homogeneous system

$$(126) \qquad \begin{cases} \dfrac{dy}{dx} + ay + bz + cu = f_1(x), \\ \dfrac{dz}{dx} + a_1 y + b_1 z + c_1 u = f_2(x), \\ \dfrac{du}{dx} + a_2 y + b_2 z + c_2 u = f_3(x). \end{cases}$$

Indeed, if we make the change of variables defined by the equations (122), C_1, C_2, C_3 being considered as new dependent variables, the system (126) is replaced by the following system,

$$(127) \qquad \begin{cases} y_1 \dfrac{dC_1}{dx} + y_2 \dfrac{dC_2}{dx} + y_3 \dfrac{dC_3}{dx} = f_1(x), \\ z_1 \dfrac{dC_1}{dx} + z_2 \dfrac{dC_2}{dx} + z_3 \dfrac{dC_3}{dx} = f_2(x), \\ u_1 \dfrac{dC_1}{dx} + u_2 \dfrac{dC_2}{dx} + u_3 \dfrac{dC_3}{dx} = f_3(x), \end{cases}$$

which is integrable by quadratures, for we derive from it

$$\frac{dC_i}{dx} = X_i. \qquad (i = 1, 2, 3)$$

Let us also observe that this transformation is unnecessary whenever we can determine directly a particular system of integrals (Y, Z, U) of the equations (126). In order to obtain the general integral of these equations, we need only add Y, Z, U, respectively, to the right-hand sides of the equations (122) which represent the general integral of the homogeneous system (121).*

---

* A method analogous to that of Cauchy (§ 39) may also be employed. Let

$$y = \phi_i(x, \alpha), \qquad z = \psi_i(x, \alpha), \qquad u = \pi_i(x, \alpha) \qquad (i = 1, 2, 3)$$

be three systems of integrals of the homogeneous equations (121), satisfying, respectively, the initial condition

$$\begin{array}{lll} \phi_1(\alpha, \alpha) = 1, & \psi_1(\alpha, \alpha) = 0, & \pi_1(\alpha, \alpha) = 0, \\ \phi_2(\alpha, \alpha) = 0, & \psi_2(\alpha, \alpha) = 1, & \pi_2(\alpha, \alpha) = 0, \\ \phi_3(\alpha, \alpha) = 0, & \psi_3(\alpha, \alpha) = 0, & \pi_3(\alpha, \alpha) = 1. \end{array}$$

When we know *one* or *two* particular systems of integrals of the equations (121), we can lower the order of the system by one or two units. Suppose, first, that we know a single system of integrals (y_1, z_1, u_1), where the function y_1 is not zero. The change of dependent variables

$$y = y_1 Y, \qquad z = z_1 Y + Z, \qquad u = u_1 Y + U$$

leads to a linear system of the same form which must have the particular system of integrals $Y = 1$, $Z = 0$, $U = 0$. Therefore the coefficients of Y in these new equations must be zero. In fact, the transformed system is

(128)
$$\begin{cases} y_1 \dfrac{dY}{dx} + bZ + cU = 0, \\ \dfrac{dZ}{dx} + z_1 \dfrac{dY}{dx} + b_1 Z + c_1 U = 0, \\ \dfrac{dU}{dx} + u_1 \dfrac{dY}{dx} + b_2 Z + c_2 U = 0. \end{cases}$$

If we replace dY/dx in the last two equations by its value derived from the first, we obtain a system of two linear homogeneous equations in the *two* dependent variables Z and U. After integrating this system Y can be obtained by a quadrature.

Suppose now that we know two *independent* systems of integrals, (y_1, z_1, u_1), (y_2, z_2, u_2). Since the three determinants

$$y_1 z_2 - y_2 z_1, \qquad y_1 u_2 - y_2 u_1, \qquad z_1 u_2 - z_2 u_1$$

do not vanish simultaneously, as we have shown above, let us suppose that $y_1 z_2 - y_2 z_1$ is different from zero. The transformation

$$y = y_1 Y + y_2 Z, \qquad z = z_1 Y + z_2 Z, \qquad u = u_1 Y + u_2 Z + U,$$

where Y, Z, U are the new dependent variables, leads to a linear system of the same form having the two particular systems of

---

It is easy to see that the functions

$$Y = \int_{x_0}^{x} [f_1(\alpha) \phi_1(x, \alpha) + f_2(\alpha) \phi_2(x, \alpha) + f_3(\alpha) \phi_3(x, \alpha)] d\alpha,$$

$$Z = \int_{x_0}^{x} [f_1(\alpha) \psi_1(x, \alpha) + f_2(\alpha) \psi_2(x, \alpha) + f_3(\alpha) \psi_3(x, \alpha)] d\alpha,$$

$$U = \int_{x_0}^{x} [f_1(\alpha) \pi_1(x, \alpha) + f_2(\alpha) \pi_2(x, \alpha) + f_3(\alpha) \pi_3(x, \alpha)] d\alpha$$

form a system of integrals of the non-homogeneous equations (126).

integrals $(Y_1 = 1, Z_1 = U_1 = 0)$, $(Y_2 = 0, Z_2 = 1, U_2 = 0)$. The coefficients of Y and Z in the equations of the new system must therefore be zero, and this new system has the form

$$\frac{dY}{dx} + AU = 0, \qquad \frac{dZ}{dx} + A_1 U = 0, \qquad \frac{dU}{dx} + A_2 U = 0,$$

as is easily verified. It is clear that this system is integrable by quadratures, since the last equation contains only U.

The preceding methods may be extended to systems of n linear equations with n dependent variables. In order to obtain the general integral of such a homogeneous system, it is sufficient to know n particular systems of integrals which form a fundamental system. If we know p independent systems of integrals ($p < n$), the integration reduces to that of a system of the same form with $n - p$ dependent variables and to a number of quadratures. Finally, the general integral of a non-homogeneous system can be obtained by quadratures if we know the general integral of the corresponding homogeneous system.

57. Adjoint systems. Given a linear homogeneous system with n dependent variables,

(129) $\qquad \dfrac{dy_i}{dx} = a_{i1} y_1 + \cdots + a_{ik} y_k + \cdots + a_{in} y_n, \qquad (i, k = 1, 2, \cdots, n)$

the linear system

(130) $\qquad \dfrac{dY_i}{dx} = -a_{1i} Y_1 - \cdots - a_{ki} Y_k - \cdots - a_{ni} Y_n,$

which is obtained from the first by replacing y_i by Y_i, and by changing the rows into columns in the determinant of the coefficients a_{ik}, after having changed the sign of each element, is called the *adjoint* of the first. It is evident from the definition itself that this relation is a *reciprocal* one between the two systems.

The integration of one of the systems (129), (130) *involves that of the other.* In fact, let $(y_1, y_2, \cdots, y_n)$ and $(Y_1, Y_2, \cdots, Y_n)$ be any two particular systems of integrals of the two adjoint systems. From the relations (129) and (130) we have

$$\frac{d}{dx}(Y_1 y_1 + Y_2 y_2 + \cdots + Y_n y_n) = \sum_i Y_i (a_{i1} y_1 + \cdots + a_{ik} y_k + \cdots + a_{in} y_n)$$
$$+ \sum_i y_i (-a_{1i} Y_1 - \cdots - a_{ki} Y_k - \cdots - a_{ni} Y_n).$$

If we permute the indices i and k in the second sum, we see immediately that the coefficient of $Y_i y_k$ on the right-hand side is

$$a_{ik} - a_{ik} = 0,$$

and the right-hand side is identically zero. We have therefore the relation between these two particular systems of integrals

(131) $\qquad Y_1 y_1 + Y_2 y_2 + \cdots + Y_n y_n = C,$

where C denotes a constant. The knowledge of a particular system of integrals $(Y_1, Y_2, \cdots, Y_n)$ of the equations (130) furnishes therefore a first integral of the system (129), which is *linear* with respect to the dependent variables $y_1, y_2, \cdots, y_n$. If we know the general integral of the adjoint system (130), the general integral of the given system (129) is represented by n relations of the form (131), where we take successively, for $Y_1, Y_2, \cdots, Y_n$, a set of n independent systems of integrals of the equations (130).

Particular attention has been paid to linear systems which are identical with their adjoint. In order to have this case, it is necessary and sufficient that the determinant of the a_{ik} be a skew symmetric determinant; that is, that we have $a_{ik} + a_{ki} = 0$, whatever may be i and k, and consequently $a_{ii} = 0$. If $(y_1, y_2, \cdots, y_n)$ and $(z_1, z_2, \cdots, z_n)$ are two particular systems of integrals, the relation (131) becomes

$$y_1 z_1 + y_2 z_2 + \cdots + y_n z_n = \text{const.};$$

and if the two systems are identical, we have also

$$y_1^2 + y_2^2 + \cdots + y_n^2 = \text{const.}$$

The integration of a linear system of the third order identical with its adjoint leads to the integration of a Riccati equation (§ 31, Ex. 2). The integration of a system of the fourth order of that kind leads to the integration of *two* Riccati equations (see Ex. 15, p. 170).

58. Linear systems with constant coefficients. If all the coefficients $a, b, c, \cdots$ of the equations

(132)
$$\begin{cases} \dfrac{dy}{dx} + ay + bz + cu = 0, \\ \dfrac{dz}{dx} + a_1 y + b_1 z + c_1 u = 0, \\ \dfrac{du}{dx} + a_2 y + b_2 z + c_2 u = 0 \end{cases}$$

are constants, the general integral can be found by the solution of an algebraic equation. For let us try to satisfy these equations by taking for y, z, u expressions of the form

(133) $\qquad y = \alpha e^{rx}, \qquad z = \beta e^{rx}, \qquad u = \gamma e^{rx},$

where α, β, γ, r are unknown parameters. Substituting these functions for y, z, u in the left-hand sides of the equations (132), and suppressing the common factor e^{rx}, we find the conditions

(134)
$$\begin{cases} (a + r)\alpha + b\beta + c\gamma = 0, \\ a_1 \alpha + (b_1 + r)\beta + c_1 \gamma = 0, \\ a_2 \alpha + b_2 \beta + (c_2 + r)\gamma = 0, \end{cases}$$

which must be satisfied by values of α, β, γ which do not all vanish. For this it is necessary and sufficient that r shall be a root of the equation of the third degree,

$$(135) \qquad F(r) = \begin{vmatrix} a+r & b & c \\ a_1 & b_1+r & c_1 \\ a_2 & b_2 & c_2+r \end{vmatrix} = 0,$$

which is called the *auxiliary equation*. Having taken for r a root of this equation, the relations (134) are consistent and we can deduce from them values for α, β, γ, at least one of which is not zero. To every root of the equation $F(r) = 0$ corresponds therefore a particular system of integrals of the form (133); there may even be several, as we shall see presently. If the auxiliary equation has three distinct roots r_1, r_2, r_3, each one furnishes a particular system of integrals. These three systems are independent, for, if they were not, we could express $e^{r_3 x}$ as a linear combination with constant coefficients of $e^{r_1 x}$ and of $e^{r_2 x}$, which would be absurd. We can therefore, in this case, obtain the general integral of the system (132) after we have solved the equation $F(r) = 0$.

It remains to treat the case in which the auxiliary equation has a multiple root. Let us denote by $f(r), \phi(r), \psi(r)$ the three cofactors of the auxiliary determinant corresponding to the elements of the same row, for example, the first. The last two equations of the system (134) are always satisfied for any value of r by taking for α, β, γ quantities proportional to these cofactors; if r is a root of $F(r) = 0$, these values of α, β, γ also satisfy the first of the equations (134). It follows from this that if r is a root of $F(r) = 0$, the functions

$$y = f(r) e^{rx}, \qquad z = \phi(r) e^{rx}, \qquad u = \psi(r) e^{rx}$$

form a particular system of integrals. Now let us suppose first that the equation $F(r) = 0$ has two roots, r_1 and r_2, whose difference is very small. Each of them furnishes a system of integrals, and the functions

$$\frac{f(r_2) e^{r_2 x} - f(r_1) e^{r_1 x}}{r_2 - r_1}, \qquad \frac{\phi(r_2) e^{r_2 x} - \phi(r_1) e^{r_1 x}}{r_2 - r_1}, \qquad \frac{\psi(r_2) e^{r_2 x} - \psi(r_1) e^{r_1 x}}{r_1 - r_1}$$

are also integrals. If we now let r_2 approach r_1 and pass to the limit, we may conclude that if r_1 is a double root of $F(r) = 0$, the two groups of functions,

$$(I) \qquad y_1 = f(r_1) e^{r_1 x}, \qquad z_1 = \phi(r_1) e^{r_1 x}, \qquad u_1 = \psi(r_1) e^{r_1 x},$$

(II) $\begin{cases} y_2 = \dfrac{\partial}{\partial r}[f(r)e^{rx}]_{r=r_1}, & z_2 = \dfrac{\partial}{\partial r}[\phi(r)e^{rx}]_{r=r_1}, \\ u_2 = \dfrac{\partial}{\partial r}[\psi(r)e^{rx}]_{r=r_1}, \end{cases}$

form two systems of integrals. Similarly (see § 44), if the equation $F(r) = 0$ has a triple root r_1, we can add to the preceding two groups the group of three functions,

(III) $\begin{cases} y_3 = \dfrac{\partial^2}{\partial r^2}[f(r)e^{rx}]_{r=r_1}, & z_3 = \dfrac{\partial^2}{\partial r^2}[\phi(r)e^{rx}]_{r=r_1}, \\ u_3 = \dfrac{\partial^2}{\partial r^2}[\psi(r)e^{rx}]_{r=r_1}, \end{cases}$

which form a third system of integrals.

Let us now consider first the case where the equation $F(r) = 0$ has a double root r_1 and a simple root r_2. If the double root r_1 does not cause all the first minors of the auxiliary determinant to vanish, we may suppose that at least one of the cofactors $f(r_1)$, $\phi(r_1)$, $\psi(r_1)$ is not zero, for we can evidently replace, in the reasoning which precedes, the first row by the second or the third. Suppose, for example, $f(r_1) \neq 0$. The two systems of integrals (I) and (II) are independent, for y_2 is equal to the product of $e^{r_1 x}$ and a binomial of the first degree $xf(r_1) + f'(r_1)$. As for the simple root r_2, it furnishes a third system of integrals which, for the same reason as above, is not a linear combination of the first two.

The reasoning fails if the double root r_1 makes all the first minors vanish, for the system (I) reduces to the trivial solution

$$y_1 = z_1 = u_1 = 0.$$

But in this case the three equations (134) reduce to a single equation when we replace in it r by r_1. If, for example, c is not zero, they reduce to the single equation $(a + r_1)\alpha + b\beta + c\gamma = 0$, and we can take the two constants α and β arbitrarily. If we take, first, $(\alpha = 1, \beta = 0)$, then $(\alpha = 0, \beta = 1)$, we obtain two independent systems of integrals of the form (133). *A double root of $F(r) = 0$, therefore, always furnishes two particular independent systems of integrals.*

Suppose, finally, that $F(r) = 0$ has the triple root $r = r_1$. If this root r_1 does not cause all the first minors of the determinant to vanish, we may suppose, for example, that $f(r_1)$ is not zero. The three particular systems of integrals (I), (II), (III) are independent, for the coefficients of $e^{r_1 x}$ in y_1, y_2, y_3 are respectively of degrees 0, 1, 2 in x.

If the triple root r_1 causes all the first minors of the determinant to vanish, we can determine first of all two independent systems of integrals of the form (133), as we have just explained in regard to the case of the double root, and we can then obtain the general integral if we can find a third system independent of these two. Developing the expressions in (III), and noting that

we find
$$f(r_1) = \phi(r_1) = \psi(r_1) = 0,$$

$$y_3 = e^{r_1 x}[2xf'(r_1) + f''(r_1)], \quad z_3 = e^{r_1 x}[2x\phi'(r_1) + \phi''(r_1)],$$
$$u_3 = e^{r_1 x}[2x\psi'(r_1) + \psi''(r_1)],$$

and this system of integrals is certainly independent of the first two unless we have at the same time $f'(r_1) = \phi'(r_1) = \psi'(r_1) = 0$.

Hence we obtain in this way a new system of integrals, unless the triple root r_1 also causes the derivatives of all the first minors to vanish. Now this cannot happen, as we see at once, unless we have

$$b = c = a_1 = c_1 = a_2 = b_2 = 0, \quad a = b_1 = c_2 = -r_1,$$

and the system (132) reduces to three identical equations,

$$\frac{dy}{dx} - r_1 y = 0, \quad \frac{dz}{dx} - r_1 z = 0, \quad \frac{du}{dx} - r_1 u = 0.$$

In this case, which may be considered as a limiting case, the three equations (134) are satisfied identically, when we replace r by r_1 in the expressions of (133), for any values whatever of the parameters α, β, γ. Summing up, *to a triple root of the auxiliary equation there always correspond three particular independent systems of integrals.*

Generalization. Similarly, a system of n linear equations with constant coefficients

(136) $$\begin{cases} \dfrac{dy_1}{dx} + a_{11}y_1 + a_{12}y_2 + \cdots + a_{1n}y_n = 0, \\ \cdots\cdots\cdots\cdots\cdots\cdots\cdots\cdots\cdots\cdots\cdots, \\ \dfrac{dy_n}{dx} + a_{n1}y_1 + a_{n2}y_2 + \cdots + a_{nn}y_n = 0, \end{cases}$$

may be integrated by finding particular systems of integrals of the form

(137) $\quad y_1 = \alpha_1 e^{rx}, \quad y_2 = \alpha_2 e^{rx}, \quad \cdots, \quad y_n = \alpha_n e^{rx},$

where $\alpha_1, \alpha_2, \cdots, \alpha_n$, r are unknown constants whose values are to be determined. We are thus led to n equations of condition

$$(138) \quad \begin{cases} (a_{11} + r)\alpha_1 + a_{12}\alpha_2 + \cdots + a_{1n}\alpha_n = 0, \\ a_{21}\alpha_1 + (a_{22} + r)\alpha_2 + \cdots + a_{2n}\alpha_n = 0, \\ \cdots \cdots \cdots \cdots \cdots \cdots \cdots \cdots \cdots \cdots \cdots, \\ a_{n1}\alpha_1 + a_{n2}\alpha_2 + \cdots + (a_{nn} + r)\alpha_n = 0, \end{cases}$$

which give for the unknown quantity r the auxiliary equation

$$(139) \quad F(r) = \begin{vmatrix} a_{11} + r & a_{12} & \cdots & a_{1n} \\ a_{21} & a_{22} + r & \cdots & a_{2n} \\ \cdots & \cdots & \cdots & \cdots \\ a_{n1} & a_{n2} & \cdots & a_{nn} + r \end{vmatrix} = 0.$$

If this equation has n distinct roots $r_1, r_2, \cdots, r_n$, we obtain by this method n particular systems of integrals of the form (137) and, consequently, the general integral. If there are multiple roots, the discussion is somewhat more complicated. Let r_1 be a p-fold root; to obtain from this root particular systems of integrals of the equations (136), we may proceed in two ways. On the one hand, applying d'Alembert's method, as in the case of three equations, we can obtain p systems of integrals corresponding to that root. These integrals will be independent only if r_1 does not make all the first minors vanish. On the other hand, if r_1 makes all the minors formed from $n - q + 1$ rows of the determinant vanish, without making all those of $n - q$ rows zero, that root furnishes q systems of integrals of the form (137), for the n equations (138) reduce to $n - q$ independent equations when we replace r by r_1. Combining these two methods, we find that they always furnish p independent systems of integrals.

Practically we can obtain all these systems by equating coefficients. In fact, by the combination just mentioned we should obtain a system of integrals depending upon p arbitrary constants, which is of the form

$$y_1 = e^{r_1 x} P_1(x), \qquad y_2 = e^{r_1 x} P_2(x), \quad \cdots, \quad y_n = e^{r_1 x} P_n(x),$$

where $P_1, P_2, \cdots, P_n$ are polynomials of degree $p - 1$ or of lower degree. If we leave the coefficients of these polynomials as unknown, and if we substitute in the given equations, we shall obtain a certain number of relations between these coefficients, which enable us to express all of them in terms of p of them, which may be taken as arbitrary constants.

59. Reduction to a canonical form. Every linear system with constant coefficients can be reduced to a simple canonical form the integration of which is immediate.

Let us write this system under a slightly different form,

$$(140) \quad \begin{cases} y_1' = a_{11} y_1 + a_{12} y_2 + \cdots + a_{1n} y_n, \\ y_2' = a_{21} y_1 + a_{22} y_2 + \cdots + a_{2n} y_n, \\ \cdots \cdots \cdots \cdots \cdots \cdots \cdots \cdots \cdots, \\ y_n' = a_{n1} y_1 + a_{n2} y_2 + \cdots + a_{nn} y_n, \end{cases}$$

where y_i' denotes dy_i/dx. If we take n dependent variables, $Y_1, Y_2, \cdots, Y_n$, linear in terms of $y_1, y_2, \cdots, y_n$,

$$(141) \quad Y_i = b_{i1} y_1 + \cdots + b_{in} y_n, \qquad (i = 1, 2, \cdots, n)$$

where the coefficients b_{ik} are constants whose determinant is different from zero, the system (140) is replaced by a system of the same form,

$$(142) \quad \begin{cases} Y_1' = A_{11}Y_1 + A_{12}Y_2 + \cdots + A_{1n}Y_n, \\ Y_2' = A_{21}Y_1 + A_{22}Y_2 + \cdots + A_{2n}Y_n, \\ \cdots \cdots \cdots \cdots \cdots \cdots \cdots \cdots \cdots \cdots \\ Y_n' = A_{n1}Y_1 + A_{n2}Y_2 + \cdots + A_{nn}Y_n, \end{cases}$$

obtained by replacing the variables $y_1, y_2, \cdots, y_n$, in the expressions for Y_i',

$$Y_i' = b_{i1}y_1' + \cdots + b_{in}y_n' = b_{i1}(a_{11}y_1 + \cdots + a_{1n}y_n) + \cdots \\ + b_{in}(a_{n1}y_1 + \cdots + a_{nn}y_n),$$

by their values given by the equations (141). If we consider the equations (140) as a linear substitution carried out on the variables $y_1, y_2, \cdots, y_n$, and $Y_1, Y_2, \cdots, Y_n$ as n new variables, the preceding calculations are precisely those which we must make in order to find the new linear substitution on the variables $Y_1, Y_2, \cdots, Y_n$, which corresponds to the linear substitution (140). Now we have seen that by suitably choosing the variables Y_i (§ 48) we can reduce every linear substitution to a simple canonical form.* In this canonical form the variables separate into a certain number of distinct groups, such that the substitution which the p variables $Y_1, Y_2, \cdots, Y_p$ of the same group undergo is of the form

$$(143) \quad Y_1' = sY_1, \quad Y_2' = s(Y_1 + Y_2), \quad \cdots, \quad Y_p' = s(Y_{p-1} + Y_p).$$

We can therefore, by a suitable change of variables of the form (141), always reduce the integration of the system (140) to the integration of a certain number of systems of the form (143), where $Y_i' = dY_i/dx$.

The integration of this system is immediate, but it is preferable to employ a somewhat different canonical form. For this purpose let us set $Y_i = s^i z_i (s \neq 0)$. The system (143) becomes

$$(144) \quad \frac{dz_1}{dx} = sz_1, \quad \frac{dz_2}{dx} = sz_2 + z_1, \quad \cdots, \quad \frac{dz_p}{dx} = sz_p + z_{p-1}.$$

This new canonical form is unchanged if we multiply all the dependent variables by a factor $e^{\lambda x}$, except for the change of s to $s + \lambda$; and it is applicable also to the case where the auxiliary equation has zero for a root.

The general integral of the system (144) is represented by the equations

$$z_i e^{-sx} = C_1 \frac{x^{i-1}}{(i-1)!} + C_2 \frac{x^{i-2}}{(i-2)!} + \cdots + C_{i-1}x + C_i \quad (i = 1, 2, \cdots, p)$$

or by equivalent equations obtained by solving for the constants C_i

$$(145) \quad z_1 e^{-sx} = C_1, \quad (z_2 - xz_1)e^{-sx} = C_2, \quad \left(z_3 - xz_2 + \frac{x^2}{2!}z_1\right)e^{-sx} = C_3,$$

$$\left\{z_i - xz_{i-1} + \frac{x^2}{2!}z_{i-2} - \cdots + (-1)^{i-1}\frac{x^{i-1}}{(i-1)!}z_1\right\}e^{-sx} = C_i. \quad (i = 1, 2, \cdots, p)$$

---

* We supposed before that the determinant of the substitution was not zero, whereas the determinant formed by the coefficients a_{ik} may be zero. But if we change y_i to $e^{\lambda x} z_i$, the coefficients $a_{11}, a_{22}, \cdots, a_{nn}$ are diminished by λ, while the a_{ik}'s, where $i \neq k$, do not change. We can therefore always choose λ in such a way that the determinant of the new coefficients shall not be zero.

60. Jacobi's equation. Let us consider again a system of three linear equations with constant coefficients, which we shall write in the form

(146)
$$\begin{cases} \dfrac{dx}{dt} = ax + by + cz, \\ \dfrac{dy}{dt} = a_1 x + b_1 y + c_1 z, \\ \dfrac{dz}{dt} = a_2 x + b_2 y + c_2 z, \end{cases}$$

where t denotes the independent variable. Let us add these three equations, after having multiplied them respectively by $y\,dz - z\,dy$, $z\,dx - x\,dz$, $x\,dy - y\,dx$. The relation obtained is

(147) $\begin{cases} (ax + by + cz)(y\,dz - z\,dy) + (a_1 x + b_1 y + c_1 z)(z\,dx - x\,dz) \\ \quad + (a_2 x + b_2 y + c_2 z)(x\,dy - y\,dx) = 0 \end{cases}$

and it is homogeneous in x, y, z. Hence it can be replaced by a relation between x/z and y/z. Indeed, if we put $x = Xz$, $y = Yz$, and divide by z^3, this relation takes the form

(148) $\begin{cases} -(aX + bY + c)\,dY + (a_1 X + b_1 Y + c_1)\,dX \\ \quad + (a_2 X + b_2 Y + c_2)(X\,dY - Y\,dX) = 0, \end{cases}$

which is exactly Jacobi's equation (pp. 11 and 32).

Let $x = f(t)$, $y = \phi(t)$, $z = \psi(t)$ be a system of integrals of the equations (146). As t varies, the point whose homogeneous coördinates are x, y, z (and whose Cartesian coördinates are $X = x/z$, $Y = y/z$) describes a plane curve Γ which is, by the preceding argument, an integral curve of Jacobi's equation (148). The integration of Jacobi's equation therefore reduces to the integration of the system (146), that is, to the solution of an algebraic equation of the third degree, as we have already seen.

If the auxiliary equation has three distinct roots s_1, s_2, s_3, the general integral of the system (146) is, according to the preceding paragraph, of the form

(I) $\qquad Pe^{-s_1 t} = C_1, \qquad Qe^{-s_2 t} = C_2, \qquad Re^{-s_3 t} = C_3,$

where P, Q, R are three linear homogeneous functions of x, y, z. It is easy to derive from these equations a homogeneous combination of degree zero which does not contain the variable t,

(α) $\qquad\qquad P^{s_2 - s_3} Q^{s_3 - s_1} R^{s_1 - s_2} = K,$

which is the same result that we obtained before by another method.

The case in which the auxiliary equation has a double or a triple root can also be easily treated. The equations representing the general integral form either two groups or a single group. In the first case these equations are of the form

(II) $\quad Pe^{-s_1 t} = C_1, \quad (Q - tP)e^{-s_1 t} = C_2, \quad Re^{-s_2 t} = C_3,$

and in the second case, of the form

(III) $\quad Pe^{-s_1 t} = C_1, \quad (Q - tP)e^{-s_1 t} = C_2, \quad \left(R - tQ + \dfrac{t^2}{2}P\right)e^{-s_1 t} = C_3,$

where P, Q, R denote in each case three linear homogeneous functions of x, y, z. From (II) we derive the following homogeneous combination of zero degree, independent of t:

(β) $\quad \dfrac{R}{P} e^{(s_1 - s_2)\frac{Q}{P}} = K,$

and from (III) the combination

(γ) $\quad \dfrac{2PR - Q^2}{P^2} = K.$

The relations (α), (β), (γ) represent the three forms possible for the general integral of Jacobi's equation.

61. Systems with periodic coefficients. Let us consider first, for simplicity, a system of three equations of the form (146), whose coefficients a, b, c, $\cdots$ are continuous functions of the variable t, each of which has the period $\omega > 0$.

Let (x_1, y_1, z_1), (x_2, y_2, z_2), (x_3, y_3, z_3) be three independent systems of integrals. Then the functions

$$X_i(t) = x_i(t + \omega), \quad Y_i(t) = y_i(t + \omega), \quad Z_i(t) = z_i(t + \omega)$$

also form a system of integrals, and we have consequently three groups of relations of the form (§ 56),

(149) $\quad \begin{cases} X_i = a_{i1} x_1 + a_{i2} x_2 + a_{i3} x_3, \\ Y_i = a_{i1} y_1 + a_{i2} y_2 + a_{i3} y_3, \\ Z_i = a_{i1} z_1 + a_{i2} z_2 + a_{i3} z_3, \end{cases} \quad (i = 1, 2, 3)$

where the a_{ik}'s are constant coefficients whose determinant H is not zero. We have, in fact, the relation

$$\begin{vmatrix} X_1 & Y_1 & Z_1 \\ X_2 & Y_2 & Z_2 \\ X_3 & Y_3 & Z_3 \end{vmatrix} = H \begin{vmatrix} x_1 & y_1 & z_1 \\ x_2 & y_2 & z_2 \\ x_3 & y_3 & z_3 \end{vmatrix},$$

or, by (125), reasoning as we have done several times (§§ 38, 56), we may write the value of H in the form

(150) $\quad H = e^{\int_0^\omega (a + b_1 + c_2)\, dt}.$

If the variable t is increased by the period ω, the three functions

$$x_1(t), \quad x_2(t), \quad x_3(t)$$

undergo a linear transformation whose determinant is different from zero, defined by the relations

(151) $\quad \begin{cases} X_1 = a_{11} x_1 + a_{12} x_2 + a_{13} x_3, \\ X_2 = a_{21} x_1 + a_{22} x_2 + a_{23} x_3, \\ X_3 = a_{31} x_1 + a_{32} x_2 + a_{33} x_3, \end{cases}$

and the two other systems of functions, (y_1, y_2, y_3), (z_1, z_2, z_3), undergo the same transformation. Now we know that it is possible to replace the three functions x_1, x_2, x_3 by three independent linear combinations with constant coefficients such that the equations (151), which define the new linear transformation, take on a simple canonical form. Taking the same linear combinations of the functions (y_1, y_2, y_3) and of (z_1, z_2, z_3), we obtain three systems of functions which are transformed by the same linear substitution of canonical form when t is changed to $t + \omega$.

The reasoning is evidently general and applies to every linear homogeneous system in n dependent variables with periodic coefficients. Let $y_1, y_2, \cdots, y_n$ be these n dependent variables. We can determine n independent systems of integrals $(y_{1i}, y_{2i}, \cdots, y_{ni})$ $(i = 1, 2, \cdots, n)$ such that the n functions

$$y_{k1}, \quad y_{k2}, \quad \cdots, \quad y_{kn}$$

undergo a linear substitution of canonical form when t changes to $t + \omega$, this linear substitution being the same for all the indices k. The consequences are the same as those which have been developed above (§ 54). All the integrals are expressible as the product of an exponential factor of the form $e^{\alpha t}$ and another factor which is either a periodic function of t or a polynomial in t whose coefficients are continuous periodic functions of t. Let

$$y_{11} = e^{\alpha t} z_1, \quad y_{21} = e^{\alpha t} z_2, \quad \cdots, \quad y_{n1} = e^{\alpha t} z_n$$

be a particular system of integrals, where $z_1, z_2, \cdots, z_n$ are polynomials in t, with periodic coefficients, of which *at least* one is of degree $p - 1$, and of which none is of a degree greater than $p - 1$. From this system of integrals we can derive $(p - 1)$ other systems of the form

$$y_{12} = e^{\alpha t} D z_1, \quad y_{22} = e^{\alpha t} D z_2, \quad \cdots, \quad y_{n2} = e^{\alpha t} D z_n,$$
$$\cdots, \quad \cdots, \quad \cdots, \quad \cdots,$$
$$y_{1,p} = e^{\alpha t} D^{p-1} z_1, \quad y_{2,p} = e^{\alpha t} D^{p-1} z_2, \quad \cdots, \quad y_{n,p} = e^{\alpha t} D^{p-1} z_n,$$

where the derivatives $D^i z_k$ are taken regarding the periodic coefficients of the powers of t as constants (§ 54). All the systems of integrals of the given equations can thus be derived from a certain number of them. The actual formation of these integrals, of which we know only the analytic form, depends, above all, on the solution of an algebraic equation of the nth degree, which is called, as before, the *characteristic equation* of the system. The coefficients of this equation can be obtained in general only by approximations, as in the case of a single differential equation of the nth order (§ 55).

62. Reducible systems. Let us consider a system of linear homogeneous equations of the form (140), whose coefficients are real, continuous, and bounded functions of the real variable t for all the values of that variable greater than a certain bound t_0, and let us suppose that we apply to this system a transformation of the form

(152) $$z_i = b_{i1} y_1 + b_{i2} y_2 + \cdots + b_{in} y_n, \qquad (i = 1, 2, \cdots, n)$$

where the coefficients b_{ik} satisfy the following conditions:

1) They are real, continuous, and bounded functions of the variable t for $t > t_0$;
2) They have derivatives satisfying the same condition;
3) The reciprocal of the determinant of the b_{ik}'s is bounded.

If we take the functions z_i for new dependent variables, it is clear that the system (140) is replaced by a linear system of the same kind as the first. We have, in fact,

$$\frac{dz_i}{dt} = b'_{i1} y_1 + b_{i1} y'_1 + \cdots,$$

or, replacing $y'_1, y'_2, \cdots, y'_n$ by their values obtained from the equations (140),

$$\frac{dz_i}{dt} = c_{i1} y_1 + c_{i2} y_2 + \cdots + c_{in} y_n,$$

where the coefficients c_{ik} have the same properties as the coefficients a_{ik}. We have now only to replace $y_1, y_2, \cdots, y_n$ in these last equations by their expressions in terms of the new dependent variables $z_1, z_2, \cdots, z_n$ obtained from the equations (152).

If it is possible to choose the coefficients b_{ik} of the transformation in such a way that the new system will be a system *with constant coefficients*, the system is said by Liapunof to be *reducible*. See page 242 of his paper cited in the footnote on page 151.

Every system whose coefficients are real, continuous, and periodic functions, with the same period ω, is reducible.

In fact, let us consider the adjoint system, which is also a system with periodic coefficients. Let s be a root of the characteristic equation and α the corresponding characteristic exponent. We shall suppose, in order to consider the most general case, that to this exponent α corresponds a group of p particular systems of integrals of the form previously considered. This group will therefore furnish (§ 57) p linear first integrals of the given system, which will be of the form

$$e^{\alpha t}(z_1 y_1 + z_2 y_2 + \cdots + z_n y_n) = C_1,$$
$$e^{\alpha t}(y_1 Dz_1 + y_2 Dz_2 + \cdots + y_n Dz_n) = C_2,$$
$$\cdots \cdots \cdots \cdots \cdots \cdots \cdots \cdots \cdots,$$
$$e^{\alpha t}(y_1 D^{p-1} z_1 + y_2 D^{p-1} z_2 + \cdots + y_n D^{p-1} z_n) = C_p,$$

where $z_1, z_2, \cdots, z_n$ are polynomials in t, of degree $p-1$ at most, with periodic coefficients, and where the derivatives D^i are taken regarding these coefficients as constants. Arranging these first integrals with respect to t, we may write them in the form

(153) $$\begin{cases} e^{\alpha t}\left[\dfrac{t^{p-1}}{(p-1)!} Y_1 - \dfrac{t^{p-2}}{(p-2)!} Y_2 + \cdots \pm Y_p\right] = C_1, \\ e^{\alpha t}\left[\dfrac{t^{p-2}}{(p-2)!} Y_1 - \dfrac{t^{p-3}}{(p-3)!} Y_2 + \cdots\right] = C_2, \\ \cdots \cdots \cdots \cdots \cdots \cdots \cdots \cdots \cdots, \\ e^{\alpha t} Y_1 = C_p, \end{cases}$$

where $Y_1, Y_2, \cdots, Y_p$ are *independent* linear combinations of $y_1, y_2, \cdots, y_n$ with periodic coefficients. For if they were not independent, we could derive from the equations (153) a relation between the *arbitrary* constants $C_1, C_2, \cdots, C_p$ and the variable t. If we take the linear combinations $Y_1, Y_2, \cdots, Y_p$ for dependent variables, the relations (153) represent precisely the general integral of the linear system of equations (§ 59),

(154) $$\frac{dY_1}{dt} = -\alpha Y_1, \quad \frac{dY_2}{dt} = -\alpha Y_2 + Y_1, \quad \cdots, \quad \frac{dY_p}{dt} = -\alpha Y_p + Y_{p-1}.$$

Proceeding similarly with all the groups of first integrals furnished by the groups of integrals of the adjoint system, we see that the given system is transformed into a linear system with constant coefficients by means of a transformation of the form

(155) $$Y_i = \phi_{i1} y_1 + \phi_{i2} y_2 + \cdots + \phi_{in} y_n,$$

where the coefficients ϕ_{ik} are periodic functions with the period ω.

The reciprocal of the determinant D of the ϕ_{ik}'s is *bounded* for $t > t_0$, for we shall show that D does not vanish for $t > t_0$. Indeed, if we consider n independent systems of integrals $(y_{1i}, \cdots, y_{ni})$ of the first system and the corresponding n systems $(Y_{1i}, \cdots, Y_{ni})$ of the transformed system, the determinant D is equal to the quotient obtained by dividing the determinant of the Y_{ki}'s by the determinant of the y_{ki}'s, and we know that these last two have finite values different from zero for all finite values of t. It follows that the absolute value of D remains greater than a certain positive minimum for all values of t between t_0 and $t_0 + \omega$.

In order to complete the proof, we may suppose that the characteristic equation of the adjoint system has no real negative roots; for, by § 54, any root is replaced by its square if we consider the period 2ω instead of the period ω. If the characteristic equation has only real positive roots, we may evidently suppose that all the functions ϕ_{ik} which appear in the equations (155) are real. Then that transformation actually satisfies all the required conditions. Moreover, all the characteristic exponents are real, and the transformed system has real coefficients. But if the characteristic equation of the adjoint system has conjugate imaginary roots, to each group of p linear combinations, such as $Y_1, Y_2, \cdots, Y_p$, in which appear imaginaries we can associate the group formed by the conjugate imaginaries. Hence, combining them in conjugate pairs, it is clear that we again obtain a system with real constant coefficients by means of a transformation of the desired form with real coefficients.

EXERCISES

1. Integrate the linear equations

$$y^{(\text{IV})} - 2y'' + y = Ae^x + Be^{-x} + C \sin x + D \cos x, \qquad y^{(\text{V})} + y'' = x,$$
$$y''' - y'' + y' - y = 2e^x - 4 \cos x,$$
$$y''' - 3y' + 2y = (ax + b)e^x + ce^{-2x},$$
$$x^2 y''' - 9xy'' + 9y' = 1 + 2x + 3x^2 \log x,$$
$$x^2 y'' - 2xy' + 2y = x^2 + px + q,$$
$$x^3 y''' - 3x^2 y'' + 7xy' - 8y = x^3 - 2x,$$
$$x^2 y'' - 3xy' + 4y = x^2 + \int_0^x \frac{dx}{\sqrt{1+x^4}},$$
$$x^3 y''' - 9x^2 y'' + 37xy' - 64y = x^4 [a + b \log x + c(\log x)^2],$$
$$x^2 y'' + 2xy' - 2y = x \cos x - \sin x,$$
$$x^2 y'' + 3xy' + y = f(x).$$

If $f(x)$ is analytic in the neighborhood of the origin, prove that this last equation has a particular integral analytic in the same neighborhood.

2. Integrate the systems of linear equations

(α) $\quad \dfrac{dy}{dt} - \dfrac{dz}{dt} + x = 0, \quad \dfrac{dz}{dt} - \dfrac{dx}{dt} + y = 0, \quad \dfrac{dx}{dt} - \dfrac{dy}{dt} + z = 0;$

(β) $\quad \dfrac{d^2x}{dt^2} + 5x + y = \cos 2t, \quad \dfrac{d^2y}{dt^2} - x + 3y = 0;$

(γ) $\quad \begin{cases} \dfrac{d^2y}{dx^2} - 5\dfrac{dy}{dx} + \dfrac{dz}{dx} + 13y + 20z = e^x, \\ \dfrac{d^2z}{dx^2} + \dfrac{dy}{dx} + 3\dfrac{dz}{dx} + 2y + 3z = e^{-x}; \end{cases}$

(δ) $\quad \dfrac{dx}{dt} + y - z = 0, \quad \dfrac{dy}{dt} - z = 0, \quad \dfrac{dz}{dt} + x - z = 0;$

(ϵ) $\quad \dfrac{dx}{dt} + x - y = 0, \quad \dfrac{dy}{dt} + y - 4z = 0, \quad \dfrac{dz}{dt} + 4z - x = 0;$

(ζ) $\quad \begin{cases} \dfrac{dy}{dx} - 3y + 8z - 4u = 0, \quad \dfrac{dz}{dx} + y - 5z + 2u = 0, \\ \dfrac{du}{dx} + 3y - 14z + 6u = 0; \end{cases}$

(η) $\quad \begin{cases} y' - (\lambda+1)y - 2z + 2(1-\lambda)u = 0, \\ z' + \lambda y + z + 2(\lambda-1)u = 0, \\ u' + \lambda y + (2\lambda-1)u = 0. \end{cases}$

3. Find the general integral of the equation

$$(2x+1)y'' + (4x-2)y' - 8y = (6x^2 + x - 3)e^x$$

from the fact that the homogeneous equation has a particular integral of the form e^{mx}, where m is a constant. [*Licence*, Caen, 1884.]

4. Prove that the differential equation

$$(x^2-1)y'' = n(n+1)y,$$

where n is a positive integer, has a polynomial $P(x)$ for integral. From this prove that the same equation has a second integral

$$P \operatorname{Log}\left(\dfrac{x+1}{x-1}\right) + Q,$$

where Q is also a polynomial. [*Licence*, Paris, 1890.]

5. The linear differential equation

$$xy'' - (x+\mu+\nu)y' + \mu y = 0,$$

where μ and ν are two positive integers, has a polynomial $y_1 = P(x)$ as an integral. Hence prove that it has a second integral $y_2 = e^x Q(x)$, where $Q(x)$ is also a polynomial. [*Licence*, Paris, 1903.]

6. Find the necessary and sufficient condition that the linear equation $y'' + py' + qy = 0$ may have two independent integrals, y_1, y_2, which satisfy the relation $y_1 y_2 = 1$. Assuming that $p = -1/x$, find the coefficient q and the general integral. [*Licence*, Paris, 1902.]

7. Derive the formula (23), p. 111, from the formula (11), p. 106, which gives the value of the determinant $\Delta(y_1, y_2, \cdots, y_n)$.

8. Bessel's equation,
$$xy'' + 2(m+1)y' + xy = 0,$$
has as a particular integral the function represented by the definite integral
$$y_1 = \int_0^1 (1-z^2)^m \cos xz \, dz,$$
provided that the real part of m is greater than -1. If m is a positive integer, that integral is of the form (see Vol. I, end of Chap. V, Ex. 20, 2d ed.; Ex. 21, 1st ed.)
$$2 \cdot 4 \cdot 6 \cdots 2m (U \sin x + V \cos x),$$
where U and V are polynomials in $1/x$ whose coefficients are all integers, and the general integral is
$$y = C(U \sin x + V \cos x) + C'(V \sin x - U \cos x).$$
[HERMITE.]

9. The integration of the system of linear equations
$$\frac{dy}{dx} + ay + bz = 0, \quad \frac{dz}{dx} + a_1 y + b_1 z = 0,$$
where a, b, a_1, b_1 are any functions of x, reduces, on putting $y = tz$, to the integration of Riccati's equation,
$$\frac{dt}{dx} + b + (a - b_1)t - a_1 t^2 = 0,$$
and to the calculation of $\int (a + b_1) dx$ (see ftn., p. 112).

10. The ratio z of two independent integrals of the linear equation
$$y'' + py' + qy = 0$$
satisfies the differential equation of the third order,
$$\frac{z'''}{z'} - \frac{3}{2}\left(\frac{z''}{z'}\right)^2 = 2q - \frac{1}{2}p^2 - p'.$$

11. Given the differential equation

(E) $\qquad x(y'' - y') - ay = 0,$

where a is constant, how must we choose the path of integration L so that the function $y(x)$ represented by the definite integral
$$y(x) = \int_{(L)} e^{xz} z^{a-1} (z-1)^{-a-1} dz$$
shall be a particular integral of (E)? Show that the equation (E) has a particular integral, which can be expressed without any sign of quadrature, when a is an integer. Deduce from it the general integral, and express it in terms of the smallest possible number of transcendentals.
[*Licence*, Paris, July, 1908.]

12. Determine the two functions $P(t)$ and $Q(t)$ so that the function y represented by the expression
$$y = (x-a)\int_b^x f(t) P(t) dt + (x-b)\int_a^x f(t) Q(t) dt$$
shall be an integral of the differential equation $y'' = f(x)$ for all possible forms of the function $f(x)$.
[*Licence*, Paris, October, 1907.]

13. The general integral of the linear equation
$$xy'' + [n + x P(x)] y' + x^{n+1} Q(x) y = 0,$$
where $P(x)$ and $Q(x)$ are analytic in the neighborhood of the origin, is single-valued in this neighborhood. The letter n denotes an integer greater than unity.

14*. Every equation of the form
$$x^{n-p} \frac{d^n y}{dx^n} + x^{n-p-1} Q_1(x) \frac{d^{n-1} y}{dx^{n-1}} + \cdots$$
$$+ x Q_{n-p-1}(x) \frac{d^{p+1} y}{dx^{p+1}} + Q_{n-p}(x) \frac{d^p y}{dx^p} + \cdots + Q_n(x) y = 0,$$
where $Q_1, Q_2, \cdots, Q_n$ are analytic functions in the neighborhood of the origin, has an analytic integral in the same neighborhood; and the value of the integral, as well as the values of its first $p-1$ derivatives, may be arbitrarily chosen for $x = 0$, provided the equation
$$(r-p) \cdots (r-n+1) + Q_1(0)(r-p) \cdots (r-n+2) + \cdots + Q_{n-p}(0) = 0$$
has no integral root greater than $p-1$.

[E. GOURSAT, *Annales de l'École Normale*, 1883, p. 265.]

Note. By an artifice analogous to the one which was used in § 50, we are led to prove the proposition for an equation of the form
$$\frac{d^p u}{dx^p} = \frac{M}{1 - \dfrac{x}{r}} \left(\frac{d^{p-1} u}{dx^{p-1}} + \cdots + \frac{du}{dx} + u \right),$$
where we have put
$$u = y + xy' + \cdots + x^{n-p} \frac{d^{n-p} y}{dx^{n-p}}.$$

15*. Let Σ be a system of four linear equations identical with its adjoint (p. 156)

(E) $\quad \dfrac{dy_i}{dx} = a_{i1} y_1 + a_{i2} y_2 + a_{i3} y_3 + a_{i4} y_4. \qquad (i = 1, 2, 3, 4) \quad a_{ih} + a_{hi} = 0$

This system has the first integral $y_1^2 + y_2^2 + y_3^2 + y_4^2 = C$. If we suppose $C = 0$, the preceding relations are satisfied by putting

$$y_1 = \rho(\eta - \xi), \qquad y_2 = \rho(1 + \xi\eta), \qquad y_3 = \rho i(1 - \xi\eta), \qquad y_4 = \rho i(\eta + \xi).$$

Substituting these expressions for y_1, y_2, y_3, y_4 in the equations (E), we obtain the system of three equations

$$2\frac{\rho'}{\rho} = (a_{21} + i a_{13})(\eta - \xi) + 2 i a_{23} + (a_{34} + i a_{24})(\eta + \xi),$$
$$2\eta' = (a_{12} + a_{43})(1 + \eta^2) + i(a_{13} + a_{24})(1 - \eta^2) + 2i(a_{32} + a_{14})\eta,$$
$$2\xi' = (a_{21} + a_{43})(1 + \xi^2) + i(a_{31} + a_{24})(1 - \xi^2) + 2i(a_{32} + a_{41})\xi,$$

of which the last two are Riccati equations. Let $\eta = f(x, C_1)$, $\xi = \phi(x, C_2)$ be the general integrals of these two equations; then the general integral of the equation in ρ is given by the equation

$$\rho^2 \frac{\partial f}{\partial C_1} \frac{\partial \phi}{\partial C_2} = C_3.$$

[E. GOURSAT, *Comptes rendus*, Vol. CVI, p. 187, and Vol. CXLVIII, p. 612.]

16. Prove the relation

$$\int_{x_0}^{x} \phi'(x)\,dx \int_{x_0}^{x} \phi'(x)\,dx \int \cdots \int_{x_0}^{x} \phi'(x)\,dx \int_{x_0}^{x} f(x)\,dx$$
$$= \frac{1}{(n-1)!} \int_{x_0}^{x} [\phi(x) - \phi(y)]^{n-1} f(y)\,dy,$$

in which the left-hand side contains n integral signs, by proving that the two sides are particular integrals of a linear differential equation of the nth order satisfying the same initial conditions.

17*. Prove that the integral $\phi(x, \alpha)$ of the linear equation $F(y) = 0$ (p. 108), considered as a function of the variable α, is an integral of the adjoint equation $G(z) = 0$, after having replaced x by α.

Note. It is seen that the integral of the equation $F(y) = 0$ which, with its first $(n-1)$ derivatives, takes on the same values for $x = x_0$ as a function $\pi(x)$ and its first $(n-1)$ derivatives, has the form

$$y = \pi(x) - \int_{x_0}^{x} F(z) \phi(x, \alpha)\,d\alpha,$$

where $z = \pi(\alpha)$. The integral on the right must depend only upon $\pi(x)$, $\pi(x_0)$, $\pi'(x_0), \cdots, \pi^{(n-1)}(x_0)$. Now we can also write (§ 42)

$$\int_{x_0}^{x} F(z) \phi(x, \alpha)\,d\alpha = \{\Psi[z, \phi(x, \alpha)]\}_{\alpha=x_0}^{\alpha=x} - \int_{x_0}^{x} z G[\phi(x, \alpha)]\,d\alpha,$$

and it is clear that the preceding condition is not satisfied unless we have $G[\phi(x, \alpha)] = 0$. It follows readily that the functions $\phi_i(x)$ defined by the equations (A) (ftn., p. 109) form a fundamental system of integrals of the adjoint equation.

CHAPTER IV

NON-LINEAR DIFFERENTIAL EQUATIONS

I. EXCEPTIONAL INITIAL VALUES

The proof which has been given for the existence of integrals that take on given initial values really supposes that the right-hand sides of the given system of equations are analytic in the neighborhood of these initial values (§ 22). Restricting ourselves to the case of a single equation, we shall examine some simple cases in which that condition is not satisfied.

63. The case where the derivative becomes infinite. Let us consider an equation of the first order,

(1) $$\frac{dy}{dx} = f(x, y),$$

where the right-hand side $f(x, y)$ becomes infinite for the pair of values $x = x_0$, $y = y_0$ in such a way that its reciprocal

$$f_1(x, y) = \frac{1}{f(x, y)}$$

is analytic in the neighborhood of this pair of values. We can write the preceding equation in the form

(2) $$\frac{dx}{dy} = \frac{1}{f(x, y)} = f_1(x, y),$$

regarding y as the independent variable and x as the dependent variable. But since the right-hand side $f_1(x, y)$ is analytic by hypothesis for $x = x_0$, $y = y_0$, Cauchy's theorem applies to the equation (2). Hence there exists an integral, *and only one*, which approaches x_0 as y approaches y_0, and that integral is analytic in the neighborhood of the point y_0. The development of $x - x_0$ in a power series according to powers of $y - y_0$ necessarily commences with a term which is at least of the second degree, since dx/dy or $f_1(x, y)$ is zero for $x = x_0$, $y = y_0$, for otherwise $f(x, y)$ itself would be analytic. Let this development of $x - x_0$ be

(3) $$x - x_0 = A_m(y - y_0)^m + A_{m+1}(y - y_0)^{m+1} + \cdots$$
$$(m \geq 2,\ A_m \neq 0)$$

From the equation (3) we derive a development for $y - y_0$ according to powers of $(x - x_0)^{1/m}$ (see II, Part I, § 99),

$$(4) \qquad y - y_0 = a_1(x - x_0)^{\frac{1}{m}} + a_2(x - x_0)^{\frac{2}{m}} + \cdots. \qquad (a_1 \neq 0)$$

It follows that *the equation* (1) *itself has this one and only this one integral of the form* (4) *which approaches* y_0 *as* x *approaches* x_0, *and the point* x_0 *is an algebraic critical point for this integral.**

64. Case where the derivative is indeterminate. The complete discussion of all the cases in which the derivative becomes indeterminate is much more complicated. Let us take first the equation studied by Briot and Bouquet,†

$$(5) \qquad xy' - by = a_{10}x + a_{20}x^2 + a_{11}xy + a_{02}y^2 + \cdots = \phi(x, y),$$

where the right-hand side is analytic in the neighborhood of the point $x = y = 0$, and let us try to determine whether there exists an analytic integral which vanishes with x. For this purpose let us substitute for y, in both sides of the equation (5), a power series

$$(6) \qquad y = c_1 x + c_2 x^2 + \cdots + c_n x^n + \cdots.$$

After the substitution the coefficient of x^n on the left-hand side is $(n - b)c_n$, while the coefficient of x^n on the right is a polynomial,

$$P_n(a_{10}, a_{20}, \cdots, a_{0n}; c_1, \cdots, c_{n-1}),$$

whose coefficients are all positive integers, and which contains only the coefficients $c_1, \cdots, c_{n-1}$, and some but not necessarily all of the coefficients a_{ih} for which $i + h \leq n$. We therefore obtain a recurrent relation for the coefficients of the series (6):

$$(7) \qquad \begin{cases} (n - b)c_n = P_n(a_{10}, a_{20}, \cdots, a_{0n}; c_1, c_2, \cdots, c_{n-1}). \\ (n = 1, 2, \cdots) \end{cases}$$

This enables us to calculate these coefficients successively, *provided that b is not equal to a positive integer*. Let us first set aside this supposition. The relation (7) gives us

$$c_1 = \frac{a_{10}}{1 - b}, \qquad c_2 = \frac{a_{20} + a_{11}c_1 + a_{02}c_1^2}{2 - b}, \qquad \cdots,$$

---

* In geometric language, we can also say that through the point (x_0, y_0) there passes one and only one *integral curve*, on which the point (x_0, y_0) is an ordinary point, and the tangent at this point is the straight line $x = x_0$. In stating the theorem we have tacitly assumed that the function $f_1(x, y)$ does not vanish for $x = x_0$ for all values of y; for in this case the integral of the equation (2), which takes on the value x_0 for $y = y_0$, reduces to $x = x_0$, and the equation (1) has no integral which approaches y_0 as x approaches x_0.

† *Journal de l'École Polytechnique*, Vol. XXI, 1856.

and so on in this way. The value of the series (6) certainly represents an integral of the equation (5) vanishing with x, provided that the series has a radius of convergence different from zero. In fact, the operations by which we have obtained the coefficients of this series are then valid (I, § 192, 2d ed.; § 186, 1st ed.).

In order to prove the convergence of this series, let us observe first that if we give to n all the integral values $1, 2, \cdots$, to infinity, the fraction $1/(n-b)$, which cannot become infinite, approaches zero. The absolute value of that fraction has therefore a maximum $1/B$, and we have for every value of the integer n, $|1/(n-b)| \leq 1/B$.

On the other hand, let

$$\Phi(x, Y) = A_{10}x + A_{20}x^2 + A_{11}xY + A_{02}Y^2 + \cdots + A_{ik}x^i Y^k + \cdots$$

be a dominant function for $\phi(x, y)$, having no constant term nor any term in Y. We might take, for example, a function of the form

$$\Phi(x, Y) = \frac{M}{\left(1 - \dfrac{x}{r}\right)\left(1 - \dfrac{Y}{\rho}\right)} - M - M\frac{Y}{\rho},$$

but it is really not necessary to specify it completely in order to carry through the proof. The auxiliary equation

(8) $\qquad BY = \Phi(x, Y)$

has, by the general theorem on implicit functions (I, § 193, 2d ed.; § 187, 1st ed.), an analytic root vanishing with x. Let

(9) $\qquad Y = C_1 x + C_2 x^2 + \cdots + C_n x^n + \cdots$

be the development of this root in a power series. In order to calculate the coefficients C_i, we can substitute this development for Y in the two sides of the relation (8). This gives the recurrent relation

(10) $\qquad BC_n = P_n(A_{10}, A_{20}, \cdots, A_{0n}; C_1, C_2, \cdots, C_{n-1}),$

where P_n denotes the polynomial which appears in the relation (7), in which a_{ik} has been replaced by A_{ik} and c_i by C_i.

But from the very way in which we have chosen the constant B and the function $\Phi(x, Y)$, we have the inequalities

$$|a_{ik}| \leq A_{ik}, \qquad \frac{1}{|n-b|} \leq \frac{1}{B}.$$

It follows that if we have

$$|c_1| < C_1, \qquad |c_2| < C_2, \qquad \cdots, \qquad |c_{n-1}| < C_{n-1},$$

we have also $|c_n| < C_n$, since all the coefficients of the polynomial P_n are positive integers. Now we have $|a_{10}| < A_{10}$, and, consequently,

$|c_i| < C_i$. Reasoning step by step, we conclude that the series (9) dominates the series (6). The latter is therefore convergent in the neighborhood of the origin. Summing up, *if the coefficient b of y in the equation* (5) *is not equal to a positive integer, that equation has one and only one analytic integral that vanishes with x.*

In order to finish the study of the analytic integrals which vanish with x, we must still examine the case where b is equal to a positive integer. Suppose first $b = 1$; the first of the relations (7) reduces to $a_{10} = 0$. If a_{10} is not zero, then there is no analytic integral fulfilling the condition. If a_{10} is zero, setting $y = \lambda x$, we are led to an equation,

$$(11) \qquad \lambda' = \psi(x, \lambda) = a_{20} + a_{11}\lambda + a_{02}\lambda^2 + \cdots,$$

where the function $\psi(x, \lambda)$ is analytic, provided that $|x| < r$, $|\lambda| < \Lambda$, r and Λ being two suitably chosen positive numbers. Now this equation (11) has an infinite number of integrals which are analytic in the neighborhood of the origin, for we can choose arbitrarily the value λ_0 of λ for $x = 0$, provided that we have $|\lambda_0| < \Lambda$. Hence in this case the equation (5) has an infinite number of analytic integrals vanishing with x.

If b is equal to a positive integer greater than unity, the coefficient of x in the development of an analytic integral vanishing for $x = 0$ must be equal to $a_{10}/(1 - b)$, and the transformation $y = a_{10}x/(1 - b) + \lambda x$ leads to an equation of the same form in which the coefficient of λ is equal to $(1 - b)$:

$$x\lambda' - (b - 1)\lambda = a'_{10}x + a'_{20}x^2 + a'_{11}\lambda x + \cdots.$$

By a succession of similar transformations we reach the case which has just been treated. Consequently, if the coefficient b is equal to a positive integer, the equation (5) has no analytic integral vanishing with x, or it has an infinite number of such integrals.

Briot and Bouquet also investigated whether there existed non-analytic integrals approaching zero with x, and proved that the equation (5) has an infinite number of such integrals when the real part of b is positive. We can easily establish this theorem by means of the method of successive approximations. Let us first point out that if the real part of b is positive, we may, without lack of generality, suppose that the real part $\mathcal{R}(b) > 1$. In fact, if we make the change of variable $x = x'^n$, where n is a positive integer, the equation (5) is replaced by an equation of the same form in which b is replaced by nb. We shall suppose, then, that we have $\mathcal{R}(b) > 1$, and that b is not an integer. As we have just seen, the equation (5) has an analytic integral y_1, which vanishes for $x = 0$. Hence, setting $y = y_1 + u$, the equation (5) becomes

$$xu' - bu = \phi(x, y_1 + u) - \phi(x, y_1) = u\psi(x, u).$$

Since the function $\phi(x, y)$ does not contain any term of the form a constant times y, the function $\psi(x, u)$ will not contain any constant term, and we can write the preceding equation in the form

$$xu' - bu = u[\alpha x + \beta u + \cdots].$$

Let us now put $u = \lambda x^b$, where λ denotes the new dependent variable. The equation takes the form

(12) $\qquad \lambda' = \lambda[\alpha + \beta\lambda x^{b-1} + \cdots] = F(\lambda, x, x^{b-1}),$

where F denotes a power series with respect to the three variables λ, x, x^{b-1}. In the plane of the variable x let us draw through the origin two rays whose inclinations are ω_0 and ω_1 ($\omega_0 < \omega_1 < \omega_0 + 2\pi$), and let us consider the circular sector A limited by these two rays and an arc of a circle with the radius r described with the origin as center. If x remains in the interior of A, and if $|\lambda|$ remains less than a positive number l, the function $F(\lambda, x, x^{b-1})$ will be analytic,* provided that the two numbers r and l are sufficiently small. Let us connect the origin with any point x of the sector A by a straight line-segment, and suppose that we take for the initial value of λ an arbitrary value λ_0 whose absolute value is less than l. We can apply to the equation (12) the method of successive approximations (§ 29), which consists in taking the successive integrals

$$\lambda_1 = \lambda_0 + \int_0^x F(\lambda_0, x, x^{b-1})\,dx, \qquad \lambda_2 = \lambda_0 + \int_0^x F(\lambda_1, x, x^{b-1})\,dx,$$

and, in general,

$$\lambda_n = \lambda_0 + \int_0^x F(\lambda_{n-1}, x, x^{b-1})\,dx,$$

all of these integrals being taken along the straight line. The fundamental hypotheses for the validity of the proof are satisfied. All the functions $\lambda_1(x)$, $\lambda_2(x)$, $\cdots$ are analytic in the interior of the sector A, and the function $\lambda_n(x)$ approaches a limit $\Lambda(x)$ if the radius r has been taken sufficiently small. Hence the equation (12) has an integral which is analytic in the interior of the sector A and which approaches the value λ_0 as x approaches zero. It follows that the equation (5) has an infinite number of non-analytic integrals in the neighborhood of the origin, each of which approaches zero as the point x approaches the origin and depends upon an arbitrary parameter λ_0. This proves Briot and Bouquet's theorem.

The condition that the real part of $b - 1$ be positive is essential. Indeed, if x approaches the origin, remaining in the sector A, its angle remains between ω_0 and ω_1, and its absolute value approaches zero. Setting $x = \rho e^{i\omega}$, $b - 1 = \mu + \nu i$, we have

(13) $\qquad x^{b-1} = e^{(\mu + \nu i)(\log\rho + i\omega)} = e^{\mu\log\rho - \nu\omega} e^{i(\nu\log\rho + \mu\omega)}.$

As ρ approaches zero, ω remaining included between the two limits ω_0 and ω_1, $\mu\log\rho - \nu\omega$ becomes infinite in absolute value in remaining negative, and the absolute value of x^{b-1} approaches zero. On the contrary, if the real part of $b - 1$ is negative, it is obvious that the absolute value of x^{b-1} becomes infinite as x approaches zero, remaining in the sector A. The function $F(\lambda, x, x^{b-1})$ is not continuous at the origin, and the previous proof no longer applies.

According to Briot and Bouquet, if the real part of b is negative, the equation (5) has no other integral than the analytic integral vanishing for $x = 0$.

---

* If x approaches the origin, remaining always in the sector A, the derivative of the function F with respect to x may become infinite if the real part of $b - 2$ is negative, but that derivative does not appear in the method of successive approximations.

But their proof, which is very similar to the one in the footnote on page 50, supposes that the variable x approaches the origin along a path of finite length that has a definite tangent at the origin, and this condition should appear in the statement of their theorem. In order to give some idea of the difficulty of the question, let us consider the function x^b, supposing that the real part μ of b is *negative*, while the coefficient ν of i is *different from zero*. The absolute value of x^b is equal to $e^{\mu \log \rho - \nu \omega}$. If we make the variable x describe a curve that approaches the origin indefinitely, $\mu \log \rho$ does approach $+\infty$, but if we make the angle ω increase in absolute value at the same time in such a way that the difference $\mu \log \rho - \nu \omega$ remains negative and increases indefinitely in absolute value, the absolute value of x^b approaches zero at the same time as $|x|$. If $\nu > 0$, we need only make the variable x describe the logarithmic spiral whose equation is $\rho = e^{\nu \omega / 2 \mu}$, for example; for we then have $|x^b| = e^{-\nu \omega / 2}$, and if the angle ω approaches $+\infty$, $|x| = \rho$ and $|x^b|$ approach zero simultaneously.

If the real part of b is negative and the real part of b/i is not zero, it follows from investigations of Picard and Poincaré that the equation (5) has an infinite number of non-analytic integrals that depend upon an arbitrary constant and approach zero as the variable x describes a path such as the preceding, along which $|x^b|$ approaches zero. The contradiction between this result and the theorem of Briot and Bouquet is only apparent, since in the two cases entirely different conditions are assumed. In particular let us notice that if the variable x takes on only real values, it cannot turn an infinite number of times around the origin; consequently there will be no other integral which approaches zero with x except the analytic integral, provided the real part of b is negative.

The results of this discussion are easy to verify with the elementary equation $xy' = ax + by$, whose general integral is $y = ax/(1-b) + Cx^b$ if $b - 1$ is not zero, and $y = ax \log x + Cx$ if $b = 1$.

65. We shall limit ourselves to a few statements concerning the general case of an equation of the form

$$(14) \qquad \frac{dy}{dx} = \frac{ax + by + cx^2 + 2\,dxy + ey^2 \cdots}{a'x + b'y + c'x^2 + 2\,d'xy + e'y^2 \cdots} = \frac{Y}{X},$$

where X and Y are power series which converge when

$$|x| < r, \qquad |y| < r.$$

We are supposing, as we may without loss of generality, that it is for $x = y = 0$ that dy/dx becomes indeterminate. Setting $y = vx$ in this equation, it becomes

$$(15) \qquad x \frac{dv}{dx} = \frac{a + bv - v(a' + b'v) + x\phi(x, v)}{a' + b'v + x\psi(x, v)},$$

where $\phi(x, v)$ and $\psi(x, v)$ are two power series which are convergent whenever $|x| < r$ and $|vx| < r$. If the equation (14) has an analytic integral vanishing with x, the coefficient of x in the development of that integral is necessarily a root of the equation

$$(16) \qquad a + bv - v(a' + b'v) = 0,$$

since the left-hand side of (15) is zero for $x = 0$. Let v_1 be a root of the equation (16). If we put $v = v_1 + u$, the two functions

$$\phi(x, v_1 + u), \qquad \psi(x, v_1 + u)$$

are still regular in the neighborhood of the values $x = 0$, $u = 0$, and the equation (15) reduces to an equation of a form already studied,

(17) $$x \frac{du}{dx} = Au + Bx + \cdots,$$

provided that $v = v_1$ does not make $a' + b'v$ vanish. Since the equation (16) is in general of the second degree, we see that we can reduce the equation (14) to the form (5) in two different ways. Hence there are in general two analytic integrals and only two vanishing for $x = 0$. But these conclusions are applicable only under the most general conditions, where the coefficients a, b, a', b' do not satisfy any special relation.

The general investigation of the integrals, analytic or not, of the equation (14), which approach zero when x approaches zero (X and Y being two regular functions which vanish for $x = y = 0$), has been the object, since the work of Briot and Bouquet, of a large number of investigations. Although it has been possible to treat more and more general cases, the question is still not exhausted. Let us notice in particular just one remarkable circumstance which we have not yet mentioned. Let us take the equation

(18) $$x^2 \frac{dy}{dx} - by = ax,$$

and let us try to find, as above, an analytic integral of this equation which vanishes for $x = 0$. If we attempt to determine the coefficients of the series (6) so that on substituting it in the equation (18) we arrive at an identity, we discover the relations

$$a + bc_1 = 0, \qquad c_1 = bc_2, \qquad 2c_2 = bc_3, \qquad \cdots, \qquad nc_n = bc_{n+1}, \cdots,$$

from which we derive

$$c_1 = -\frac{a}{b}, \qquad c_2 = -\frac{a}{b^2}, \qquad c_3 = -\frac{2a}{b^3}, \qquad \cdots, \qquad c_{n+1} = -\frac{n!\,a}{b^{n+1}}.$$

We thus obtain a unique value for each coefficient, but *the series which we obtain is divergent except for $x = 0$.* The origin is an essentially singular point for all the integrals, as is verified by direct integration. Similarly, the point $x = 0$ is an essentially singular point for all the integrals of the equation $xy' + y^2 = 0$; and all these integrals approach zero with $|x|$.

If we assign only real values to the variables x and y, and wish to *construct* the integral curves of the equation (14) (X and Y being, for example, two polynomials in x and y with real coefficients), it is very important to know the form of these integral curves in the neighborhood of a point common to the two curves $X = 0$, $Y = 0$. We shall study from this point of view the simple equation

(19)
$$\frac{dy}{dx} = \frac{ax + by}{a'x + b'y},$$

which can be integrated by elementary methods (§ 3) by the substitution $y = tx$. We can integrate it in a more symmetric way by replacing the equation (19) by the system

(20)
$$\frac{dx}{a'x + b'y} = \frac{dy}{ax + by} = dt,$$

where t is an auxiliary variable introduced for the sake of symmetry. We have seen above (§ 59) that this system can be reduced to a simple canonical form by replacing x and y by two linear homogeneous combinations X and Y of these variables. In this case the characteristic equation is

$$s^2 - (a' + b)s + ba' - ab' = 0.$$

This equation cannot have zero for a root, since we suppose that $ab' - ba'$ is not zero. Several cases are now to be distinguished according to the nature of the roots:

1) If the characteristic equation has two real and distinct roots s_1, s_2, we can reduce the system (20) to the form

$$\frac{dX}{s_1 X} = \frac{dY}{s_2 Y} = dt,$$

and the given equation consequently becomes

$$Y dX = \frac{s_1}{s_2} X dY.$$

The general integral is given by the equation

$$Y = C X^{\frac{s_2}{s_1}}.$$

If s_1 and s_2 have the same sign, Y approaches zero with X, and all the integral curves pass through the origin, which is a node. If s_2/s_1 is negative, there exist only two integral curves passing through the origin, the straight lines $X = 0$, $Y = 0$; hence the origin is a saddleback.

2) If the characteristic equation has two conjugate imaginary roots $\alpha + \beta i$, $\alpha - \beta i$ ($\beta \neq 0$), we can reduce the system (20) to the form

$$\frac{d(X + Yi)}{(\alpha + \beta i)(X + Yi)} = \frac{d(X - Yi)}{(\alpha - \beta i)(X - Yi)} = dt,$$

where X and Y are linear homogeneous combinations of x and y with *real coefficients*. We can then write these equations in the form

$$\frac{dX}{\alpha X - \beta Y} = \frac{dY}{\beta X + \alpha Y} = dt,$$

from which we derive

$$\frac{X dX + Y dY}{\alpha(X^2 + Y^2)} = \frac{X dY - Y dX}{\beta(X^2 + Y^2)}.$$

The general integral of the equation (19) is therefore represented by the equation

$$\sqrt{X^2 + Y^2} = Ce^{\frac{\alpha}{\beta} \arctan \frac{Y}{X}}.$$

If α is not zero, all these curves have the form of spirals which approach the origin as an asymptotic point. The origin is said to be a *focus*.

If α is zero, the general integral is represented by concentric conics. The origin is called a *center*, but this case must be considered as exceptional, since it occurs only when α satisfies an *equality*.

3) If the characteristic equation has a double root s, that root is real and different from zero, and the system (20) reduces to the form

$$\frac{dX}{sX} = \frac{dY}{s(X+Y)} = dt.$$

The equation (19) itself becomes $dY/dX = 1 + Y/X$, and the general integral is $Y = CX + X \log X$. In order to construct these curves, we can express X and Y in terms of an auxiliary variable by putting $X = e^\theta$, which gives $Y = Ce^\theta + \theta e^\theta$. When θ approaches $-\infty$, X and Y, and consequently x and y, approach zero, and the origin is again a focus.

The preceding classification is due to Poincaré, who has extended the discussion to equations of the general form (14) whose coefficients are real.

II. A STUDY OF SOME EQUATIONS OF THE FIRST ORDER

66. Singular points of integrals. The developments in series by which we have established the existence of analytic integrals of a system of differential equations enable us to calculate these integrals only in the interior of the circle of convergence; but the knowledge of these developments suffices, as we have noticed in general (see II, Part I, § 86), to *virtually* determine these functions in the whole domain of their existence. Let us consider, for definiteness, an algebraic differential equation of the first order,

(21) $$F(x, y, y') = 0,$$

where F is a polynomial in x, y, y'. Let (x_0, y_0) be a pair of values for which the equation $F(x_0, y_0, y') = 0$ has a simple root y_0'. When x and y approach x_0 and y_0 respectively, the equation (21) has one and only one root approaching y_0', and that root, $y' = f(x, y)$, is a regular function in the neighborhood of the point (x_0, y_0). The equation (21) has therefore an analytic integral which reduces to y_0 for $x = x_0$, and whose derivative takes on the value y_0' for $x = x_0$. This integral is defined by its power-series development only in the interior of a circle C_0 about x_0 as center, whose radius is in general finite. But this function, whose analytic extension may be followed outside of the circle C_0, satisfies the equation (21) in its whole domain of existence. Let us observe that we may make use of the equation (21) itself to calculate the coefficients of the different series which we use in the method of analytic extension. If at a point x_1 in the circle C_0 the integral considered is equal to y_1, its derivative is equal to one of the roots y_1' of the equation $F(x_1, y_1, y') = 0$, and we shall be able to derive the values of the other derivatives at the point x_1 by successive differentiations of (21).

It follows that every differential equation of the first order defines an infinite number of analytic functions, depending upon one arbitrary constant. These are in general transcendental functions which cannot be expressed in terms of the classic transcendentals, and the same thing is true a fortiori of the functions defined by algebraic differential equations of the second, or higher, order. The study of the properties of these new transcendentals and their classification constitutes the object of the analytic theory of differential equations.

We may be guided in this study by two different motives: we may seek the necessary and sufficient conditions that equations of a given sort may be integrated by means of functions already known; or, on the other hand, we may propose to ourselves the problem of discovering the algebraic differential equations that define transcendentals not reducible to the classic transcendentals, and possessing certain remarkable properties, such as being single-valued and analytic, or analytic except for poles, etc. Whatever may be the object that we have specially in view, the investigation of the possible singularities of the integrals is an essential question. While the singular points of the integrals of a linear equation are fixed, the singular points of the integrals of non-linear equations vary in general with their initial values. For example, the integral of the equation $x + yy' = 0$ which takes on the value y_0 for $x = 0$ is $y = \sqrt{y_0^2 - x^2}$. This function has the two critical points $+ y_0, - y_0$, which depend upon the initial value. Similarly, the integral of the equation $y' = y^2$ which takes on the value y_0 for $x = 0$ is $y_0/(1 - xy_0)$; this solution has the pole $x = 1/y_0$. We are therefore led to distinguish two classes of singular points for a differential equation: the *fixed* singular points which do not depend upon the chosen initial values (not being necessarily singular points for all the integrals), and the *movable* singular points, poles, or critical points which depend upon the initial values. A differential equation may have both kinds of singular points.

67. Functions defined by a differential equation $y' = R(x, y)$. We shall study in particular the differential equation

$$(22) \qquad \frac{dy}{dx} = R(x, y) = \frac{P(x, y)}{Q(x, y)},$$

where $P(x, y)$ and $Q(x, y)$ are two polynomials in x and y which have no common polynomial divisor. The pair of simultaneous equations $P(x, y) = 0$, $Q(x, y) = 0$ has a certain number of solutions $(a_1, b_1), \cdots, (a_n, b_n)$. Let us mark the points $a_1, a_2, \cdots, a_n$ in the x-plane.

The transformation $y = 1/z$ reduces the equation (22) to an equation of the same form,

$$(23) \qquad \frac{dz}{dx} = R_1(x, z) = \frac{P_1(x, z)}{Q_1(x, z)},$$

and the pair of equations $P_1(x, z) = 0$, $Q_1(x, z) = 0$ has a certain number of systems of solutions $(a'_1, b'_1), \cdots, (a'_m, b'_m)$. Let us mark the points $a'_1, a'_2, \cdots, a'_m$ in the x-plane. The points a_i, a'_k are in general singular points for some of the integrals of the equation (22), but they are known a priori; that is, they are the *fixed* singular points.

Let now (x_0, y_0) be any pair of values such that $Q(x_0, y_0)$ is not zero. Then by Cauchy's fundamental theorem the equation (22) has an analytic integral, in the neighborhood of the point x_0, which takes on the value y_0 for $x = x_0$. Suppose that we make the variable x describe any path L proceeding from the point x_0 and not passing through any of the points a_i, a'_k. We can continue the analytic extension of this integral along L so long as we do not encounter any singular point. But it may happen that we are stopped by the presence of such a point; let α be the first singular point which we encounter. The integral considered is analytic in the neighborhood of every point X of the path L included between x_0 and α, but the circle of convergence of the power series which represents it, and whose center is at X, never contains the point α in its interior, however small $|X - \alpha|$ may be. The equation $Q(\alpha, y) = 0$ has a certain number of roots $\beta_1, \beta_2, \cdots, \beta_N$. Let us mark the points β_i in the y-plane. The equation $Q(\alpha, y) = 0$ has only a finite number of roots, for otherwise the polynomial $Q(x, y)$ would be divisible by $(x - \alpha)$ and the point α would be included among the points a_i, a'_k. For the same reason it is seen that the two equations $P(\alpha, y) = 0$, $Q(\alpha, y) = 0$ have no common root.

There are now several possible cases to be examined. Let Y be the value of the integral in terms of X; we cannot suppose that Y approaches a finite value β different from $\beta_1, \beta_2, \cdots, \beta_N$ as X approaches α, for $R(x, y)$ is a regular function for $x = \alpha, y = \beta$. Now, by Cauchy's fundamental theorem, there would then exist a single integral approaching β as x approaches α, and that integral would be analytic at the point α, contrary to hypothesis. Let us suppose next that Y approaches the value β_i when $|X - \alpha|$ approaches zero. The function $R(x, y)$ is infinite for $x = \alpha, y = \beta_i$, but its reciprocal is a regular function, since we cannot have $P(\alpha, \beta_i) = 0$. We have seen in § 63 that the equation (22) has one and only one integral which approaches β_i as $|X - \alpha|$ approaches zero, and which has the point α as an algebraic critical point. Similarly, if $|Y|$ becomes infinite as $|X - \alpha|$ approaches zero, the equation (23) has an integral which approaches zero with $|X - \alpha|$. We cannot have simultaneously $P_1(\alpha, 0) = 0$, $Q_1(\alpha, 0) = 0$, since the point α is not contained among the points a'_i. If $Q_1(\alpha, 0)$ is not zero, z is analytic in the neighborhood of the point α, which is a pole for the integral considered. If $Q_1(\alpha, 0) = 0$, the point α is an algebraic critical point for z and thus also for y.

We have not yet exhausted all the possibilities. Might it not happen, for example, that Y does not approach any limit, although $|Y|$ does not become infinite as $|X - \alpha|$ approaches zero? Painlevé has shown, in the following way, that this is not possible. Previously this had been assumed without adequate proof. About the point α as a center let us describe a circle C with a very small radius r. The roots of the equation $Q(X, y) = 0$ which approach respectively $\beta_1, \beta_2, \cdots, \beta_N$ as $|X - \alpha|$ approaches zero remain respectively contained in the interior of the circles $\gamma_1, \gamma_2, \cdots, \gamma_N$ about the points $\beta_1, \beta_2, \cdots, \beta_N$ as centers with radii $\rho_1, \rho_2, \cdots, \rho_N$. We can take the radius r so small that all these radii ρ_i are themselves smaller than any given positive number ϵ. Let us consider at the same time a circle Γ with a very large radius R, described in the plane of the variable y about the origin as center, and let (E) be the portion of the y-plane exterior to the circles γ_i and interior to the circle Γ. We shall show that when $|X - \alpha|$ approaches zero, the corresponding point Y finally remains constantly in the interior of one of the circles γ_i or exterior to the circle Γ. If this were not the case, we should always find on the path L certain points X such that $|X - \alpha|$ is less than any given number and for which Y would be in the region (E). Suppose now that we have $|X - \alpha| < r/2$, for example, while only values of Y in the region (E) are considered. *We shall show that there is a positive minimum for the radius of the circle of convergence of that integral of the equation* (22) *which is equal to* Y *for* $x = X$. In fact, there is evidently a maximum for $|R(X, Y)|$ when the points X, Y remain respectively in the preceding regions, while there is a positive minimum for the numbers a and b (see § 22). Let η be the minimum of the radius of this circle of convergence. We could find, by hypothesis, a point X' on the path L whose distance from the point α would be less than η and such that the corresponding point Y' would be in the region (E). Since the circle of convergence of the series, which represents the integral which takes on the value Y' for $x = X'$, has a radius at least equal to η, the point α would be in the interior of this circle, which is evidently impossible, since α is a singular point.

The point Y, therefore, finally remains constantly in the interior of one of the circles γ_i or outside of the circle Γ as $|X - \alpha|$ approaches zero. Since the radius ρ_i can be taken as small as we please and the radius R as large as we please, this means that Y approaches one of the values β_i unless $|Y|$ becomes infinite. We have just examined what happens in these two cases. It follows that the point α is either a pole or an algebraic critical point. Hence we can avoid the singular point by replacing the portion of the path L near the point α by an arc of a circle of infinitesimal radius described about this point as center, and we shall be able to continue the analytic extension beyond this point until we meet a new singular point. We shall show that on a path L *of finite length* there are never more than a *finite* number of poles or of algebraic critical points. In fact, with each of the points a_i, a'_k as centers let us describe a very small circle in the plane of the x's, and let us describe also a circle of very large radius about the origin as center, so that all of the path L shall lie in the region (E') of the x-plane bounded by these circumferences. Let x_1 be any point of (E'). Then the integral whose absolute value becomes infinite as $|x - x_1|$ approaches zero is equal to a polynomial in $(x - x_1)^{-1}$ plus a power series in $(x - x_1)$ which converges in a circle of radius ρ_1. Similarly, the different integrals which have the point x_1 as an algebraic critical point are represented by series arranged according to fractional powers of $x - x_1$. Let ρ_2 be the smallest of the radii of convergence of these different series. It is clear that these numbers ρ_1 and ρ_2 vary continuously with the position of the point x_1; hence they have a minimum $\lambda > 0$, and the distance between two neighboring singular points on the path L is of necessity greater than λ.* We can therefore encounter on this path only a finite number of poles or of algebraic critical points. Consequently, *the only movable singular points of the integrals of the equation* (22) *are poles or algebraic critical points.* These integrals cannot have movable essentially singular points, and consequently they can have no natural boundaries.

The preceding arguments can be extended without difficulty to equations of the form (22), where $P(x, y)$, $Q(x, y)$ are polynomials in y whose coefficients are analytic functions of x. We have only to add to the points a_i, a'_k, which are defined as above, the singular points of all these coefficients. If the path described by the variable x remains in a region not containing any of the points a_i, a'_k or any of the singular points of the coefficients of the various powers of y, the only singular points which the integrals can have are poles or algebraic critical points.

As an application, let us consider the question of finding the equations of the form (22) which have no *movable critical points*. In order that this may be true, the denominator must not contain y. In fact, let α be any value of x, and β a corresponding value of y, for which $Q(\alpha, \beta) = 0$, while the numerator $P(\alpha, \beta)$ is not zero. The integral of the equation (22) which approaches β when $|x - \alpha|$ approaches zero has this point as a critical point, and it is clear that it is not a critical point for all the integrals. Hence the desired equation must be of the form

$$\frac{dy}{dx} = P_m y^m + P_{m-1} y^{m-1} + \cdots,$$

where P_m, P_{m-1}, $\cdots$ are functions of x. Moreover, the equation obtained by putting $y = 1/z$ must be of the same form, so that m

---

* It should be noticed that an integral can have an infinite number of critical points, and even an infinite number of them in the neighborhood of any value of x. Consider, for example, the equation

$$2yy' = R(x),$$

where $R(x)$ is a rational function; the general integral of this equation is

$$y^2 = y_0^2 + \int_{x_0}^{x} R(x) \, dx.$$

Let us suppose that the definite integral $\int R(x) \, dx$ has the four periods 1, α, i, βi, where α and β are two real irrational numbers. In the interior of a circle c described with any point x_1 as center and with an arbitrary radius, it is easy to prove (see II, Part I, § 53, *Note*) that we can find an infinite number of roots of y^2 by suitably choosing the paths of integration, and each of these roots is a critical point. But a path of finite length described by the variable never contains an infinite number of them.

cannot be greater than 2. It follows that the most general equation which satisfies the given conditions is a Riccati equation. It is easily verified that the condition is satisfied by any Riccati equation. If we take, for example, the expression in (26) (§ 40) which represents the general integral, it is clear that that integral can have for singular points, besides the singular points of the functions y_1, y_2, only poles resulting from the roots of the denominator $y_1 + Cy_2$, that is, poles which vary with the constant C.

Similarly, we may consider the question of determining the equations of the form (22) whose integrals have no *movable poles*. Let m and n be the degrees of $P(x, y)$ and of $Q(x, y)$ with respect to y; the equation obtained by putting $y = 1/z$ is of the form

(24) $$\frac{dz}{dx} = z^{n+2-m} \frac{P_1(x, z)}{Q_1(x, z)},$$

where P_1 and Q_1 are two new polynomials in z. Let $x = \alpha$ be any value of x not contained among the fixed singular points. The equation (24) has an integral which approaches zero as $|x - \alpha|$ approaches zero. It would seem from this that the equation (22) always has an integral whose absolute value becomes infinite as $|x - \alpha|$ approaches zero, but this conclusion is incorrect if the integral of the equation (24) reduces to $z = 0$. It is necessary and sufficient for this that $m < n + 2$; hence this is the condition that there shall be no movable poles. It follows that the only type of equation which has no movable singular point of any kind is the linear equation.

Application. The preceding result enables us to determine whether the general integral of a differential equation of the first order *is a rational function of the constant of integration*, when that constant is suitably chosen. Let

(A) $$y = R(x, C) = \frac{P(x, C)}{Q(x, C)}$$

be a rational function of the parameter C, where the coefficients of the two polynomials in C, $P(x, C)$ and $Q(x, C)$, are any functions of x. It is clear that the derivative y' is also a rational function of C,

$$y' = R'(x, C).$$

The elimination of the parameter C leads to a relation of the form

(E) $$F(y, y'; x) = 0,$$

where F is a polynomial in y, y' whose coefficients may be any functions of x. From the manner in which this equation is obtained we see that it is of deficiency zero in y and y', regarding x as a parameter.

Conversely, let us consider a differential equation of the first order (E), in which the left-hand side is a polynomial in y and y' whose coefficients are any analytic functions of x. In order that such an equation have a general integral of the form (A), it must first of all be of deficiency zero in y and y'. When this is the case, we can express y and y' as rational functions of a parameter u,

$$y = r(x, u), \qquad y' = r_1(x, u),$$

in such a way that we have, conversely, $u = s(x, y, y')$, where the functions r and r_1 are rational in u, and where s is a rational function of y and y'. Hence the given differential equation (E) may be replaced by the equation

$$\frac{\partial r}{\partial x} + \frac{\partial r}{\partial u}\frac{du}{dx} = r_1(x, u),$$

which is of the form

(E$_1$) $$\frac{du}{dx} = F(x, u),$$

where F is a rational function of u. If the general integral of the equation (E) is $y = R(x, C)$, the general integral of the equation (E$_1$) is, according to the above,

$$u = s[x, R(x, C), R'(x, C)];$$

that is, it is a rational function of C. But the only singular points of such an expression which vary with C are evidently poles. The only movable singular points that the equation (E$_1$) can have are therefore poles; consequently the equation (E$_1$) must be a Riccati equation.*

Let us consider, for example, the equation

$$y'^2 = (Py + Q)^2 (y-a)(y-b),$$

where P and Q are functions of x, and where a and b are two constants. This relation is of deficiency *zero* in y and y', and in order to express y and y' as rational functions of a parameter, we need only set $(y-b)/(y-a) = t^2$, which gives

$$y'(1-t^2)^2 = (b-a)[P(bt - at^3) + Q(t - t^3)],$$

and the equation (E$_1$) is the Riccati equation

$$2\frac{dt}{dx} = P(b - at^2) + Q(1 - t^2).$$

68. Single-valued functions deduced from the equation $(y')^m = R(y)$.
Let us now consider the integrals of the differential equation

(25) $$(y')^m = R(y) = \frac{P(y)}{Q(y)},$$

where m is a positive integer and where $P(y)$ and $Q(y)$ are two mutually prime polynomials in y with constant coefficients. We shall now propose to determine all the equations of this kind whose general integrals are single-valued and in general analytic. Let x_0 be any value of x, and y_0 an arbitrary value of y which does not cause either of the polynomials $P(y)$, $Q(y)$ to vanish. The equation (25), after y has been replaced by y_0, has m distinct roots in y'. Let us choose one of these roots, y_0'. The equation (25) has an analytic

---

* The converse is immediate. If (E$_1$) is a Riccati equation, the general integral u is a *linear* function of the arbitrary constant C, and consequently $y = r(x, u)$ is a *rational* function of C.

integral in the neighborhood of the point x_0, which takes on the value y_0 at this point and whose derivative is equal to y_0' for $x = x_0$. We can extend this integral analytically along the whole of any path starting from the point x_0, so long as we do not encounter any singular points. Let a be the first singular point encountered, let X be a point of the path L lying between x_0 and a, and let Y be the corresponding value of the integral at the point X. The arguments of the preceding paragraph, which apply here without essential modification, show that as $|X - a|$ approaches zero, Y approaches a root of one of the equations

$$P(y) = 0, \qquad Q(y) = 0,$$

or else $|Y|$ becomes infinite; but it can never happen that Y does not approach any limit.

Let us examine the different possible cases. Suppose first that b is a q-fold root of the denominator $Q(y) = 0$. From the equation (25) we have

(26) $\qquad dx = (y-b)^{\frac{q}{m}}[c_0 + c_1(y-b) + \cdots] dy. \qquad (c_0 \neq 0)$

If y describes a path from y_0 to b in the plane of the y's, the variable x describes a path starting from x_0 and ending at a point which we shall call a in the finite portion of the x-plane. Conversely, if x goes from x_0 to a along this path, y goes from y_0 to b. Putting $y - b = t^m$, we derive from the equation (26) a development of $x - a$ in powers of t beginning with a term of degree $m + q$. Conversely, we have for t a development according to fractional powers of $x - a$ beginning with a term in $(x-a)^{1/(m+q)}$, and therefore a development of $y - b$ of the form

$$y - b = (x-a)^{\frac{m}{m+q}}[\alpha_0 + \alpha_1(x-a)^{\frac{1}{m+q}} + \cdots]. \qquad (\alpha_0 \neq 0)$$

Since q is positive, $m + q$ is greater than m, and the point $x = a$ is an algebraic critical point for the integral considered. In order that the general integral of the equation (25) may be a single-valued function, it is necessary that the polynomial $Q(y)$ shall reduce to a constant, or that the equation shall be of the form

(27) $\qquad\qquad (y')^m = P(y),$

where $P(y)$ is a polynomial. Since the equation $z'^m = (-1)^m z^{2m} P(1/z)$, obtained by the transformation $y = 1/z$, must also be of the same form, *the degree of the polynomial $P(y)$ cannot be greater than $2m$.*

Finally, we may suppose that $P(y)$ is of degree $2m$. In fact, if $P(y)$ is of degree $2m - q$, putting $y = a + 1/z$, where a is not a root of $P(y)$, we are led to the equation

$$z'^m = (-1)^m [z^{2m} P(a) + z^{2m-1} P'(a) + \cdots],$$

where the right-hand side is now a polynomial of degree $2m$.

Conversely, given an equation of the form (27), where $P(y)$ is a polynomial of degree $2m$, if b is a root of that polynomial the substitution $y = b + 1/z$ leads to an equation in z of the same kind, where the right-hand side is a polynomial of degree less than $2m$.

Let us suppose, then, that $P(y)$ is a polynomial of degree $2m$, and let α be the first singular point that we find on the path L starting from x_0. If $|Y|$ becomes infinite when X approaches α, the equation in z, obtained by putting $y = 1/z$, has an analytic integral which is zero for $X = \alpha$. The point α is therefore a pole for y.

There remains the case where Y approaches a root b of $P(y)$ as X approaches α. This can happen only in case the order of multiplicity of that root is less than m. For let us suppose that $P(y)$ is divisible by $(y - b)^q$, where $q \geqq m$. From the equation (27) we obtain, according to the initial conditions,

$$X - x_0 = \int_{y_0}^{Y} \frac{\phi(y) \, dy}{(y - b)^{\frac{q}{m}}},$$

where $\phi(y)$ is regular in the neighborhood of the point b, and we see that $|X|$ becomes infinite as Y approaches b. It follows that $q < m$, and the given equation can again be written in the form

(28) $\qquad dx = (y - b)^{-\frac{q}{m}} [c_0 + c_1(y - b) + \cdots] dy, \qquad (c_0 \neq 0)$

whence we may derive a development of $x - \alpha$ in powers of $(y - b)^{1/m}$ beginning with a term of degree $m - q$. Conversely, we may derive from (28) a development for $y - b$ according to fractional powers of $x - \alpha$ beginning with a term in $(x - \alpha)^{m/(m-q)}$. The point α is therefore, in general, an algebraic critical point. In order that it may be an ordinary point, $m/(m - q)$ must be an integer i; that is, we must have $q = m(1 - 1/i)$, where i is an integer greater than *unity*. This condition is also sufficient, for if it is satisfied, we may derive from the equation (28) a development of the form

$$x - \alpha = k_1 (y - b)^{\frac{1}{i}} + k_2 (y - b)^{\frac{1}{i} + 1} + \cdots, \qquad (k_1 \neq 0)$$

and, conversely, we have for $(y-b)^{1/i}$ a development according to integral powers of $x-a$.

In order that the integrals of the equation (27) shall have no critical points, where $P(y)$ is a polynomial of degree $2m$ it is necessary and sufficient, by what precedes, that *the order of multiplicity of each root of $P(y)=0$ be equal to or greater than m, or of the form $m(1-1/i)$, where i is an integer greater than unity*. If all these conditions are satisfied, the general integral of the equation (27) is a single-valued function whose singular points in the finite portion of the plane can only be poles.

To complete the discussion, we shall distinguish several cases:

First case. There is one linear factor in $P(y)$ whose exponent is greater than m (evidently there can be only one). If there are also p linear factors distinct from this one, the sum of the exponents of these factors is less than m:

$$m\left(1-\frac{1}{i_1}\right)+\cdots+m\left(1-\frac{1}{i_p}\right)<m.$$

Hence we have $p-1<1/i_1+\cdots+1/i_p$ and, since $i_1, i_2, \cdots, i_p$ are each greater than unity, $p-1<p/2$, or $p<2$. We have therefore $p=1$, and, extracting the mth root of the two sides, we may write the equation (27) in the form

(I) $\qquad y' = A(y-a)^{1+\frac{1}{i}}(y-b)^{1-\frac{1}{i}}.$

The case where $i=1$ should not be excluded, for it corresponds to an hypothesis which we have not examined — that of a single linear factor in $P(y)$.

Second case. The equation $P(y)=0$ has an m-fold root. If it has two, the equation (27) becomes, after extracting the mth root of the two sides,

(II) $\qquad y' = A(y-a)(y-b).$

If the equation $P(y)=0$ has only one root of multiplicity m, it has p ($p \geqq 2$) roots whose order of multiplicity is less than m, and we have a relation of the form

$$m\left(1-\frac{1}{i_1}\right)+\cdots+m\left(1-\frac{1}{i_p}\right)=m,$$

or

$$p-1=\frac{1}{i_1}+\cdots+\frac{1}{i_p}\leqq\frac{p}{2},$$

whence we derive $p \leqq 2$.

Since p is greater than unity, we have necessarily $p = 2$, $i_1 = i_2 = 2$; the number m is an even number, and the equation (27) reduces to the form

(III) $\qquad y'^2 = A(y-a)^2(y-b)(y-c),$

a, b, c being three different numbers.

Third case. The equation $P(y) = 0$ has only roots whose order of multiplicity is less than m. Let p be the number of these roots; the sum of the orders of multiplicity being equal to $2m$, we have a relation of the form

$$m\left(1-\frac{1}{i_1}\right) + m\left(1-\frac{1}{i_2}\right) + \cdots + m\left(1-\frac{1}{i_p}\right) = 2m,$$

or

$$p - 2 = \frac{1}{i_1} + \cdots + \frac{1}{i_p} \leqq \frac{p}{2}.$$

Hence $p \leqq 4$, and since $p > 2$, we can have only $p = 4$ or $p = 3$. If $p = 4$, the sum $1/i_1 + 1/i_2 + 1/i_3 + 1/i_4$ must be equal to 2; and since each of the denominators is equal to at least 2, we must have

$$i_1 = i_2 = i_3 = i_4 = 2.$$

If $p = 3$, it is a question of finding three integers, i_1, i_2, i_3, each greater than unity, such that the sum of their reciprocals is equal to 1. If none of these numbers is equal to 2, each must be equal to 3. If one of them, i_1, is equal to 2, the sum of the reciprocals of the other two must be equal to $1/2$; if the two are equal, each of them is equal to 4. If they are unequal, the smaller must be less than 4; it is therefore equal to 3, and the greater is then equal to 6. We have, then, in all only four possible combinations, and the equation (27) may be reduced to one of the following forms:

(IV) $\qquad y'^2 = A(y-a)(y-b)(y-c)(y-d),$
(V) $\qquad y'^3 = A(y-a)^2(y-b)^2(y-c)^2,$
(VI) $\qquad y'^4 = A(y-a)^3(y-b)^3(y-c)^2,$
(VII) $\qquad y'^6 = A(y-a)^5(y-b)^4(y-c)^3,$

where a, b, c, d are different numbers. If, in the equation (27), the polynomial $P(y)$ is of degree $2m$, and if the general integral is a single-valued function, the equation (27) has one of the forms which we have just obtained. Conversely, every integral of any one of these equations is a single-valued function, since on any path described by the variable we cannot encounter any other singular points than poles.

It is convenient to add to the types which we have just enumerated, in order to have all the equations of the form (25) whose general integral is single-valued, the types which are obtained by a transformation of the form $y - a = 1/z$, where a is a root of the polynomial $P(y)$. The new types of equations thus obtained are

(I)′ $\quad y' = A(y-a)^{1-\frac{1}{i}},$

(I)″ $\quad y' = A(y-a)^{1+\frac{1}{i}},$

(II)′ $\quad y' = A(y-a),$

(III)′ $\quad y'^2 = A(y-a)^2(y-b),$

(III)″ $\quad y'^2 = A(y-b)(y-c),$

(IV)′ $\quad y'^2 = A(y-a)(y-b)(y-c),$

(V)′ $\quad y'^3 = A(y-a)^2(y-b)^2,$

(VI)′ $\quad y'^4 = A(y-a)^3(y-b)^3,$

(VI)″ $\quad y'^4 = A(y-a)^3(y-b)^2,$

(VII)′ $\quad y'^6 = A(y-a)^5(y-b)^4,$

(VII)″ $\quad y'^6 = A(y-a)^5(y-b)^3,$

(VII)‴ $\quad y'^6 = A(y-a)^4(y-b)^3.$

The equations (I), (I)′, (I)″, which are transformable one into the other, have a rational function for their general integral, as we see immediately from the equation (I)′, for example. It is easy to show that the equations (II), (II)′, (III), (III)′, (III)″ have a simply periodic function for their general integral. Finally, the general integral of the equations (IV) and (IV)′ is an elliptic function. There remain, then, as new types of differential equations of the form (25) whose general integral is single-valued, only the equations (V), (VI), (VII), and those which reduce to these forms. These equations separate into three groups, and it is sufficient to integrate one equation from each of the groups, for example, the equations (V)′, (VI)″, (VII)‴.

If, in the equation (VI)″, we put $y = a + z^2$ and extract the square root of the two sides, it becomes

$$4z'^2 = A^{\frac{1}{2}} z(z^2 + a - b),$$

and the general integral in z is an elliptic function. Similarly, if in the equation (VII)‴ we put $y = a + z^3$ and extract the cube root of the two sides, it becomes

$$9z'^2 = A^{\frac{1}{3}}(z^3 + a - b),$$

which is an equation of the form (IV)′.

In order to integrate the equation (V)', we observe that that relation between y and y' is of deficiency one. We can therefore express y and y' rationally in terms of the coördinates of a point of a cubic or in terms of a parameter t and the square root of a polynomial of the third degree. In fact, if we put $y' = At^2$, we derive from the equation (V)'

$$y = \frac{a+b}{2} + \frac{1}{2}\sqrt{(a-b)^2 + 4At^2},$$

and the relation $dy = y' dx$ leads to the new equation

$$3\frac{dt}{dx} = \sqrt{(a-b)^2 + 4At^3}.$$

The general integral of this equation, $t = f(x + C)$, is an elliptic function. Hence the general integral of the equation (V)' is of the form

$$y = \frac{a+b}{2} + \frac{3}{2}f'(x+C).$$

It follows that the general integral of every equation of the form (25), if it is a single-valued function, is a rational function of x or of the exponential function e^{ax}, or is an elliptic function.

Except in the cases which have just been enumerated, the general integral of the equation (25) is never a single-valued function. For example, the inverse function of a hyperelliptic integral of the first kind cannot be a single-valued function. In fact, let $P(y)$ be a polynomial prime to its derivative and of degree greater than 4. The differential equation $y'^2 = P(y)$ cannot have a single-valued integral. Let (x_0, y_0) be the initial values of the two variables x and y. As $|y|$ becomes infinite, x approaches a finite value α; and, conversely, when x goes from x_0 to α, $|y|$ becomes infinite. The point $x = \alpha$ is an algebraic critical point, as we have just seen, for the integral of the equation $z'^2 = z^4 P(1/z)$ which approaches zero when x approaches α, since the degree of $P(y)$ is greater than 4.*

---

* In one of their papers Briot and Bouquet set for themselves the problem of determining all the equations $F(y, y') = 0$, where F is a polynomial, whose general integral is a single-valued function (*Journal de l'École Polytechnique*, Vol. XXI). From the conditions found by them Hermite proved that the relation between y and y' is of deficiency zero or one (*Cours lithographié de l'École Polytechnique*, 1873); hence we can apply the method of § 11 in integrating them. If the relation is of deficiency zero, we can express y and y' as rational functions of a parameter t. In order that the integral of the given equation be a single-valued function, the variable z, which is obtained by a quadrature, must be a linear function of t, $z = (at+b)/(ct+d)$, or else the logarithm of such a function, $z = A \operatorname{Log}[(at+b)/(ct+d)]$. If the relation is of deficiency one, we can express y and y' as elliptic functions of a parameter u, and $dx/du = (1/y') dy/du$ must reduce to a constant. The problem of Briot and Bouquet

69. Existence of elliptic functions deduced from Euler's equation. The reasoning of the preceding paragraph proves, in particular, that the general integral of the equation $y'^2 = R(y)$, where $R(y)$ is a polynomial of the third or of the fourth degree, prime to its derivative, is a single-valued function analytic except for poles in the whole plane. On the other hand, the inverse function, which is an elliptic integral of the first kind, has two periods whose ratio is imaginary (see II, Part I, § 56). This single-valued function is therefore doubly periodic, and we thus demonstrate the existence of elliptic functions by means of the integral calculus alone.

The preceding proof of the single-valued property of the inverse function of the elliptic integral of the first kind is distinct from the one which has been given in § 78 of Vol. II, Part I, in which we make use of the properties of the function $p(u)$. We shall also show in brief how we can take as our point of departure the theory of the integration of Euler's equation, which will give an idea of the method pursued by the originators of the theory.

Let us first consider the differential equation

$$(29) \qquad \frac{dx}{\sqrt{1-x^2}} + \frac{dy}{\sqrt{1-y^2}} = 0,$$

whose general integral is $x\sqrt{1-y^2} + y\sqrt{1-x^2} = C$ (§ 14). It is also clear that the general integral is given by the equation

$$\arcsin x + \arcsin y = C',$$

and therefore that we have between the two a relation of the form

$$\arcsin x + \arcsin y = F(x\sqrt{1-y^2} + y\sqrt{1-x^2}).$$

In order to determine the function F, let us suppose $y = 0$; there results the definite relation

$$(30) \qquad \arcsin x + \arcsin y = \arcsin(x\sqrt{1-y^2} + y\sqrt{1-x^2}).$$

This relation is equivalent to the addition formula. For let us take two angles u and v determined by the conditions

$$x = \sin u, \qquad \sqrt{1-x^2} = \cos u, \qquad y = \sin v, \qquad \sqrt{1-y^2} = \cos v,$$

where the radicals have the same values as in the preceding equations. The relation (30) gives

$$x\sqrt{1-y^2} + y\sqrt{1-x^2} = \sin(u+v),$$

or

$$\sin(u+v) = \sin u \cos v + \sin v \cos u.$$

However, to see in this work only an ingenious proof of the addition formula for the sine function would be to overlook entirely its broad significance. Indeed, we shall show that it leads to a very simple proof of the existence of a single-valued *integral* function which satisfies the differential equation

$$(31) \qquad y'^2 = 1 - y^2,$$

---

has been generalized by Fuchs, who formulated the necessary and sufficient conditions that the general integral of an equation of the first order $F(x, y, y') = 0$, algebraic in y and y', may have only fixed critical points. Poincaré has since shown that when these conditions are satisfied, we are led to quadratures or to Riccati equations (*Acta mathematica*, Vol. VII).

and which reduces to zero for $x = 0$, while y' is equal to $+1$ for $x = 0$. Cauchy's general theorem shows, indeed, that there exists an analytic function $\phi(x)$ satisfying these conditions and analytic in the neighborhood of the origin, but it does not give the radius of convergence of the power series which represents $\phi(x)$. Let R be this radius of convergence. The circle C of radius R about the origin as center is the greatest circle described about the origin as center within which the function $\phi(x)$ is analytic. The derivative $\phi'(x)$ is analytic in the same circle, and we have $\phi'^2(x) = 1 - \phi^2(x)$. Let us now return to the equation (29), and let us make the change of variables $x = \phi(u)$, $y = \phi(v)$, where u and v are the two new variables and ϕ the function which has just been defined. If we choose the determination of the radicals in a suitable way, we have also

$$\sqrt{1-x^2} = \phi'(u), \qquad \sqrt{1-y^2} = \phi'(v),$$

and the equation (29) becomes $du + dv = 0$. The general integral of this equation can therefore be written in two different ways:

$$u + v = C, \qquad x\sqrt{1-y^2} + y\sqrt{1-x^2} = C'$$

or

$$\phi(u)\phi'(v) + \phi'(u)\phi(v) = C'.$$

Hence it follows, as before, that we have a relation between $u + v$ and $\phi(u)\phi'(v) + \phi'(u)\phi(v)$. Putting $v = 0$, the relation is determined, and we have

(32) $$\phi(u+v) = \phi(u)\phi'(v) + \phi'(u)\phi(v).$$

This relation holds, provided $|u| < R$, $|v| < R$, $|u+v| < R$, which will surely be true if we have $|u| < R/2$, $|v| < R/2$. Let us put $v = u$ and $|u| < R/2$; then the equation (32) becomes

(33) $$\phi(2u) = 2\phi(u)\phi'(u).$$

Let $\phi_1(u)$ be the function $2\phi(u/2)\phi'(u/2)$. This function $\phi_1(u)$ is analytic in the circle of radius $2R$ about the origin as center, and, by (33), it is *identical* with the analytic function $\phi(u)$ in the circle C of radius R. These two functions, $\phi(u)$, $\phi_1(u)$, form therefore only a single analytic function, which is analytic outside of the circle C. It is therefore impossible that the radius R of this circle of convergence has a finite value; consequently the function $\phi(u)$ is an *integral function* of u.

Let us now consider the differential equation

(34) $$y'^2 = (1-y^2)(1-k^2y^2),$$

adopting for the right-hand side Legendre's normal form, and let us study the integral $\lambda(x)$ of this equation which is zero for $x = 0$ and whose derivative is equal to $+1$ for $x = 0$. This function $\lambda(x)$ is analytic in the neighborhood of the origin. Let C be the greatest circle about the origin as center in the interior of which the function $\lambda(x)$ is analytic except for poles, and let R be its radius. If the nearest singular point of $\lambda(x)$ to the origin were not a pole, we should take for C the circle through this singular point, and the function $\lambda(x)$ would then be analytic in this circle.

Let us now consider Euler's equation

(35) $$\frac{dx_1}{\sqrt{(1-x_1^2)(1-k^2x_1^2)}} + \frac{dx_2}{\sqrt{(1-x_2^2)(1-k^2x_2^2)}} = 0.$$

Multiplying the numerator and the denominator of the right-hand side of the equation (66) (p. 29) by the conjugate of the denominator, and suppressing the common factor $x_1^2 - x_2^2$, we have for its general integral

$$(36) \quad \frac{x_2 \sqrt{(1-x_1^2)(1-k^2 x_1^2)} + x_1 \sqrt{(1-x_2^2)(1-k^2 x_2^2)}}{1 - k^2 x_1^2 x_2^2} = -C.$$

On the other hand, choosing the sign of the radicals suitably, the change of variables $x_1 = \lambda(u)$, $x_2 = \lambda(v)$ reduces the equation (35) to the equation $du + dv = 0$, whose general integral is $u + v = C'$. If we make the same substitution in the formula (36), we have a relation of the form

$$\frac{\lambda(u)\lambda'(v) + \lambda(v)\lambda'(u)}{1 - k^2 \lambda^2(u) \lambda^2(v)} = F(u+v).$$

We can determine the form of the function F, as before, by supposing $v = 0$, which gives $F(u) = \lambda(u)$; and we have the definite relation

$$(37) \quad \lambda(u+v) = \frac{\lambda(u)\lambda'(v) + \lambda(v)\lambda'(u)}{1 - k^2 \lambda^2(u) \lambda^2(v)}.$$

Putting $v = u$, we find

$$(38) \quad \lambda(2u) = \frac{2\lambda(u)\lambda'(u)}{1 - k^2 \lambda^4(u)},$$

a formula which holds true whenever $|u| < R/2$.

Let us consider the function

$$\Phi(u) = \frac{2\lambda\left(\frac{u}{2}\right)\lambda'\left(\frac{u}{2}\right)}{1 - k^2 \lambda^4\left(\frac{u}{2}\right)}.$$

This function is analytic except for poles in the circle of radius $2R$ about the origin as center, since it is the quotient of two such functions. Moreover, it coincides with $\lambda(u)$ in the interior of the circle C, by the relation (38). Hence the two functions $\lambda(u)$ and $\Phi(u)$ form a single analytic function, and $\lambda(u)$ is analytic except for poles in a larger circle than C. It is therefore impossible to suppose that the radius R of this circle has a finite value, and consequently the function $\lambda(u)$ is analytic except for poles in the entire plane.

The equation (37) constitutes the formula for the addition of the arguments of the function $\lambda(u)$. When k approaches zero, we find again at the limit the addition formula for $\sin u$. The function $\sin u$ can, in fact, be considered as a degenerate case of $\lambda(u)$, obtained by letting k approach zero.

70. Equations of higher order. The study of the properties of the functions defined by differential equations of higher order presents much more serious difficulties than those which arise in studying equations of the first order. These difficulties result in a great measure from the possible presence of movable essential singularities. These singularities may be at the same time essentially singular points and transcendental critical points, as in the following example, due to Painlevé. The function

$$(39) \quad y = \mathrm{p}[\mathrm{Log}(Ax+B); g_2, g_3],$$

where A and B are two arbitrary constants, is the general integral of the equation of the second order,

$$(40) \qquad y'' = y'^2 \left(\frac{6y^2 - \dfrac{g_2}{2}}{4y^3 - g_2 y - g_3} - \frac{1}{\sqrt{4y^3 - g_2 y - g_3}} \right).$$

In the neighborhood of every value of x different from $-B/A$ this function (39) is analytic or analytic except for poles.

When x turns around the point $-B/A$, the function has an infinite number of different values, unless $2\pi i$ is an aliquot part of one of the periods of the function $p(u; g_2, g_3)$. On the other hand, when the variable x describes any sort of curve such that $|Ax + B|$ approaches zero, the point which represents the quantity $u = \text{Log}\,(Ax + B)$ describes a curve with an infinite branch. This curve therefore crosses an infinite number of parallelograms of periods of the function $p(u)$, and consequently y does not approach any finite or infinite limit. Thus, although the general integral of the equation (40) presents neither fixed critical points nor *movable algebraic critical points*, we cannot conclude from this that it is single-valued. This is on account of the presence of the movable transcendental critical point, $x = -B/A$.

Beginning with equations of the third order, the movable transcendental singular points may form lines. It is easy to see how an analytic function having natural boundaries may not have any critical points in its whole domain of existence without being on that account single-valued. Consider, for example, a function $f(x)$ analytic in the ring included between the two circles C and C' described about the center a and having C and C' as natural boundaries (II, Part I, § 87). The function $F(x) = f(x) + \text{Log}\,(x - a)$ has the same circles C and C' as natural boundaries, and it is analytic at every point between C and C'. Nevertheless it has an infinite number of determinations for every value of x in this domain.

For a long time these difficulties arrested the progress of this theory, but Painlevé, in recent investigations, has obtained algebraic differential equations of the second order which are integrable by means of essentially new single-valued transcendentals. We shall cite only one of the equations given by Painlevé,

$$y'' = \alpha y^2 + \beta x,$$

where α and β are constants ($\alpha\beta \neq 0$). The general integral of this equation is a transcendental function analytic except for poles.* (*Bulletin de la Société Mathématique*, Vol. XXVIII.)

---

* Starting with linear equations, it is easy to form systems of differential equations which generalize Riccati's equation, and whose integrals have no other movable singular points than poles. Consider, for example, a system of three linear equations of the first order,

(α) $\quad y' + ay + bz + cu = 0, \qquad z' + a_1 y + b_1 z + c_1 u = 0, \qquad u' + a_2 y + b_2 z + c_2 u = 0.$

If we put $y = uY$, $z = uZ$, Y and Z are integrals of the system of equations

(β) $\quad \begin{cases} Y' + aY + bZ + c - Y(a_2 Y + b_2 Z + c_2) = 0, \\ Z' + a_1 Y + b_1 Z + c_1 - Z(a_2 Y + b_2 Z + c_2) = 0, \end{cases}$

and it is clear that the only movable singular points of the integrals are poles. But

III. SINGULAR INTEGRALS

71. Singular integrals of an equation of the first order. We have already remarked on several occasions (§§ 10, 14) that a differential equation of the first order may have certain integrals which it would be impossible to obtain by assigning a particular value to the arbitrary constant which appears in the general integral. This result appears to contradict the theorem of § 26, from which we deduced a precise definition of the general integral. This leads us to consider again Cauchy's fundamental theorem and to determine by a closer examination whether or not the hypotheses of that theorem are necessarily fulfilled for all integrals. Let us consider, for definiteness, an equation of the first order,

$$(41) \qquad F(x, y, y') = 0,$$

where F is an irreducible polynomial in x, y, y' of the mth degree in y'. To every system of values (x_0, y_0) the equation

$$(41') \qquad F(x_0, y_0, y') = 0$$

determines in general m corresponding distinct and finite values $y_1', y_2', \cdots, y_m'$ for y'. Let us suppose first that this is actually true at a given point (x_0, y_0). Then, as $x - x_0$ and $y - y_0$ approach zero, the m roots of the equation (41) approach respectively $y_1', y_2', \cdots, y_m'$, and each of them is an analytic function in the neighborhood of the point (x_0, y_0). The root which approaches y_i', for example, is represented by a power-series development of the form

$$(42) \qquad y' = y_i' + \alpha_i(x - x_0) + \beta_i(y - y_0) + \cdots.$$

We can apply Cauchy's theorem to the equation (42), and we conclude from it that that equation has one and only one integral which approaches y_0 as $|x - x_0|$ approaches zero. This integral is analytic, and the development of $y - y_0$ begins with the term $y_i'(x - x_0)$. To each root of the equation (41') corresponds thus an integral of the given equation.

It is to be noticed that this is not the most general system of differential equations of the form

$$(\gamma) \qquad Y' = R(x, Y, Z), \qquad Z' = R_1(x, Y, Z),$$

where R and R_1 are rational functions of Y and Z, which possess this property. In fact, let $Y = \phi(Y_1, Z_1)$, $Z = \psi(Y_1, Z_1)$ be relations defining a *Cremona transformation*, such that we can derive from them the inverse relations $Y_1 = \phi_1(Y, Z)$, $Z_1 = \psi_1(Y, Z)$, where $\phi, \psi, \phi_1, \psi_1$ are rational functions. If we apply this transformation to the system (β), we are led to a system having the same property, which is surely of the form (γ) but not in general of the form (β).

The equation (41) has therefore m and only m integrals which take on the value y_0 for $x = x_0$, and these m integrals are analytic in the neighborhood of the point x_0. In geometric language we may say that through the point M_0 whose coördinates are (x_0, y_0) there pass m integral curves with m distinct tangents, and that the point M_0 is an ordinary point on each of them. Besides, all the integrals of the equation (42) which for $x = x_0$ take on values differing only slightly from y_0 satisfy a relation of the form $\Phi(x, y; x_0, y_0 + C) = 0$ (§ 26), and the integral considered corresponds to the value $C = 0$ of the arbitrary constant.

If for $x = x_0$, $y = y_0$ a root of the equation (41') is infinite, it will suffice to regard y as the independent variable and x as the dependent variable. The equation (41) is replaced by an equation of the same form, $F_1(x, y, x'_y) = 0$, which for $x = x_0$, $y = y_0$ has a zero root $x' = 0$. If this is a simple root, we derive from it a development for $x - x_0$ in powers of $y - y_0$ beginning with a term of at least the second degree. Conversely, the point x_0 is an algebraic critical point for the integral which approaches y_0 when $|x - x_0|$ approaches zero (II, Part I, § 100). Through the point (x_0, y_0) there passes an integral curve whose tangent at that point is the straight line $x = x_0$.

The coördinates (x_0, y_0) of a point for which the equation (41) has a multiple root satisfy the relation

(43) $$R(x, y) = 0,$$

which is obtained by eliminating y' from the two relations $F = 0$, $\partial F / \partial y' = 0$. The equation (43) represents a certain curve (γ), and for all the points of this curve the equation (41) has one or several multiple roots. Let (x_0, y_0) be the coördinates of an ordinary point M_0 taken on this algebraic curve. We shall suppose, in order to treat the simplest possible case, that the equation

$$F(x_0, y_0, y') = 0$$

has a double root y'_0 but no other multiple root finite in value. If this double root were infinite, it would suffice to interchange x and y in order to pass to the case where it is zero. When $|x - x_0|$ and $|y - y_0|$ are very small, the equation (41) has two roots which differ very little from y'_0. These roots are not, in general, analytic functions of the variables x and y in the neighborhood of the point (x_0, y_0), but their sum and their product are analytic functions,[*] so that these two

---

[*] The proofs of these properties are analogous to the proofs of the corresponding theorems on implicit functions of a single variable (II, Part I, § 98).

roots of the equation (41), which approach y_0' as $|x-x_0|$ and $|y-y_0|$ approach zero, are also roots of an equation of the second degree,

(44) $$y'^2 - 2P(x,y)y' + Q(x,y) = 0,$$

where $P(x, y)$ and $Q(x, y)$ are analytic functions in the neighborhood of (x_0, y_0). From the equation (44) we find

(45) $$y' = P(x,y) \pm \sqrt{P^2(x,y) - Q(x,y)},$$

and these two roots are equal for all the points of the curve (γ_1) whose equation is $P^2 - Q = 0$. This curve (γ_1) is necessarily part of the curve (γ), and since it passes through the point (x_0, y_0), it coincides with (γ) in the neighborhood of (x_0, y_0). In order to study the corresponding integral curve, we shall suppose that the origin has been transformed to the point M_0, which amounts to putting $x_0 = y_0 = 0$. Since the origin is a simple point of the curve (γ), if we have chosen the axes of coördinates in such a way that the tangent at the origin is not the axis Oy itself, the equation $P^2 - Q = 0$ has an analytic root $y = y_1(x)$ which approaches zero as x approaches zero.

In general, the slope of the tangent to the curve (γ) at the origin is different from the double root $y_0' = P(0,0)$ of the equation (45) for $x = y = 0$. Let us first assume this point, which is almost self-evident, and return to it later. Then, if we put $y = y_1 + z$ in the equation (45), it becomes

$$y_1' + z' = P(x, y_1 + z) \pm \sqrt{z\Phi(x,z)},$$

where $\Phi(x, z)$ is a power series in x and z. It is clear that z must be a factor under the radical after the substitution $y = y_1 + z$, since y_1 is a root of the equation $P^2 - Q = 0$. If we arrange $\Phi(x, z)$ in powers of z, we have a development of the form

$$\psi_0(x) + z\psi_1(x) + z^2\psi_2(x) + \cdots,$$

where ψ_0, ψ_1, ψ_2 are regular functions of x in the neighborhood of the origin. The function $\psi_0(x)$ cannot be zero for $x = 0$, for otherwise the development of $z\Phi(x, z)$ would contain no terms of the first degree in x, z; whence the development of $P^2 - Q$ would contain no terms of the first degree in x, y, contrary to hypothesis. Similarly, if we replace y_1 by its development in the difference $P(x, y_1 + z) - y_1'$, we have, after arranging in powers of z,

$$P(x, y_1 + z) - y_1' = \phi_0(x) + z\phi_1(x) + \cdots,$$

where the first function $\phi_0(x)$ does not vanish for $x = 0$, since by supposition the derivative y_1' is different from $P(0, 0)$ at the origin.

The equation (45) therefore reduces to an equation of the form

(46) $\quad z' = \phi_0(x) + z\phi_1(x) + \cdots \pm \sqrt{z}\sqrt{\psi_0(x) + z\psi_1(x) + \cdots}$,

where neither of the functions $\phi_0(x)$ and $\psi_0(x)$ vanishes for $x = 0$.

In this last equation let us put $z = u^2$. Selecting a determination of the radical on the right, we find

(47) $\quad 2u\dfrac{du}{dx} = \phi_0(x) + u^2\phi_1(x) + \cdots + u\sqrt{\psi_0(x) + u^2\psi_1(x) + \cdots}$.

The right-hand side is analytic in the neighborhood of the point $x = 0$, $u = 0$, since $\psi_0(0)$ is not zero. Moreover, this right-hand side is not zero for $x = 0$, $u = 0$, since $\phi_0(0)$ is not zero. The derivative du/dx is infinite for $x = u = 0$. Hence the equation (47) has one and only one integral which approaches zero as x approaches zero (§ 63), and for which the origin is an algebraic critical point.

It follows that the given equation (44) has an integral $y = y_1 + u^2$ which approaches zero as x approaches zero. The adoption of the opposite determination of the radical in the equation (47) would amount to changing u to $-u$ in that equation, and we should obtain the same function $y_1 + u^2$. The origin is an algebraic critical point for this integral. Let a_0 be the term independent of x and of u in the development of the right-hand side of the equation (47), and let b_0 be the coefficient of u in the same development. Developing x in powers of u, we find

$$x = \dfrac{u^2}{a_0} - \dfrac{2b_0}{3a_0^2}u^3 + \cdots.$$

Conversely, we derive from this a series for u in powers of $x^{1/2}$,

$$u = \sqrt{a_0}\,x^{\frac{1}{2}} + \dfrac{b_0}{3}x + \cdots,$$

and the development of $y_1 + u^2$ contains a term in $x^{3/2}$. The origin is therefore a cusp for the integral curve which passes through this point, and we can say now that *the curve (γ), represented by the equation* (43), *is, in general, the locus of the cusps of the integral curves.*

Through a point of the curve (γ) there passes, in general, an integral curve that has a cusp of the first kind at this point, and the tangent at the cusp has for its slope the double root y_0'. If the equation (41) is of higher degree than 2, there pass through the same point other integral curves, corresponding to the simple roots of the equation $F(x_0, y_0, y') = 0$, for which this point is an ordinary point.

The discussion is entirely different when for every point (x_0, y_0) of the curve (γ) the corresponding double root y_0' of the equation (41) is equal to the slope of the tangent to the curve (γ) at this point. In this case we see first of all that the curve (γ) is an integral curve of the equation (41). Moreover, it is an integral which is entirely unaccounted for in Cauchy's fundamental theorem, whatever may be the point chosen on the curve to fix the initial values of x and y. For if we take the point (x_0, y_0), the equation

$$F(x, y, y') = 0$$

has two roots which approach y_0' as $|x - x_0|$ and $|y - y_0|$ approach zero; but these two roots are not in general regular functions of the variables x and y in the neighborhood of the values x_0, y_0, and we cannot apply Cauchy's theorem. The integral thus obtained is said to be a *singular integral*. The investigation of singular integrals does not offer any theoretical difficulties, since it is evidently sufficient to determine whether the curve represented by the equation (43) satisfies the differential equation (41), and this necessitates only an elimination. It may happen that the equation (43) represents two distinct curves, one of which is a singular integral curve and the other the locus of the cusps of the integral curves.

If the curve (γ) is a singular integral, through each point of that curve there passes in general another integral curve tangent to (γ). Let us take for origin any point of (γ). We know in advance an integral y_1 of the equation (45), namely, the singular integral for which we have simultaneously

(48) $\qquad y_1' = P(x, y_1), \qquad P^2(x, y_1) = Q(x, y_1).$

Putting $y = y_1 + z$, as above, the equation takes the form (46), but in this case the function $\phi_0(x)$ is zero, since $z = 0$ must be an integral of this new equation. Retaining the other hypotheses, the function $\psi_0(x)$ is not zero for $x = 0$, and if we next put $z = u^2$ in the equation (46), we are led to an equation all of whose terms are divisible by u. Dividing by u, there remains a differential equation

(49) $\quad 2u' = u[\phi_1(x) + u^2\phi_2(x) + \cdots] \pm \sqrt{\psi_0(x) + u^2\psi_1(x) + \cdots},$

to which we can apply Cauchy's general theorem. Since the function $\psi_0(x)$ is not zero for $x = 0$, the two determinations of the radical are analytic for $x = 0$, $u = 0$. The equation (49) has therefore two analytic integrals in the neighborhood of the origin which vanish for $x = 0$, and it is easily seen that these two integrals are deducible one from the other by changing u to $-u$. It follows that the equation in y has another integral curve

$$y = y_1 + u^2,$$

which is tangent to the curve (γ) at the origin. But there is an essential difference between these two integrals. In fact, we can apply the general theorems of § 26 to the equation (49), and the integral of this equation which is zero for $x = 0$ belongs to a family of integrals which depend upon one arbitrary constant. The same thing is therefore true of the integral curve which is tangent to the singular integral curve at the origin, whereas the singular integral itself is in general an *isolated* solution. This fact is easily explained, since we cannot apply to this integral the reasoning which proves the existence of a general integral (§ 21) from which we could obtain the former by giving a particular value to the constant which appears in the latter.

The singular integral is therefore in general the envelope of the other integral curves. Lagrange had already noticed that the envelope of the curves represented by the general integral of a differential equation of the first order is also an integral of the same equation, which is almost self-evident, since at any point of the enveloping curve the slope of the tangent is the same for the envelope and for the particular curve enveloped at that point. We can also find in this way the rule which enables us to deduce the singular integral from the differential equation itself. In fact, let us first take a point M very near the envelope. Through this point M there pass two integral curves very close to each other. Moreover, the slopes of the tangents to these two curves differ from each other very little. When the point M approaches the envelope, these tangents approach coincidence, and the equation (41) has a double root in y' (see I, § 208, 2d ed.; § 202, 1st ed.).

Summing up, we see that for an equation of the first order two entirely distinct cases may present themselves, according as the curve (γ) is a singular integral curve or the locus of the cusps of the integral curves. It is natural to ask which of these two cases ought to be considered as the *normal* case. A little attention will show that it is the second. In fact, the curve (γ) is also the envelope of the curves represented by the equation $F(x, y, a) = 0$, where a is the variable parameter. If the differential equation (41) had a singular integral, whatever the polynomial F might be, we should be led to assert a consequence which is manifestly absurd — that is, that at every point of the envelope of a family of algebraic curves the slope of the tangent is equal to the value of the parameter for the corresponding curve of the family tangent to the envelope at that point. If this condition is satisfied by a family of curves, it suffices to change the parameter (putting, for example, $a = a' + \epsilon$) in order that this condition shall cease to hold. We see, therefore, that if we start from an equation of the first order in which the coefficients of F are taken at random, rather than from an equation furnished by the elimination of an arbitrary constant, the cases where there exists a singular integral must be considered as *exceptional*. If this result formerly appeared paradoxical to some mathematicians, that was no doubt because, up to the time of Cauchy's work, the equations studied had been principally those whose general integral is represented by algebraic curves. As a family of algebraic curves has in general an envelope, it appeared quite natural to extend the conclusion to the integral curves of any differential equation of the first order. We have just seen that this induction was not justified.* Moreover, even in the case where a family of plane curves depending upon a variable parameter has an envelope, the method which enables us to find that envelope gives also, as we have seen (I, §§ 207, 208, 2d ed.; §§ 201, 202, 1st ed.), the locus of singular points.

72. General comments. *Example* 1. Let us take the equation

(50) $$y'^2 + 2xy' - y = 0.$$

The two values of y' are equal for all the points of the parabola $y + x^2 = 0$, and the double root is equal to $-x$, while the slope of the tangent to the parabola is $-2x$. This curve is therefore not a singular integral curve. We shall show that it is the locus of the cusps of the integral curves. The equation (50) is a Lagrange equation. Applying to it the general method of § 9, we find that the coördinates x and y of a point of an integral curve are expressed in terms of a parameter p by means of the equations

(51) $$x = \frac{C}{p^2} - \frac{2p}{3}, \qquad y = \frac{2C}{p} - \frac{p^2}{3}.$$

---

* In the theory of envelopes we suppose tacitly that in the neighborhood of a system of solutions (x_0, y_0, a_0) of the two equations $f(x, y, a) = 0$, $\partial f/\partial a = 0$ the functions f and $\partial f/\partial a$, together with their partial derivatives, are continuous, so that we can apply to the functions x and y of a defined by these two equations the reasoning which we apply to implicit functions. Now, given a differential equation of the first order, we know certainly that it has an infinite number of integrals depending upon an arbitrary constant and represented in a certain region by an equation $\phi(x, y, C) = 0$, but there is nothing to prove a priori that this function $\phi(x, y, C)$ satisfies the conditions which we have just mentioned. We may even assert that it is not true in general.

It follows that these integrals are represented by unicursal curves of the fourth degree. For the values of the parameter which are roots of the equation $p^3 + 3\,C = 0$ we have $dx/dp = dy/dp = 0$. Each of these curves has therefore three cusps, and the locus of these points can be found by eliminating p and C from the equations (51) and the relation $p^3 = -\,3\,C$, which gives the parabola $y + x^2 = 0$.

Example 2. Let us again consider Euler's equation $Xy'^2 = Y$. The two values of y' are equal for all the points of any one of the eight straight lines represented by the equation $XY = 0$. These eight lines represent the singular solutions, and form the envelope of the curves represented by the general integral.

Example 3. We can use the following method to determine whether singular solutions exist. From what we have seen, such an integral, if it exists, satisfies the equations

$$F(x, y, y') = 0, \qquad \frac{\partial F}{\partial y'} = 0,$$

and consequently also the equation $\partial F/\partial x + \partial F/\partial y\, y' = 0$ obtained by differentiating the first. Conversely, suppose that for all the points of a curve (γ) the three equations

(52) $\qquad F(x, y, m) = 0, \qquad \dfrac{\partial F}{\partial m} = 0, \qquad \dfrac{\partial F}{\partial x} + \dfrac{\partial F}{\partial y}\, m = 0$

have a common solution in m. Along the curve (γ), x, y, and m are three functions of a single variable satisfying the three relations (52). We have therefore the relation between their differentials,

$$\frac{\partial F}{\partial x}\, dx + \frac{\partial F}{\partial y}\, dy + \frac{\partial F}{\partial m}\, dm = 0,$$

which, by (52), takes the form

$$\frac{\partial F}{\partial y}\left(m - \frac{dy}{dx}\right) = 0.$$

If $\partial F/\partial y$ is not zero at all the points of the curve (γ), we have therefore $y' = m$, and this curve is a singular integral curve.* If $\partial F/\partial y = 0$, we must also have $\partial F/\partial x = 0$, and a direct verification is necessary to determine whether the curve (γ) is an integral curve.

This remark applies in particular to Clairaut's equation

$$F(x, y, y') = f(y', y - xy') = 0.$$

---

* See an article by Darboux in the *Bulletin des Sciences mathématiques*, Vol. IV, 1873, pp. 158-176.

Putting, for the sake of brevity, $u = y - xy'$, the three equations which are to be compatible are here

$$f(y', y - xy') = 0, \quad \frac{\partial f}{\partial y'} - x\frac{\partial f}{\partial u} = 0, \quad \frac{\partial f}{\partial u}(-y') + y'\frac{\partial f}{\partial u} = 0,$$

and they reduce to only two equations. A singular integral is therefore obtained by eliminating y' from these two relations.

Example 4. Consider the equation

$$x^2 + y^2 - 2x(x + yy') + \frac{1}{m^2}(x + yy')^2 + K = 0,$$

whose general integral is represented by the circles which have double contact with the conic

$$x^2(1 - m^2) + y^2 + K = 0,$$

and which have their centers on the x-axis. This conic represents a singular solution. Moreover, the two values of y' become infinite for every point of the axis of x. This straight line is not, however, a locus of the cusps. Through any point of it there pass two integral curves tangent to each other, the common tangent being parallel to the axis of y.

Example 5. In order that a curve C represent a singular integral, it is not enough to require that at all the points of that curve the equation (41) shall have a double root. It is also necessary that that double root shall be precisely the slope of the tangent to C. Let us consider, for example, the cissoids represented by the equation $(y - 2a)^2(x - a) - x^3 = 0$. The straight line $x = 0$ is the locus of the cusps of these curves, and it represents also a particular integral obtained by supposing $a = 0$. At every point of this integral curve the corresponding differential equation has the double root $y' = 0$ and an infinite root. It is therefore not a singular integral curve.

Example 6. Let S be a surface having convex regions and also regions where its curvature is negative. These regions are separated by a curve Γ, the locus of the parabolic points, at every point of which the differential equation of the asymptotic lines (I, § 243, 2d ed.; § 242, 1st ed.),

$$D du^2 + 2 D' du\, dv + D'' dv^2 = 0,$$

has a double root in dv/du. This double root furnishes the direction of the single asymptotic tangent. If the tangent to Γ does not coincide with this asymptotic tangent (which is the general case), the curve Γ is the locus of the cusps of the asymptotic lines; but if the asymptotic tangent at each point M of Γ coincides with the tangent to Γ, the curve is the envelope of the asymptotic lines. This curve Γ, therefore, is at the same time an asymptotic line and a line of curvature, since the tangent is also an axis of the indicatrix. The normals to the surface S along Γ form, therefore, a developable surface, and since the normal to S is the binormal to the curve Γ, it follows that Γ is a plane curve (I, § 235, 2d ed.; § 231, 1st ed.) and the given surface S is tangent to the plane P of the curve Γ along the entire length of that curve.

Let us consider, for example, a surface of revolution. In order that one of the principal radii of curvature at a point M of this surface be infinite, the radius of curvature of the meridian must be infinite or the tangent to this meridian must be perpendicular to the axis. In the first case the curve Γ is a parallel each point of which is a point of inflection for the meridian, the asymptotic tangent is perpendicular to the tangent to Γ, and this parallel is a locus of the cusps of the asymptotic lines. On the other hand, in the second case the curve Γ is a parallel in all of whose points the surface is tangent to the plane of this parallel, as in an anchor ring. It is also the envelope of the asymptotic lines. All these results are easy to verify directly from the differential equation of the asymptotic lines in polar coördinates.

73. Geometric interpretation. The preceding discussion may be presented in a somewhat different form, which we shall rapidly indicate. We shall continue to employ geometric language, although the reasoning can be extended without difficulty to the domain of complex variables.

We have already pointed out (§ 8) that the integration of a differential equation of the first order $F(x, y, y') = 0$ is equivalent to the determination of the curves Γ which lie on the surface S whose equation is

$$F(x, y, z) = 0 \tag{53}$$

and for which $dy - z\,dx = 0$. The projection c on the xy-plane of a curve Γ of the surface S satisfying the preceding conditions is an integral curve of the given differential equation, and conversely. We shall suppose in the discussion that this surface S has no other singularities than the double curves along which two sheets of the surface cross with distinct tangent planes. Instead of studying the curves c in the xy-plane, we shall study the curves Γ on the surface S.

Let us consider first a point $M_0(x_0, y_0, z_0)$ of the surface S not on a double curve nor where the tangent plane is parallel to the z-axis. The tangent to the curve Γ which passes through M_0 lies in the tangent plane at this point,

$$(X - x_0)\left(\frac{\partial F}{\partial x}\right)_0 + (Y - y_0)\left(\frac{\partial F}{\partial y}\right)_0 + (Z - z_0)\left(\frac{\partial F}{\partial z}\right)_0 = 0, \tag{54}$$

and also, since we must have $dy - z\,dx = 0$, in the plane

$$Y - y_0 - z_0(X - x_0) = 0. \tag{55}$$

These two planes are distinct, since $(\partial F/\partial z)_0$ is not zero; hence they intersect in a straight line *not parallel to Oz*. Through the point M_0 there passes, therefore, one and only one curve Γ whose tangent is not parallel to the z-axis. The projection c of this curve on the xy-plane passes through the point m_0, the projection of M_0, and m_0 is an ordinary point for c. If the point M_0 belongs to a double curve of S, the preceding reasoning applies to each of the two sheets, provided that none of the tangent planes at M_0 are parallel to Oz. Through the point M_0 there pass, therefore, two curves Γ corresponding to the two sheets of the surface S. It remains to find out what happens if the point M_0 lies on the curve D of S, the locus of the points for which we have simultaneously $F = 0$, $\partial F/\partial z = 0$. We shall suppose that this curve D is not a double curve. It is, then, the locus of the points of S where the tangent plane is parallel to Oz, and one at least of the partial derivatives $\partial F/\partial x$, $\partial F/\partial y$ is different from zero at the point M_0. Hence the two planes (54) and (55) are parallel to the z-axis, and their intersection is parallel to Oz unless these two planes coincide, that is, unless we have

$$\left(\frac{\partial F}{\partial x}\right)_0 + z_0\left(\frac{\partial F}{\partial y}\right)_0 = 0. \tag{56}$$

Let us first discard the case in which this happens. The tangent to the curve Γ which passes through M_0 is parallel to Oz, but this curve itself does not present any singularity at the point M_0. To assure ourselves of this, we shall replace the system of the two equations

(57) $\qquad F(x, y, z) = 0, \qquad dy = z\,dx$

by the system of the two simultaneous equations

(58) $$\frac{dx}{\dfrac{\partial F}{\partial z}} = \frac{dy}{z\dfrac{\partial F}{\partial z}} = \frac{-dz}{\dfrac{\partial F}{\partial x} + z\dfrac{\partial F}{\partial y}},$$

with the initial conditions $x = x_0$, $y = y_0$, $z = z_0$. The two systems are equivalent. In fact, from the equations (58) we derive the integrable combination $dF = 0$. Hence we have $F(x, y, z) = F(x_0, y_0, z_0) = 0$. Now, since

$$\left(\frac{\partial F}{\partial x}\right)_0 + z_0 \left(\frac{\partial F}{\partial y}\right)_0$$

does not vanish by hypothesis, we derive from the equations (58) the developments of $x - x_0$ and of $y - y_0$ in powers of $z - z_0$ beginning with terms of at least the second degree,

$$x - x_0 = \alpha_2 (z - z_0)^2 \cdots, \qquad y - y_0 = \beta_2 (z - z_0)^2 \cdots.$$

The point M_0 is therefore an ordinary point for the curve Γ which passes through this point, but the projection m_0 of M_0 on the plane xOy is a cusp (in general of the first kind) for the curve c, the projection of Γ. This results, moreover, from a general property, which is easily verified, that the projection of a space curve on a plane, in a direction parallel to the tangent at a point M of the curve, has a cusp at the point m, the projection of M (I, Exercise 13, p. 582, 2d ed.). If d denotes the projection of the curve D on the xy-plane, it follows that the curve d is the locus of the cusps of the integral curves c, as we have shown before. The preceding method has the advantage of showing us how this singularity disappears when we pass from the plane to the surface S.

The result is quite different when the relation (56) is satisfied at all the points of the curve D. The two planes (54) and (55) are then coincident, and we have the case in which there exists a singular integral. Through every point of D there pass in general two curves Γ, the curve D itself and the second curve whose projection on the xy-plane is tangent to the singular integral curve d.

74. Singular integrals of systems of differential equations. The theory of the singular integrals may be extended to systems of differential equations of the first order, and therefore also to equations of higher order. We shall study only a system of two equations of the first order (which covers also the case of a single equation of the second order), and we shall employ a process which is the reverse of the preceding — that is, we shall consider first of all a system obtained by the elimination of the constants.* Let

(59) $\qquad F(x, y, z;\ a, b) = 0, \qquad \Phi(x, y, z;\ a, b) = 0$

---

* See E. GOURSAT, *Sur les solutions singulières des équations différentielles simultanées* (*American Journal of Mathematics*, Vol. XI).

be the equations of a family of plane or skew curves which depend upon two arbitrary parameters a and b. Such a family is called a *congruence* of curves. Let us suppose, for simplicity, that the functions F and Φ are polynomials. The curves of the congruence are then algebraic. We shall first generalize the theorems established for the congruences of straight lines (I, § 255). If we establish a relation between a and b of arbitrary form $b = \phi(a)$, we obtain an infinite number of curves Γ depending upon a single arbitrary parameter a. In general these curves do not have an envelope. In order that an envelope exist, it is necessary that the four equations (59) and (60) shall have a system of common solutions in x, y, z (I, § 215, 2d ed.; § 223, 1st ed.):

$$(60) \qquad \frac{\partial F}{\partial a} + \frac{\partial F}{\partial b}\frac{db}{da} = 0, \qquad \frac{\partial \Phi}{\partial a} + \frac{\partial \Phi}{\partial b}\frac{db}{da} = 0.$$

The elimination of x, y, z from these four equations leads to a relation between a, b, and db/da,

$$(61) \qquad \Pi\left(a, b, \frac{db}{da}\right) = 0;$$

that is, to a differential equation of the first order. If we have taken for $b = \phi(a)$ an integral of this equation, the curves Γ will generate a surface Σ and will be tangent to a curve C lying on Σ. We shall call this curve C the *edge of regression* of Σ, as in the case of line congruences. If the equation (61) is of degree m in db/da, every curve Γ of the congruence belongs, in general, to m surfaces similar to Σ, and it touches the corresponding edge of regression on each of these surfaces in a definite point. Thus there exist m remarkable particular points on each curve Γ of the congruence, which we call the *focal points*. These focal points can be obtained without integrating the differential equation (61), for we need only solve the four equations (59) and (60) for x, y, z, db/da. We find first the relation (61), which gives db/da, and, eliminating db/da from the two equations (60), we have a new relation,

$$(62) \qquad \frac{D(F, \Phi)}{D(a, b)} = \frac{\partial F}{\partial a}\frac{\partial \Phi}{\partial b} - \frac{\partial F}{\partial b}\frac{\partial \Phi}{\partial a} = 0,$$

which, together with the two equations (59) of the curve Γ, enable us to calculate the coördinates of the focal points.

The locus of the focal points is the *focal surface* of the congruence. We obtain the equation of this surface by eliminating a and b from the three relations (59) and (62). The focal surface is also the locus of the edges of regression C of the surfaces Σ. In fact, any point of the curve C is a focal point for the curve of the congruence which is tangent to C at that point. It follows that every curve Γ of the congruence is tangent to m sheets of the focal surface at the m corresponding focal points, since at each of these points it is tangent to a curve C lying on the focal surface. All these properties are exactly analogous to the properties of congruences of straight lines. In general, if F and Φ are any polynomials, the m sheets of the focal surface are represented by a single equation, but it may also happen that this equation breaks up into several distinct equations. In certain particular cases it may also happen that some of the sheets of the focal surface reduce to curves. In such a case the corresponding edge of regression C reduces to a point.

The conclusion which we can derive from these properties with respect to differential equations is as follows: The curves Γ are the integral curves of a system of differential equations which is obtained by eliminating the constants a and b from the equations (59) and the equations obtained by differentiating them:

(63) $\quad \dfrac{\partial F}{\partial x} + \dfrac{\partial F}{\partial y} y' + \dfrac{\partial F}{\partial z} z' = 0, \quad \dfrac{\partial \Phi}{\partial x} + \dfrac{\partial \Phi}{\partial y} y' + \dfrac{\partial \Phi}{\partial z} z' = 0.$

Let

(64) $\quad \mathscr{F}(x, y, z, y', z') = 0, \quad \mathscr{F}_1(x, y, z, y', z') = 0$

be the system of differential equations thus obtained. The equations (59) represent the general integral of this system, since by hypothesis we can choose the constants a and b in such a way that the curve Γ passes through any point (x_0, y_0, z_0) of space. If through this point there pass n curves Γ, the equations (59) determine n systems of values for a and b. The equations (63) determine y' and z', and we see that for the point (x_0, y_0, z_0) the equations (64) determine n systems of values for y' and z'. But the edges of regression C are also integral curves of the equations (64), since in a point of C the values of x, y, z, y', z' are the same for C and for the curve Γ tangent to C at that point. The equations (64) have, therefore, besides the integrals represented by curves Γ, an infinite number of other integrals, not included in the equations (59), which are obtained by integrating the equation of the first order (61); these are the *singular integrals* of the system.

On closer examination we see that the existence of the focal surfaces does not in reality require that the curves Γ shall be algebraic. It is sufficient that, in the neighborhood of a system of solutions $(x_0, y_0, z_0, a_0, b_0)$ of the three equations

(65) $\quad F(x, y, z, a, b) = 0, \quad \Phi(x, y, z, a, b) = 0, \quad \dfrac{D(F, \Phi)}{D(a, b)} = 0,$

the implicit functions x, y, z of the parameters a and b, defined by these three equations, which reduce to x_0, y_0, z_0, for $a = a_0, b = b_0$, shall be continuous and have continuous derivatives in the neighborhood. In fact, let

(66) $\quad x = f_1(a, b), \quad y = f_2(a, b), \quad z = f_3(a, b)$

be these three functions. The sheet of the focal surface which passes through the point (x_0, y_0, z_0) is represented in the neighborhood of this point by the equations (66), where the values of the parameters a and b are near a_0 and b_0. It is easy to derive from this the equation of the plane tangent to the focal surface. In fact, when the point x, y, z describes any curve on this surface, x, y, z, a, b are functions of a single independent variable which satisfy the equations (65); hence the differentials of these functions satisfy the two relations

$$\dfrac{\partial F}{\partial x}\delta x + \dfrac{\partial F}{\partial y}\delta y + \dfrac{\partial F}{\partial z}\delta z + \dfrac{\partial F}{\partial a}\delta a + \dfrac{\partial F}{\partial b}\delta b = 0,$$

$$\dfrac{\partial \Phi}{\partial x}\delta x + \dfrac{\partial \Phi}{\partial y}\delta y + \dfrac{\partial \Phi}{\partial z}\delta z + \dfrac{\partial \Phi}{\partial a}\delta a + \dfrac{\partial \Phi}{\partial b}\delta b = 0.$$

Making use of the last of the relations (65), we can eliminate δa and δb, and we find the new relation

(67) $\quad \dfrac{D(F, \Phi)}{D(x, b)}\delta x + \dfrac{D(F, \Phi)}{D(y, b)}\delta y + \dfrac{D(F, \Phi)}{D(z, b)}\delta z = 0.$

We have only to replace δx, δy, δz by $X - x_0$, $Y - y_0$, $Z - z_0$, respectively, in order to have the equation of the plane tangent to the focal surface. It is easy to show that this plane passes through the tangent to the curve Γ. The properties of the focal surface suppose, therefore, only that we can apply the theory of implicit functions to the equations (65), and in particular that the functions F, Φ, together with their partial derivatives, are continuous in the neighborhood of a system of solutions x_0, y_0, z_0, a_0, b_0. This is certainly true when F and Φ are polynomials, but it is clear that it is also true for many other functions. Let us also observe that if the curves Γ have singular points, the locus of these singular points forms a part of the focal surface. This is shown as in the case of the analogous proposition relative to plane curves (I, § 207, 2d ed.; § 201, 1st ed.).

Let us now examine the question from the opposite point of view. Given a system of two differential equations of the first order, such as the system (64), let us propose to determine whether this system has singular integrals. We shall suppose that $\mathcal{F}$ and $\mathcal{F}_1$ are polynomials. Let M_0 be any point (x_0, y_0, z_0) of space. If x, y, z are replaced by x_0, y_0, z_0, respectively, in the equations (64), these equations have in general a certain number of systems of solutions. Let y_0', z_0' be one of these systems. Let us assume first that, for this system of solutions, the Jacobian $D(\mathcal{F}, \mathcal{F}_1)/D(y', z')$ is not zero. From the equations (64), y' and z' can be found as regular functions in the neighborhood of the point (x_0, y_0, z_0),

$$y' = y_0' + \alpha(x - x_0) + \cdots, \qquad z' = z_0' + \alpha_1(x - x_0) + \cdots,$$

which reduce to y_0' and z_0', respectively, for $x = x_0$, $y = y_0$, $z = z_0$. The equations (64) have therefore an integral curve passing through the point M_0 tangent to the straight line whose equations are $Y - y_0 = y_0'(X - x_0)$, $Z - z_0 = z_0'(X - x_0)$. Moreover, this curve forms part of a family of integral curves depending upon two arbitrary parameters (§ 26). This conclusion does not hold if we have $D(\mathcal{F}, \mathcal{F}_1)/D(y_0', z_0') = 0$; but this can occur only if the coördinates (x_0, y_0, z_0) satisfy the relation

(68) $$R(x, y, z) = 0,$$

which is obtained by eliminating y' and z' from the three equations

(69) $$\mathcal{F} = 0, \qquad \mathcal{F}_1 = 0, \qquad \frac{D(\mathcal{F}, \mathcal{F}_1)}{D(y', z')} = 0.$$

The equation (68) represents a surface S, and, from what we have just seen, an integral curve which does not lie on the surface S cannot be a singular integral curve.

If the point M_0 is on the surface S, the three equations (69) have for this point a system of common solutions, $y' = y_0'$, $z' = z_0'$. If the straight line D represented by the equations

(70) $$\frac{X - x_0}{1} = \frac{Y - y_0}{y_0'} = \frac{Z - z_0}{z_0'}$$

is not tangent to S (which is the general case), there is an integral curve passing through the point M_0 and tangent to the straight line D. It has been shown that the point M_0 is in general a cusp for that curve. What is essential for us is that this integral curve cannot be on the surface, since its tangent is not in the tangent plane. In order that singular integrals may exist, in each point of

S the corresponding straight line D must therefore be situated in the plane tangent to the surface. This condition is sufficient, for then through each point of S there passes a curve lying on the surface and tangent to the line D. These curves are determined by a differential equation of the first order, and they are indeed singular integral curves, for at each of their points the values of y' and of z' form a *multiple* system of solutions of the equations (64).

Example 1. Consider the simultaneous system of equations

(71) $\qquad y - xy' = 0, \qquad x^2 z'^2 = x^2 + y^2 - 1.$

The two values of z' are equal for all the points of the cylinder $x^2 + y^2 - 1 = 0$, and the direction corresponding to that double root is the perpendicular dropped from the point (x, y) on the z-axis. Since this perpendicular is not in the tangent plane to the cylinder, there cannot be any singular integrals. In this example it is easy to verify that the cylinder is the locus of the cusps of the integral curves, for the general integral of the system (71) is represented by the equations

$$y = C_1 x, \qquad z = \sqrt{x^2 + y^2 - 1} - \arctan \sqrt{x^2 + y^2 - 1} + C_2.$$

Example 2. Every system of differential equations of the form

(72) $\qquad F(y - xy', z - xz', y', z') = 0, \qquad \Phi(y - xy', z - xz', y', z') = 0,$

which may be considered as a generalization of Clairaut's equation, is easily integrated by observing that the preceding relations lead to the equations

$$\left(\frac{\partial F}{\partial y'} - x \frac{\partial F}{\partial u}\right) y'' + \left(\frac{\partial F}{\partial z'} - x \frac{\partial F}{\partial v}\right) z'' = 0,$$

$$\left(\frac{\partial \Phi}{\partial y'} - x \frac{\partial \Phi}{\partial u}\right) y'' + \left(\frac{\partial \Phi}{\partial z'} - x \frac{\partial \Phi}{\partial v}\right) z'' = 0,$$

where $u = y - xy'$, $v = z - xz'$. These last equations are satisfied by assuming that $y'' = 0$, $z'' = 0$, or by supposing that we have

(73) $\qquad \left(\frac{\partial F}{\partial y'} - x \frac{\partial F}{\partial u}\right)\left(\frac{\partial \Phi}{\partial z'} - x \frac{\partial \Phi}{\partial v}\right) - \left(\frac{\partial F}{\partial z'} - x \frac{\partial F}{\partial v}\right)\left(\frac{\partial \Phi}{\partial y'} - x \frac{\partial \Phi}{\partial u}\right) = 0.$

Under the first supposition, y' and z' are constants a and b; whence we see that the curves which correspond to the general integral are *the straight lines of the congruence* represented by the two equations

$$F(y - ax, z - bx, a, b) = 0, \qquad \Phi(y - ax, z - bx, a, b) = 0.$$

There are also singular integrals, since the straight lines of the congruence are tangent to the two sheets of a focal surface. These singular integrals are the edges of regression of the developables of the congruence, and are obtained by the integration of a differential equation of the first order. The equation of the focal surface is obtained by eliminating y' and z' from the relations (72) and (73).

EXERCISES

1. Examine the following differential equations for singular solutions:

$$y'^2 + \left(x + \frac{x^3}{2}\right) y' - (1 + x^2) y - \frac{x^4}{16} = 0. \qquad [\text{Serret.}]$$
$$xy^2 y'^2 - y^3 y' + a^2 x = 0. \qquad [\text{Schlömilch.}]$$
$$y'^2 - 2x\sqrt{y}\, y' + 4y\sqrt{y} = 0. \qquad [\text{Boole.}]$$
$$(xy' - y)^2 - 2xy(1 + y'^2) = 0. \qquad [\text{Hoüel.}]$$
$$2xy(1 + y'^2) - (xy' + y)^2 = 0. \qquad [\text{Moigno.}]$$

2*. The equation $\Pi(x, y) = 0$, obtained by eliminating y' between the two relations $F(x, y, y') = 0$, $\partial F/\partial x + \partial F/\partial y\, y' = 0$, represents the locus of the points of inflection of the integral curves.

Deduce from this the theorem of § 72, in regard to the locus of the cusps of the integral curves, by means of a transformation of reciprocal polars.

[Darboux, *Bulletin des Sciences mathématiques*, Vol. IV, 1873.]

3. Determine the singular integrals of the system of differential equations
$$y = xy' + y'^2 + z', \qquad z = z'x + y'z'. \qquad \text{[Serret.]}$$

4. Determine whether the differential equation of the second order,
$$(1 + x^2) y''^2 - \left(2xy' + \frac{x^2}{2}\right) y'' + y'^2 + xy' - y = 0,$$
has singular integrals, and find any that exist. [Lagrange.]

[Replace this equation by a system of two equations of the first order.]

5*. Given a differential equation of the second order,
$$F(x, y, y', y'') = 0,$$
by eliminating y'' between this equation and the relation $\partial F/\partial y'' = 0$ we obtain a differential equation of the first order $P(x, y, y') = 0$, whose integrals have *in general* the following property: Through each point M of one of these integral curves C there passes an integral curve of the equation $F = 0$, which has a cusp of the second kind at M, and whose cuspidal tangent is the tangent to the curve C at this point. [*American Journal of Mathematics*, Vol. XI, p. 364.]

6. Establish the properties of e^x by starting with the general integral of the differential equation $dx/x + dy/y = 0$, written in the algebraic form $xy = C$.

Consider the same question for the function $\tan x$, finding first the general integral in algebraic form of the differential equation
$$\frac{dx}{1 + x^2} + \frac{dy}{1 + y^2} = 0.$$

7*. Let $y' = R(x, y)$, where $R(x, y)$ is a rational function of y whose coefficients are analytic functions of x, be a differential equation of the first order having a general integral of the form

(1) $$\frac{\phi_0(x) y^n + \phi_1(x) y^{n-1} + \cdots + \phi_n(x)}{\psi_0(x) y^n + \psi_1(x) y^{n-1} + \cdots + \psi_n(x)} = F(x, y) = C.$$

Prove that this equation can be reduced to a Riccati equation by a substitution of the form $u = R_1(x, y)$, where R_1 is a rational function of y. [Painlevé.]

Note. It will be noticed that the equation (1) can be written in the form
$$y^n + [A_1(x) + B_1(x)u] y^{n-1} + \cdots + [A_{n-1}(x) + B_{n-1}(x)u] y + u = 0,$$
where $u = (\phi_n - C\psi_n)/(\phi_0 - C\psi_0)$, and that u satisfies a Riccati equation, while the functions A_i, B_i are known.

8. If we seek to determine the function $f(\alpha)$ so that the envelope of the straight lines $x \cos \alpha + y \sin \alpha = f(\alpha)$ shall be a given curve C, we are led to a differential equation whose general integral is represented by the straight lines which pass through a fixed point of C. The true solution is furnished by the singular integral.

CHAPTER V

PARTIAL DIFFERENTIAL EQUATIONS OF THE FIRST ORDER

This chapter is devoted to the theory of partial differential equations of the first order. We shall consider for the most part the reduction of the integration of an equation of this type to that of a system of ordinary differential equations. Although this reduction is not, in many cases, of any practical utility, it nevertheless possesses great theoretical interest, for it enables us to determine just how difficult the problem is. Although not all the arguments require that the integrals considered shall be analytic, we shall restrict ourselves to that case unless the contrary is particularly stated.

I. LINEAR EQUATIONS OF THE FIRST ORDER

75. General method. We have already seen that the integration of the homogeneous equation

$$(1) \qquad X_1 \frac{\partial f}{\partial x_1} + X_2 \frac{\partial f}{\partial x_2} + \cdots + X_n \frac{\partial f}{\partial x_n} = 0,$$

where $X_1, X_2, \cdots, X_n$ are functions of $x_1, x_2, \cdots, x_n$, and the integration of the system of differential equations

$$(2) \qquad \frac{dx_1}{X_1} = \frac{dx_2}{X_2} = \cdots = \frac{dx_n}{X_n}$$

are equivalent problems (§ 31). If $f_1, f_2, \cdots, f_{n-1}$ are $(n-1)$ independent first integrals of the system (2), the general integral of the equation (1) is an arbitrary function,

$$\Phi(f_1, f_2, \cdots, f_{n-1}),$$

of these $(n-1)$ integrals.

We can obtain the integral satisfying the Cauchy condition as follows: Suppose that the coefficients X_i are analytic in the neighborhood of a particular system of values $x_1^0, x_2^0, \cdots, x_n^0$, and that the first coefficient $(X_1)_0$ does not vanish at that point. Solving equation (1) with respect to $\partial f/\partial x_1$, we can apply to it the general theorem of § 25. Hence there exists an analytic integral in the neighborhood mentioned, which reduces, for $x_1 = x_1^0$, to a given analytic function

$\phi(x_2, x_3, \cdots, x_n)$ of the $(n-1)$ variables $x_2, x_3, \cdots, x_n$. In order to obtain this integral, let us write the system (2) in the form

(3) $$\frac{dx_2}{dx_1} = \frac{X_2}{X_1}, \quad \cdots, \quad \frac{dx_n}{dx_1} = \frac{X_n}{X_1},$$

where the right-hand sides are analytic in the neighborhood of the point $(x_1^0, x_2^0, \cdots, x_n^0)$. There exists a system of analytic integrals reducing to given values $C_2, C_3, \cdots, C_n$ for $x_1 = x_1^0$, provided that each of the absolute values $|C_i - x_i^0|$ is less than a certain limit, and these integrals are analytic functions of x_1 and of the parameters $C_2, C_3, \cdots, C_n$ (§ 26), which are represented by developments of the form

(4) $\quad x_i = C_i + (x_1 - x_1^0) P_i(x_1, C_2, C_3, \cdots, C_n). \quad (i = 2, 3, \cdots, n)$

Solving these $(n-1)$ equations for the C_i's, we obtain a system of $(n-1)$ first integrals of the equations (2), represented by the developments

(5) $\quad C_i = x_i + (x_1 - x_1^0) Q_i(x_1, x_2, \cdots, x_n), \quad (i = 2, 3, \cdots, n)$

where the Q_i's are analytic functions. It is clear that the function $\phi(C_2, C_3, \cdots, C_n)$ of these $(n-1)$ first integrals is analytic in the neighborhood of the point $(x_1^0, \cdots, x_n^0)$ and reduces to $\phi(x_2, x_3, \cdots, x_n)$ for $x_1 = x_1^0$.

Let us now consider any linear equation

(6) $$P_1 \frac{\partial z}{\partial x_1} + P_2 \frac{\partial z}{\partial x_2} + \cdots + P_n \frac{\partial z}{\partial x_n} - R = 0,$$

where $P_1, P_2, \cdots, P_n, R$ may depend both upon the independent variables $x_1, x_2, \cdots, x_n$ and upon the dependent variable z. We shall reduce this equation to the form (1) by means of a device very often used in the study of partial differential equations. Instead of trying to find the unknown function z directly, we shall try to define it by means of an equation not solved for z,

(7) $$V(z, x_1, x_2, \cdots, x_n) = 0,$$

where the function V of the $(n+1)$ variables $z, x_1, x_2, \cdots, x_n$ is now the unknown function. From this relation we derive, by differentiation,

$$\frac{\partial V}{\partial x_1} + \frac{\partial V}{\partial z} \frac{\partial z}{\partial x_1} = 0, \quad \cdots, \quad \frac{\partial V}{\partial x_n} + \frac{\partial V}{\partial z} \frac{\partial z}{\partial x_n} = 0;$$

and, replacing $\partial z/\partial x_1, \cdots, \partial z/\partial x_n$ by the values derived from the preceding relations, the equation (6) becomes

(8) $$F(V) = P_1 \frac{\partial V}{\partial x_1} + \cdots + P_n \frac{\partial V}{\partial x_n} + R \frac{\partial V}{\partial z} = 0.$$

The new equation is of the form (1), and its integration is equivalent to that of the system

$$\text{(9)} \qquad \frac{dx_1}{P_1} = \frac{dx_2}{P_2} = \cdots = \frac{dx_n}{P_n} = \frac{dz}{R};$$

hence we may state the following proposition:

If $u_1, u_2, \ldots, u_n$ are n independent first integrals of the system (9), every function z of the n variables $x_1, x_2, \ldots, x_n$, defined by a relation of the form

$$\text{(10)} \qquad \Phi(u_1, u_2, \ldots, u_n) = 0,$$

where Φ indicates an arbitrary function of $u_1, u_2, \ldots, u_n$, is an integral of the equation (6).

We cannot conclude from this that we obtain all the integrals of the equation (6) in this way. In fact, in order that the implicit function defined by the relation (7) be an integral, it is not necessary that we have identically $F(V) = 0$; it is sufficient that the equation $F(V) = 0$ be a consequence of the equation $V = 0$. If, for example, we take for V an integral of an equation of the form $F(V) = KV$, where K indicates a constant different from zero, the relation $V = 0$ still defines an integral of the equation (6). It is quite in order, therefore, to determine whether or not the relation (10) gives all the integrals of the given equation. In order to prove that this is really the case, with certain exceptions which we shall state, let us suppose that in the n functions $u_1, u_2, \ldots, u_n$ we replace z by an integral of the equation (6). The resulting expressions are n functions $U_1, U_2, \ldots, U_n$ of the n variables $x_1, x_2, \ldots, x_n$. If we prove that the Jacobian of these n functions is identically zero, it will follow that we have a relation of the form

$$\psi(U_1, U_2, \ldots, U_n) = 0,$$

and consequently that the integral considered satisfies a relation of the form (10) in which the function Φ is replaced by ψ. This Jacobian is of the form

$$\Delta = \begin{vmatrix} \frac{\partial u_1}{\partial x_1} + p_1 \frac{\partial u_1}{\partial z} & \frac{\partial u_1}{\partial x_2} + p_2 \frac{\partial u_1}{\partial z} & \cdots & \frac{\partial u_1}{\partial x_n} + p_n \frac{\partial u_1}{\partial z} \\ \cdots & \cdots & \cdots & \cdots \\ \frac{\partial u_n}{\partial x_1} + p_1 \frac{\partial u_n}{\partial z} & \cdots & \cdots & \frac{\partial u_n}{\partial x_n} + p_n \frac{\partial u_n}{\partial z} \end{vmatrix}. \qquad \left(p_i = \frac{\partial z}{\partial x_i}\right)$$

Noting that certain determinants in the development of Δ have two columns identical and therefore vanish, we may write

$$(11) \quad \Delta = \frac{D(u_1, u_2, \cdots, u_n)}{D(x_1, x_2, \cdots, x_n)} + \sum_{i=1}^{i=n} P_i \frac{D(u_1, u_2, \cdots, u_n)}{D(x_1, \cdots, x_{i-1}, z, x_{i+1}, \cdots, x_n)}.$$

But, since $u_1, u_2, \cdots, u_n$ are n first integrals of the system (9), we have

$$P_1 \frac{\partial u_i}{\partial x_1} + P_2 \frac{\partial u_i}{\partial x_2} + \cdots + P_n \frac{\partial u_i}{\partial x_n} + R \frac{\partial u_i}{\partial z} = 0; \quad (i = 1, 2, \cdots, n)$$

hence, by the theory of linear homogeneous equations, we have

$$(12) \quad \frac{R}{\dfrac{D(u_1, u_2, \cdots, u_n)}{D(x_1, x_2, \cdots, x_n)}} = \frac{-P_i}{\dfrac{D(u_1, u_2, \cdots, u_n)}{D(x_1, \cdots, x_{i-1}, z, x_{i+1}, \cdots, x_n)}} = M,$$

$$(i = 1, 2, \cdots, n)$$

where M is a function of $x_1, x_2, \cdots, x_n, z$ which we can always calculate when we know the first integrals $u_1, u_2, \cdots, u_n$. Substituting in (11) the values of the determinants deduced from (12), we find

$$(12') \quad M\Delta = R - P_1 p_1 - P_2 p_2 - \cdots - P_n p_n.$$

If z is an integral of the equation (6), the right-hand side is zero; hence this integral satisfies either the condition $\Delta = 0$ or else $M = 0$. In the first case, as we have just shown, this integral is defined by a relation of the form (10). As for the relation $M = 0$, it can define only one or more completely determined implicit functions. Hence, except for certain exceptional integrals which do not depend upon any arbitrary constant, all the integrals of the equation (6) satisfy a relation of the form (10). We shall hereafter say that the relation (10) represents the *general integral* of the equation (6).

To see if an integral can satisfy the relation $M = 0$, let us consider any point of that integral, $(x_1^0, x_2^0, \cdots, x_n^0, z_0)$, and let us suppose that all the coefficients $P_1, P_2, \cdots, P_n, R$ are analytic in the neighborhood of this system of values without being all zero simultaneously for $x_i = x_i^0, z = z_0$. Let us assume, for example, that P_1 is not zero for this system of values. We can then solve the equation (8) for $\partial V/\partial x_1$, and, by Cauchy's theorems (§ 25), we can take for $u_1, u_2, \cdots, u_n$ functions analytic in the neighborhood of this system of values. Now one of the equations (12) can be written in the form

$$-P_1 = M \frac{D(u_1, u_2, \cdots, u_n)}{D(z, x_2, \cdots, x_n)}.$$

Since the determinant on the right is analytic, and since P_1 is not zero for $x_i = x_i^0, z = z_0$, it follows that this system of values cannot make M zero. Since the point $(x_1^0, \cdots, x_n^0, z_0)$ is any point of the integral, we see that there cannot exist integrals satisfying the relation $M = 0$ except in the two following cases:

1) There exists a function $V(x_1, x_2, \cdots, x_n, z)$ such that every system of values of the variables x_i, z that makes the function V vanish, also causes P_1, $P_2, \cdots, P_n$, and R to vanish. All these coefficients are therefore divisible by the same factor, and it is clear that by equating this factor to zero we obtain an integral. This trivial case is of slight interest.

2) The reasoning would again be faulty if the integral defined by the relation $V = 0$ were such that, in the neighborhood of every system of values satisfying that relation, some of the coefficients P_i, R ceased to be analytic. This case can actually occur, as we shall show presently.

76. Geometric interpretation. The preceding general method is susceptible of a simple geometric interpretation in the case of an equation in three variables, which we shall write in the customary notation,

$$(13) \qquad Pp + Qq = R, \qquad p = \frac{\partial z}{\partial x}, \qquad q = \frac{\partial z}{\partial y},$$

where P, Q, R are functions of the three variables x, y, z. Let S be any *integral surface*. Since the equation of the plane tangent to this surface is

$$Z - z = p(X - x) + q(Y - y),$$

the relation (13) expresses the fact that this tangent plane passes through the straight line D represented by the equations

$$(14) \qquad \frac{X - x}{P} = \frac{Y - y}{Q} = \frac{Z - z}{R}.$$

Hence the problem of the integration of the equation (13) may be stated in geometric language as follows:

To each point M of space, whose coördinates are (x, y, z), there corresponds a straight line D through that point, represented by the equations (14). A surface S is to be determined so that the tangent plane at each of its points passes through the straight line associated with that point.

The surfaces possessing this property constitute the general integral of the linear equation (13). The three functions P, Q, R determine the law according to which the straight line D moves when the point M changes its position. These three functions are usually analytic functions of x, y, z, but it is sufficient for the argument that they satisfy the conditions stated in our previous study of differential equations (§§ 27 ff.).

The preceding statement leads us to seek the curves Γ which are in each of their points tangent to the corresponding straight line D. We shall call these the *characteristic curves*. We shall first show that *every integral surface is generated by characteristic curves*. Consider, in fact, such a surface S. In each point M of that surface the corresponding straight line D lies in the tangent plane. We can therefore propose to determine the curves on that surface which are tangent at each of their points to the corresponding straight line D. These curves may be obtained by the integration of a differential equation of the first order (§ 17). Through each point of S there passes in general one and only one curve, possessing this property, which lies entirely on the surface. It is clear that these curves are characteristic curves, which proves the proposition.

The converse is almost self-evident. If a surface is a locus of characteristic curves, the tangent plane at any one of its points contains the tangent to the characteristic curve lying upon the surface and passing through that point — that is, the straight line D. The given problem is therefore reduced to the determination of the characteristic curves.

The differential equations of these curves, by their very definition, are of the form

(15)
$$\frac{dx}{P} = \frac{dy}{Q} = \frac{dz}{R}.$$

Through each point of space there passes, therefore, in general one and only one characteristic curve tangent to the corresponding straight line D. Suppose that we have integrated these equations (15). Let u and v be two independent first integrals of this system. The general integral is represented by the equations

(16) $\qquad u(x, y, z) = a, \qquad v(x, y, z) = b,$

where a and b are two arbitrary constants. The characteristic curves, which depend upon two parameters, therefore form a *congruence*. In order to obtain a surface generated by the curves of this congruence, we must establish between the two parameters a and b an arbitrary relation, say $\phi(a, b) = 0$, and the corresponding integral surface will have for its equation $\phi(u, v) = 0$. This is exactly the result to which the general method of the preceding paragraph would lead us, for u and v are here two independent integrals of the equation

$$P\frac{\partial u}{\partial x} + Q\frac{\partial u}{\partial y} + R\frac{\partial u}{\partial z} = 0.$$

Example 1. Consider the equation $px + qy = mz$. The differential equations of the characteristic curves,

$$\frac{dx}{x} = \frac{dy}{y} = \frac{dz}{mz},$$

have the two first integrals $y/x = a$, $z/x^m = b$, and the general equation of the integral surfaces is $z = x^m f(y/x)$. If $m = 1$, the characteristic curves are straight lines passing through the origin, and the integral surfaces are cones having their vertices at the origin. If $m = 0$, the characteristic curves are straight lines parallel to the xy-plane and meeting the z-axis. The integral surfaces are conoids.

Example 2. Consider the equation $py - qx + a = 0$. The differential equations of the characteristic curves,

$$\frac{dx}{y} = \frac{dy}{-x} = \frac{dz}{-a},$$

give the two integrable combinations

$$x\,dx + y\,dy = 0, \qquad dz - a\frac{x\,dy - y\,dx}{x^2 + y^2} = 0,$$

and the characteristic curves are represented by the equations

$$x^2 + y^2 = C_1, \qquad z - a \arctan\frac{y}{x} = C_2.$$

These are helices with the pitch $2\pi a$ lying upon cylinders of revolution having Oz for axis, and the general integral is represented by helicoids (the axes of coördinates being supposed rectangular). In the particular case where $a = 0$, the characteristic curves are circles having their centers on the z-axis and their planes parallel to the xy-plane. The integral surfaces are surfaces of revolution about the z-axis.

Example 3. *Orthogonal trajectories.* Let

(17) $$F(x, y, z) = C$$

be the equation of a family of surfaces Σ which depend upon an arbitrary parameter C in such a way that through every point of space (or at least of a portion of space) there passes one and only one of these surfaces. Let us consider the problem of finding another surface S, represented by the equation

$$z = \phi(x, y),$$

which cuts orthogonally at each of its points the surface Σ through that point. Since the direction cosines of the normals to the two surfaces are respectively proportional to $\partial F/\partial x$, $\partial F/\partial y$, $\partial F/\partial z$ for Σ, and to p, q, -1 for S, the condition of orthogonality leads to the linear equation

(18) $$p\frac{\partial F}{\partial x} + q\frac{\partial F}{\partial y} - \frac{\partial F}{\partial z} = 0.$$

The characteristic curves, whose differential equations are

(19) $$\frac{dx}{\frac{\partial F}{\partial x}} = \frac{dy}{\frac{\partial F}{\partial y}} = \frac{dz}{\frac{\partial F}{\partial z}},$$

are the curves tangent at each of their points to the normal to the surface Σ through that point.

Suppose, for example, that we have $F(x, y, z) = zf(x, y)$, where $f(x, y)$ is a homogeneous function of the mth degree. The differential equations of the characteristic curves are here

$$\frac{dx}{f'_x} = \frac{dy}{f'_y} = \frac{z\,dz}{f}.$$

By Euler's relation, we have the integrable combination

$$x\,dx + y\,dy - mz\,dz = 0,$$

from which we derive the first integral $x^2 + y^2 - mz^2 = a$. On the other hand, dy/dx is a homogeneous function of degree zero in the variables x, y. Hence we can obtain a new first integral by a quadrature (§ 3).

Example 4. It is sometimes possible to determine the characteristic curves without any calculation, merely from their geometric definition. Let it be required, for example, to determine the surfaces S such that *the tangent plane at any point M of one of these surfaces meets a fixed straight line Δ in a point T, equally distant from the point M and from a fixed point O on the straight line Δ.*

Let M be a point in space; there exists on the straight line Δ one and only one point T such that $TO = TM$, and this point is the intersection of Δ with the plane perpendicular to the segment OM at its middle point. Let D be the straight line through the two points M and T. The tangent plane to every surface satisfying the given condition and passing through the point M therefore contains this straight line D. Consequently these surfaces are obtained by the integration of a linear equation. Since the tangents to the characteristic curves all meet the straight line Δ, these curves are plane curves, lying in planes passing through Δ. The characteristic curves lying in one of these planes are the integral curves of a differential equation of the first order, and it is easy to see, from their definition, that they are circles tangent to the straight line Δ at O. The required surfaces are therefore generated by the circles tangent at O to the straight line Δ.

We can dispose of the arbitrary function $\phi(u, v)$ in such a way that the integral surface passes through a given curve Γ; we shall obtain that surface by taking the locus of the characteristic curves passing through the different points of the given curve. If Γ is represented by the system of two equations

(20) $\qquad \Phi(x, y, z) = 0, \qquad \Phi_1(x, y, z) = 0,$

the whole question reduces to finding the relation which must hold between the parameters a and b in order that a characteristic curve shall meet the curve Γ. It is clear that that relation may be found by eliminating x, y, z between the equations (20) and the equations $u = a$, $v = b$ of the characteristic curve. The problem has only one solution, unless the curve Γ is itself a characteristic curve. In this singular case it suffices, in order to obtain an integral surface passing through Γ, to consider the surface generated by a family of characteristic curves which depend upon an arbitrary parameter, and of which the curve Γ is a member.

77. Congruences of characteristic curves.

To every linear equation of the form (13) there corresponds a *congruence of characteristic curves* formed by the characteristic curves of that equation. Conversely, every congruence of curves, that is, every family of curves depending upon two arbitrary parameters a and b, is the congruence of characteristic curves for an equation of the form (13).* Suppose, in fact, that the equations which define that congruence are solved for the two parameters a and b:

$$u(x, y, z) = a, \qquad v(x, y, z) = b.$$

Every surface S generated by the curves of this congruence, associated according to an arbitrary law, is represented by an equation of the form $v = \pi(u)$. Taking the partial derivatives with respect to x and to y, we find

$$\frac{\partial v}{\partial x} + \frac{\partial v}{\partial z} p = \pi'(u)\left(\frac{\partial u}{\partial x} + \frac{\partial u}{\partial z} p\right), \qquad \frac{\partial v}{\partial y} + \frac{\partial v}{\partial z} q = \pi'(u)\left(\frac{\partial u}{\partial y} + \frac{\partial u}{\partial z} q\right).$$

The elimination of $\pi'(u)$ leads to a linear equation

$$\frac{D(u, v)}{D(z, y)} p + \frac{D(u, v)}{D(x, z)} q + \frac{D(u, v)}{D(x, y)} = 0,$$

for which the given congruence is evidently the congruence of characteristic curves.

Let us now consider the general case of a congruence defined by two equations of any form whatever,

(21) $\qquad U(x, y, z, a, b) = 0, \qquad V(x, y, z, a, b) = 0.$

If we set up an arbitrary relation $\phi(a, b) = 0$ between the two parameters a and b, we shall have the equation of a surface S generated by the curves Γ of the congruence by eliminating a and b from the equations (21) and the relation $\phi = 0$. All these surfaces again satisfy, whatever may be the function ϕ, the same partial differential equation of the first order. To obtain this equation we may proceed as follows: The three equations

(22) $\qquad\qquad U = 0, \qquad V = 0, \qquad \phi(a, b) = 0$

define three implicit functions z, a, b of the independent variables x and y, and the last contains only a and b. Hence we have

(23) $\qquad\qquad\qquad \dfrac{D(a, b)}{D(x, y)} = 0.$

---

* We suppose, in addition, that through any point of space (or of a portion of space) passes one of these curves, which would not happen if they were all situated the same surface.

On the other hand, if we differentiate the first two of the equations (22) with respect to x and to y, we can derive from the resulting relations expressions for $\partial a/\partial x$, $\partial b/\partial x$, $\partial a/\partial y$, $\partial b/\partial y$ in terms of x, y, z, p, q, a, b, and, by replacing these derivatives in the determinant (23) by their values, we obtain a new relation,

$$\Phi(x, y, z, p, q, a, b) = 0.$$

We need only eliminate a and b from this relation and the two relations (21) in order to obtain an equation containing only x, y, z, p, q,

(24) $$F(x, y, z, p, q) = 0,$$

which applies to all the surfaces generated by the curves of the congruence. It is easy to show, from the very way in which this equation has been obtained, that it breaks up into a system of linear equations in p and q. The same fact results from its meaning. Let us suppose, for definiteness, that through a point M of space there pass m curves of the congruence, and let $D_1, D_2, \cdots, D_m$ be the m tangents to these curves at the point M. Every surface through the point M generated by the curves of the congruence must contain one of the m curves of this congruence which pass through M; hence the tangent plane at the point M must pass through one of the straight lines $D_1, D_2, \cdots, D_m$. Let P_i, Q_i, R_i be proportional to the direction cosines of the straight line D_i. Every surface generated by the curves of the congruence must therefore satisfy one of the m equations,

(25) $$E_i = P_i p + Q_i q - R_i = 0, \qquad (i = 1, 2, \cdots, m)$$

and the left-hand side of the equation (24) is identical, except for a factor independent of p and of q, with the product of the m linear factors $E_1, E_2, \cdots, E_m$. It should be noticed also that it would be impossible, in general, to separate these m factors analytically.

Similarly, certain problems of geometry may lead to partial differential equations of the first order which decompose into a product of linear factors. Let us consider again, for example, the problem of the orthogonal trajectories to a family of surfaces whose equation $F(x, y, z, C) = 0$ is of degree m in the arbitrary parameter C. To obtain the partial differential equation of orthogonal surfaces, we must again eliminate C between the relation $F = 0$ and the condition

$$p \frac{\partial F}{\partial x} + q \frac{\partial F}{\partial y} - \frac{\partial F}{\partial z} = 0.$$

Through a point M of space there pass, by hypothesis, m surfaces of the given family. Let $D_1, D_2, \cdots, D_m$ be the normals to these surfaces. The tangent plane to an orthogonal surface through M must contain one of these straight lines. Hence the partial differential equation decomposes into a system of m equations which are linear in p and q.

Conversely, given any equation of this type, to each point of space there correspond m straight lines $D_1, D_2, \cdots, D_m$, and the plane tangent to any integral surface contains one of these straight lines. If we give the name *characteristic curve* to every curve which, at each of its points, is tangent to one of the corresponding m straight lines, the reasoning employed above shows again that every integral surface is a locus of characteristic curves. To obtain the differential equations of these curves, we are not compelled to carry out the decomposition of the left-hand side of the equation into linear factors. Indeed, expressing the fact that the left-hand side is divisible by the factor $Pp + Qq - R$, we obtain equations of condition homogeneous in P, Q, R, which furnish m systems of values for the ratios of these coefficients for each point (x, y, z). Replacing P, Q, R in these conditions by the proportional quantities dx, dy, dz, we obtain the differential equations of the characteristic curves, and the integration of the partial differential equation is reduced to the integration of a system of ordinary differential equations.

The preceding theory explains very simply how a linear equation may have integrals which are not included in the general integral. Consider a partial differential equation of the form

(26) $$F(x, y, z, p, q) = 0,$$

whose left-hand side is the product of a certain number of linear factors in p and q that are not analytically distinct, and let

(27) $$\Phi\left(x, y, z, \frac{dy}{dx}, \frac{dz}{dx}\right) = 0, \qquad \Psi\left(x, y, z, \frac{dy}{dx}, \frac{dz}{dx}\right) = 0$$

be the differential equations of the characteristic curves of this system. The curves which represent the general integral of this system form a congruence, which is the congruence of the characteristic curves of the equation (26), and the general integral is represented by the surfaces generated by the curves of this congruence associated according to an arbitrary law. But it may happen that the equations (27) have singular integrals. This will happen if the congruence of the characteristic curves has a focal surface (Σ). Then through each point of this surface there passes a curve of the congruence of characteristics tangent to this surface. The plane tangent to (Σ) contains, therefore, one of the straight lines D_i relative to the point of contact, and consequently (Σ) is an integral surface of the equation (26). Moreover, it is not a member, at least in general, of the surfaces which represent the general integral; that is, it is a singular integral surface.

Consider, for example, the equation

(28) $$p(x^2 - z^2) + q(xy \pm z\sqrt{x^2 + y^2 - z^2}) = 0,$$

which in reality is equivalent to two linear equations. We can write the differential equations of the characteristic curves in the form

$$\frac{dz}{dx} = 0, \quad \left(y - x\frac{dy}{dx}\right)^2 = z^2\left[1 + \left(\frac{dy}{dx}\right)^2\right].$$

The integration is immediate, and the congruence of characteristic curves is formed by the straight lines

$$z = C, \quad y = C_1 x \pm z\sqrt{1 + C_1^2},$$

which are parallel to the xy-plane and tangent to the cone $x^2 + y^2 = z^2$. The general integral is represented by the conoid surfaces generated by these straight lines, and there is a singular integral, the cone itself.

The coefficient of q in the equation (28) is not analytic in the neighborhood of any point (x_0, y_0, z_0) of this cone, which confirms a previous remark (§ 75).

II. TOTAL DIFFERENTIAL EQUATIONS

78. The equation $dz = A\,dx + B\,dy$. The existence of integrals of a *completely integrable* system of total differential equations was established in § 24. The integration of such a system reduces to the integration of several systems of ordinary differential equations with a single independent variable. The method, which we shall develop only in the simplest case, is extensible to the general case.

Let the equation be

(29) $$dz = A(x, y, z)\,dx + B(x, y, z)\,dy,$$

where z is an unknown function of the two independent variables x and y. This equation is equivalent to two distinct relations

(30) $$\frac{\partial z}{\partial x} = A(x, y, z), \quad \frac{\partial z}{\partial y} = B(x, y, z).$$

Every integral common to these two equations satisfies also the two new equations

$$\frac{\partial^2 z}{\partial x\,\partial y} = \frac{\partial A}{\partial y} + \frac{\partial A}{\partial z}B, \quad \frac{\partial^2 z}{\partial y\,\partial x} = \frac{\partial B}{\partial x} + \frac{\partial B}{\partial z}A,$$

and consequently the relation

(31) $$\frac{\partial A}{\partial y} + \frac{\partial A}{\partial z}B = \frac{\partial B}{\partial x} + \frac{\partial B}{\partial z}A.$$

If this relation does not reduce to an identity, there can be no integrals of the given equation (29), except possibly one or more of the implicit functions defined by the equation (31). Hence in this case we can always determine by substitution whether the equations (30) have a common integral. On the other hand, in order that these equations may have an infinite number of integrals depending upon an arbitrary constant, the relation (31) must be satisfied identically. If it is, the equation (29) is said to be *completely integrable.*

In order to obtain all its integrals, let us first disregard the second of the equations (30), and consider only the first. If we regard y as a parameter, this equation is a differential equation of the first order between the independent variable x and the dependent variable z; hence it has an infinite number of integrals $z = \phi(x, y, C)$ that depend upon an arbitrary constant C. We may replace this constant C by any function $u(y)$ of the variable y, since the expression for $\partial z/\partial x$ remains the same when we replace C by a function of y. The solution of the problem therefore depends upon the determination of this function $u(y)$ in such a way that the derivative of the function $z = \phi[x, y, u(y)]$ with respect to y shall be equal to $B(x, y, \phi)$. This leads to the equation

$$\frac{\partial \phi}{\partial y} + \frac{\partial \phi}{\partial u}\frac{du}{dy} = B[x, y, \phi(x, y, u)],$$

or

(32) $$\frac{du}{dy} = \frac{B[x, y, \phi(x, y, u)] - \dfrac{\partial \phi}{\partial y}}{\dfrac{\partial \phi}{\partial u}}.$$

We shall show that the right-hand side of this equation depends only upon the variables y and u. It is sufficient to show that the derivative with respect to x is identically zero, that is, that we have

(33) $$\frac{\partial \phi}{\partial u}\left(\frac{\partial B}{\partial x} + \frac{\partial B}{\partial \phi}\frac{\partial \phi}{\partial x} - \frac{\partial^2 \phi}{\partial x\, \partial y}\right) - \left[B(x, y, \phi) - \frac{\partial \phi}{\partial y}\right]\frac{\partial^2 \phi}{\partial u\, \partial x} = 0.$$

From the very manner in which the function $\phi(x, y, u)$ has been obtained, we have the relation

(34) $$\frac{\partial \phi}{\partial x} = A(x, y, \phi),$$

which is satisfied for all values of x, y, and u. It follows that we may write

$$\frac{\partial^2 \phi}{\partial x\, \partial y} = \frac{\partial A}{\partial y} + \frac{\partial A}{\partial \phi}\frac{\partial \phi}{\partial y},$$

$$\frac{\partial^2 \phi}{\partial x\, \partial u} = \frac{\partial A}{\partial \phi}\frac{\partial \phi}{\partial u}.$$

Replacing $\partial\phi/\partial x$, $\partial^2\phi/\partial x\,\partial y$, $\partial^2\phi/\partial u\,\partial x$ by the preceding values, the relation to be verified reduces to the form

$$\frac{\partial\phi}{\partial u}\left(\frac{\partial B}{\partial x}+\frac{\partial B}{\partial\phi}A-\frac{\partial A}{\partial y}-\frac{\partial A}{\partial\phi}B\right)=0.$$

The second factor is identically zero by the condition of integrability (31). The equation (32) is therefore of the form

(35) $$\frac{du}{dy}=F(y,u).$$

Let $u=\psi(y,C)$ be the general integral of this equation, where C is a constant independent both of x and of y. Then if we replace u by $\psi(y,C)$ in the function $\phi(x,y,u)$, we obtain the general integral of the completely integrable equation (29), and we see that *the integration of this equation reduces to the successive integrations of two ordinary differential equations* (34) *and* (35).

Example. Consider the total differential equation

(36) $$dz=\frac{1+yz}{1+xy}dx+\frac{x(z-x)}{1+xy}dy,$$

which is equivalent to the system

(36′) $$\frac{\partial z}{\partial x}=\frac{1+yz}{1+xy},\quad \frac{\partial z}{\partial y}=\frac{x(z-x)}{1+xy}.$$

The condition of integrability is verified, and the first of the equations (36′), which is linear in z and $\partial z/\partial x$, has for its general integral

$$z=-\frac{1}{y}+u(y)(1+xy),$$

where $u(y)$ is an arbitrary function of y. Substituting this value of z in the second of the equations (36′), it becomes $du/dy+1/y^2=0$, whence we derive $u(y)=1/y+C$. Hence the general integral of the equation (36) is

(37) $$z=x+C(1+xy),$$

where C indicates an arbitrary constant.

The preceding problem can also be interpreted geometrically. In order to simplify the statement, we shall again call an *integral surface* any surface represented by an equation $z=f(x,y)$, where the function $f(x,y)$ is an integral of the equation (29). The two conditions (30), or

$$p=A(x,y,z),\quad q=B(x,y,z),$$

express the fact that the tangent plane to the integral surface S at a point (x, y, z) of that surface coincides with the plane P whose equation is

(38) $$Z - z = A(X - x) + B(Y - y),$$

so that the problem of the integration of the equation (29) is equivalent to the following geometric problem:

To each point of space (x, y, z) there corresponds a plane P through that point, which is represented by the equation (38). It is required to find the surfaces S whose tangent plane at each point (x, y, z) is the plane P associated with that point.

The proposition is analogous to that of § 76. But in the present case the problem does not always have a solution. If the condition of integrability (31) is satisfied, there exists, in general, one and only one integral of the equation (29) which takes on a given value z_0 when x and y take on given values x_0 and y_0. Through every point in space there passes, therefore, in general, one and only one integral surface.

Let us consider, for example, a family of skew curves Γ which depend upon two arbitrary parameters a and b, and which are represented by a system of two equations

(39) $$u(x, y, z) = a, \qquad v(x, y, z) = b$$

such that through every point of space (or of a region of space) there passes one and only one curve of this family. There does not always exist a family of surfaces S which has these curves Γ for orthogonal trajectories. In fact, the tangent plane to the surface S passing through a point would have to coincide with the normal plane to the curve Γ passing through the same point. We are therefore led to a particular case of the preceding problem, which proves that the curves of an arbitrarily assigned congruence of curves are not, in general, the orthogonal trajectories of any family of surfaces. The plane tangent to the surface S through the point (x, y, z) must be perpendicular to the planes tangent to the two surfaces (39) which pass through the tangent to the curve Γ. Hence we have, in rectangular coördinates, the two conditions

$$\frac{\partial u}{\partial x} p + \frac{\partial u}{\partial y} q - \frac{\partial u}{\partial z} = 0, \qquad \frac{\partial v}{\partial x} p + \frac{\partial v}{\partial y} q - \frac{\partial v}{\partial z} = 0.$$

From these equations the values of p and q are found to be

$$p = A(x, y, z), \qquad q = B(x, y, z),$$

and the condition (31) must be satisfied identically in order that the problem have a solution.

Let us take, for example, the family of curves
$$X = aZ, \qquad Y = bZ,$$
where X is a function of x alone, and Y and Z are respectively functions of y alone and of z alone. The preceding method gives the following values for p and q,
$$p = -\frac{XZ'}{X'Z}, \qquad q = -\frac{YZ'}{Y'Z},$$
and the total differential equation can be written in the form
$$\frac{Z\,dz}{Z'} + \frac{X\,dx}{X'} + \frac{Y\,dy}{Y'} = 0.$$
It is clear that this equation is completely integrable, and the general integral is obtained by quadratures
$$\int \frac{Z}{Z'}\,dz + \int \frac{X}{X'}\,dx + \int \frac{Y}{Y'}\,dy = C.$$

79. Mayer's method. The preceding method requires two successive integrations. We can replace these two integrations by a single integration, as follows: Let us suppose, for definiteness, that the coefficients $A(x, y, z)$ and $B(x, y, z)$ are analytic in the neighborhood of the point (x_0, y_0, z_0). Then there exists one and only one integral surface S_0 through the point (x_0, y_0, z_0) if the condition (31) is satisfied. Mayer's method for obtaining this surface reduces to determining first the sections cut from that surface by the planes parallel to the z-axis through the point (x_0, y_0, z_0). Let Γ be the intersection of S_0 with the plane

(40) $$y - y_0 = m(x - x_0),$$

where m has any given value. Along this curve Γ we have $dy = m\,dx$, and, replacing y and dy in the equation (29) by the preceding values, we obtain the relation

(41) $$dz = \{A[x, y_0 + m(x - x_0), z] + mB[x, y_0 + m(x - x_0), z]\}\,dx,$$

which is also satisfied along the whole length of the curve Γ. Now this is a relation containing only the two variables x and z; that is, it is a differential equation of the first order, the integration of which determines the curve Γ. Let

(42) $$z = \phi(x\,;\,x_0, y_0, z_0, m)$$

be the integral of this equation which reduces to z_0 for $x = x_0$. The curve Γ is represented by the two equations (40) and (42). Since the required surface S_0 is the locus of the curves Γ as the parameter m varies, the equation of this surface is obtained by eliminating m from the equations (40) and (42). To accomplish this it is sufficient to replace m in the equation (42) by $(y - y_0)/(x - x_0)$. This method presents an evident analogy with the one which has been indicated for the integration of the total differentials $P(x, y)\,dx + Q(x, y)\,dy$ (I, § 152). We might generalize it still further by replacing the planes parallel to the z-axis by cylinders passing through a given point (x_0, y_0, z_0) and having their generators parallel to Oz.

For example, let us again take the equation (36), and let us suppose $x_0 = y_0 = 0$. Substituting $y = mx$, $dy = m\,dx$, that equation becomes

$$\frac{dz}{dx} = \frac{2mxz}{1+mx^2} + \frac{1-mx^2}{1+mx^2}.$$

This is a linear equation which is readily integrated, and the integral which reduces to z_0 for $x = 0$ has the form

$$z = x + z_0(1 + mx^2).$$

Hence the surface S_0 has the equation $z = x + z_0(1 + xy)$, which is the result obtained by the first method.

80. The equation $P\,dx + Q\,dy + R\,dz = 0$. The problem of the integration of a total differential equation can be put in a more general and more symmetrical form. Let $P(x, y, z)$, $Q(x, y, z)$, $R(x, y, z)$ be three functions of the variables x, y, z. To integrate the equation

(43) $\qquad P(x, y, z)\,dx + Q(x, y, z)\,dy + R(x, y, z)\,dz = 0$

is to find a relation $F(x, y, z) = 0$ between x, y, z such that these three variables and their differentials dx, dy, dz satisfy the given relation. If the function F contains the variable z, we may regard x and y in it as two independent variables and z as a function of these two variables, and we see that that function must satisfy the equation

$$dz = -\frac{P}{R}\,dx - \frac{Q}{R}\,dy,$$

which is of the form (29). Replacing A by $-P/R$ and B by $-Q/R$, and carrying out the differentiations, the condition of integrability (31) becomes

(44) $\qquad P\left(\dfrac{\partial Q}{\partial z} - \dfrac{\partial R}{\partial y}\right) + Q\left(\dfrac{\partial R}{\partial x} - \dfrac{\partial P}{\partial z}\right) + R\left(\dfrac{\partial P}{\partial y} - \dfrac{\partial Q}{\partial x}\right) = 0.$

This condition remains the same when we permute x, y, z and P, Q, R circularly. Hence we should have obtained the same relation if, instead of regarding z as the dependent variable, we had taken one of the variables x or y for the unknown dependent variable. The problem of the integration of the equation (43), therefore, does not differ essentially from the problem already treated; but when we write a total differential equation in this way, it is not necessary to specify which of the variables have been chosen as the independent variables.

The condition (44) arises in a question which is closely connected with the preceding. Given an expression

$$P(x, y)\,dx + Q(x, y)\,dy,$$

we have seen (§§ 12, 26) that there always exist an infinite number of factors $\mu(x, y)$ such that the product $\mu(P\,dx + Q\,dy)$ is the total differential of a function of the two variables x and y. When we pass from two to three variables, this does not remain true in general. Let us consider, in fact, three functions, P, Q, R, of the variables x, y, z. In order that the product $\mu(P\,dx + Q\,dy + R\,dz)$ be an exact differential, the factor $\mu(x, y, z)$ must satisfy the three conditions

$$\frac{\partial(\mu Q)}{\partial z} = \frac{\partial(\mu R)}{\partial y}, \quad \frac{\partial(\mu R)}{\partial x} = \frac{\partial(\mu P)}{\partial z}, \quad \frac{\partial(\mu P)}{\partial y} = \frac{\partial(\mu Q)}{\partial x}.$$

If we add these three equations, after having multiplied them by P, Q, R respectively, and then divide by μ, we find again the condition of integrability (44). This condition is therefore *necessary* in order that the trinomial $P\,dx + Q\,dy + R\,dz$ have an integrating factor. It is also *sufficient*. For if it is satisfied, the equation (43) is completely integrable. Let

(45) $$F(x, y, z) = C$$

be the general integral of this equation. The values of $\partial z/\partial x$ and of $\partial z/\partial y$ derived from the equation (45) must be identical with the values $-P/R$ and $-Q/R$ obtained from the equation (43), since we can choose the arbitrary constant C so that the integral surface passes through any point of space. For this we must have

$$\frac{F'_x}{P} = \frac{F'_y}{Q} = \frac{F'_z}{R} = \mu,$$

or

$$dF = \mu(P\,dx + Q\,dy + R\,dz).$$

The factor μ, which is equal to the common value of the preceding ratios, is therefore an integrating factor. Repeating the reasoning of § 12, we see, in a similar manner, that there are in this case an infinite number of integrating factors, which are of the form $\mu\pi(F)$, where π is an arbitrary function.

The condition of integrability (44) is *invariant* with respect to every change of variables. Consider, in fact, a transformation defined by the equations

(46) $\quad x = f(u, v, w), \quad y = \phi(u, v, w), \quad z = \psi(u, v, w),$

where the Jacobian of the functions f, ϕ, ψ with respect to u, v, w is not identically zero. This transformation carries the trinomial $P\,dx + Q\,dy + R\,dz$ into an expression of the same form, $P_1\,du + Q_1\,dv + R_1\,dw$, where P_1, Q_1, R_1 are functions of u, v, w. If now the relation (44) is satisfied, the analogous relation

(47) $\quad P_1\left(\dfrac{\partial Q_1}{\partial w} - \dfrac{\partial R_1}{\partial v}\right) + Q_1\left(\dfrac{\partial R_1}{\partial u} - \dfrac{\partial P_1}{\partial w}\right) + R_1\left(\dfrac{\partial P_1}{\partial v} - \dfrac{\partial Q_1}{\partial u}\right) = 0$

is also satisfied identically. We might verify this by a direct calculation (I, Chap. III, Ex. 19, 2d ed.; I, Chap. II, Ex. 19, 1st ed.), but it also results from the meaning of the condition. In fact, if the relation (44) is satisfied, there exist two functions $\mu(x, y, z)$ and $F(x, y, z)$ such that

$$\mu(P\,dx + Q\,dy + R\,dz) = dF.$$

If we carry out the change of variables defined by the equations (46), the functions μ and F change into two functions $\mu_1(u, v, w)$, $F_1(u, v, w)$ of the new variables, and we have identically $dF = dF_1$. Hence the preceding identity becomes

$$\mu_1(P_1\,du + Q_1\,dv + R_1\,dw) = dF_1,$$

and the trinomial $P_1\,du + Q_1\,dv + R_1\,dw$ has an integrating factor. This shows that P_1, Q_1, R_1 satisfy also the relation (47).

This remark enables us to present the method of integration of § 78 under a more general form. For let us suppose that the trinomial $P\,dx + Q\,dy + R\,dz$ has been converted by a transformation into a binomial of the form $P_1\,du + Q_1\,dv$, containing now only two differentials du and dv. In the relation (47) we must suppose $R_1 = 0$, and that relation reduces to

$$P_1 \frac{\partial Q_1}{\partial w} = Q_1 \frac{\partial P_1}{\partial w},$$

which shows that the ratio of the two coefficients P_1 and Q_1 is independent of w. The integration of the given total differential equation is therefore reduced to the integration of an equation of the form $dv + \pi(u, v)\,du = 0$, that is, to an ordinary differential equation.

Every trinomial $P\,dx + Q\,dy + R\,dz$ can be reduced to a binomial $P_1\,du + Q_1\,dv$ in an infinite number of ways. For example, we can proceed as follows: We determine first two functions, $\mu(x, y, z)$ and $F(x, y, z)$, such that, whatever dx and dy may be,

$$\frac{\partial F}{\partial x}dx + \frac{\partial F}{\partial y}dy = \mu[P(x, y, z)\,dx + Q(x, y, z)\,dy].$$

This amounts in reality to integrating the differential equation $P\,dx + Q\,dy = 0$, regarding z as a parameter. Again, we may write the preceding equation in the form

$$dF + \left(\mu R - \frac{\partial F}{\partial z}\right)dz = \mu(P\,dx + Q\,dy + R\,dz).$$

Then if we select a new system of independent variables, of which $F(x, y, z)$ and z are two, we see that $P\,dx + Q\,dy + R\,dz$ is actually replaced by an expression in which there appear only the two differentials dF and dz. This procedure can be varied in many ways. It is clear, for example, that we can begin by integrating either of the two equations

$$Q\,dy + R\,dz = 0, \qquad P\,dx + R\,dz = 0;$$

this last method is in reality identical with the method of § 78.

We can also connect with the preceding remark an elegant method due to Joseph Bertrand. Assuming that the equation (43) is completely integrable, let us begin by integrating the linear partial differential equation

$$(48) \quad X(f) = \left(\frac{\partial Q}{\partial z} - \frac{\partial R}{\partial y}\right)\frac{\partial f}{\partial x} + \left(\frac{\partial R}{\partial x} - \frac{\partial P}{\partial z}\right)\frac{\partial f}{\partial y} + \left(\frac{\partial P}{\partial y} - \frac{\partial Q}{\partial x}\right)\frac{\partial f}{\partial z} = 0.$$

Let u and v be two independent integrals of this equation. If between the two relations

$$X(u) = 0, \quad X(v) = 0$$

and the condition of integrability (44) we eliminate the three differences

$$\frac{\partial Q}{\partial z} - \frac{\partial R}{\partial y}, \quad \frac{\partial R}{\partial x} - \frac{\partial P}{\partial z}, \quad \frac{\partial P}{\partial y} - \frac{\partial Q}{\partial x},$$

we obtain the equality

$$\begin{vmatrix} \frac{\partial u}{\partial x} & \frac{\partial u}{\partial y} & \frac{\partial u}{\partial z} \\ \frac{\partial v}{\partial x} & \frac{\partial v}{\partial y} & \frac{\partial v}{\partial z} \\ P & Q & R \end{vmatrix} = 0.$$

There exist, therefore, two functions λ and μ for which we have

$$(49) \quad P = \lambda\frac{\partial u}{\partial x} + \mu\frac{\partial v}{\partial x}, \quad Q = \lambda\frac{\partial u}{\partial y} + \mu\frac{\partial v}{\partial y}, \quad R = \lambda\frac{\partial u}{\partial z} + \mu\frac{\partial v}{\partial z},$$

and we can write the given equation in the form

$$\lambda\, du + \mu\, dv = 0.$$

Now we have seen that the ratio λ/μ can depend only upon the variables u and v; hence this equation is a differential equation in u and v.

This method appears to be more complicated than the preceding, since the integration of the equation (48) requires first the integration of a system of two differential equations of the first order. But it is more symmetric, and it may be preferable if the given equation is itself symmetric in x, y, and z.

Consider, for example, the equation

$$(y^2 + yz + z^2)\,dx + (z^2 + zx + x^2)\,dy + (x^2 + xy + y^2)\,dz = 0.$$

The condition (44) is satisfied, and the linear equation (48) is here

$$(z - y)\frac{\partial f}{\partial x} + (x - z)\frac{\partial f}{\partial y} + (y - x)\frac{\partial f}{\partial z} = 0.$$

The corresponding system of differential equations,

$$\frac{dx}{z - y} = \frac{dy}{x - z} = \frac{dz}{y - x},$$

gives easily the two integrable combinations

$$d(x + y + z) = 0, \quad x\,dx + y\,dy + z\,dz = 0.$$

Hence we may take

$$u = x + y + z, \quad v = x^2 + y^2 + z^2,$$

and the values of the factors λ and μ derived from the equations (49) are

$$\lambda = x^2 + y^2 + z^2 + xy + yz + zx = \frac{u^2 + v}{2}, \quad \mu = -\frac{x + y + z}{2} = -\frac{u}{2}.$$

The transformed equation in u and v is therefore
$$(u^2 + v)\,du - u\,dv = 0,$$
or
$$du = \frac{u\,dv - v\,du}{u^2} = d\left(\frac{v}{u}\right).$$

It follows that the general integral is $u - v/u = C$, or, returning to the variables x, y, z,
$$\frac{xy + yz + zx}{x + y + z} = C.$$

81. The parenthesis (u, v) and the bracket $[u, v]$. Any total differential equation is really equivalent to two simultaneous equations
$$p = A(x, y, z), \quad q = B(x, y, z).$$
Let us now consider any two equations,

(50) $\quad F(x, y, z, p, q) = 0, \quad \Phi(x, y, z, p, q) = 0,$

in the two independent variables x and y, the unknown dependent function z, and its two partial derivatives p and q.

If we can solve these two equations for p and q, we obtain two equations, $p = f(x, y, z)$, $q = \phi(x, y, z)$, of a form which has already been studied, and it will be possible to determine whether these two relations are compatible. But we can determine whether the condition of integrability is satisfied without first solving the equations (50) for p and q. We have only to apply the rules for the calculation of the derivatives of implicit functions. Let us consider, in fact, the relations (50) as defining two implicit functions, $p = f(x, y, z)$, $q = \phi(x, y, z)$, of the three independent variables x, y, z. Differentiating with respect to x, we find

$$\frac{\partial F}{\partial x} + \frac{\partial F}{\partial p}\frac{\partial p}{\partial x} + \frac{\partial F}{\partial q}\frac{\partial q}{\partial x} = 0, \quad \frac{\partial \Phi}{\partial x} + \frac{\partial \Phi}{\partial p}\frac{\partial p}{\partial x} + \frac{\partial \Phi}{\partial q}\frac{\partial q}{\partial x} = 0,$$

and consequently
$$\frac{D(F, \Phi)}{D(p, q)}\frac{\partial q}{\partial x} + \frac{D(F, \Phi)}{D(p, x)} = 0.$$

Similarly, we have
$$\frac{D(F, \Phi)}{D(p, q)}\frac{\partial p}{\partial y} + \frac{D(F, \Phi)}{D(y, q)} = 0, \quad \frac{D(F, \Phi)}{D(p, q)}\frac{\partial p}{\partial z} + \frac{D(F, \Phi)}{D(z, q)} = 0,$$
$$\frac{D(F, \Phi)}{D(p, q)}\frac{\partial q}{\partial z} + \frac{D(F, \Phi)}{D(p, z)} = 0.$$

Substituting the values of $\partial p/\partial y$, $\partial p/\partial z$, $\partial q/\partial x$, $\partial q/\partial z$ in the condition of integrability
$$\frac{\partial p}{\partial y} + \frac{\partial p}{\partial z}q = \frac{\partial q}{\partial x} + \frac{\partial q}{\partial z}p,$$

that condition becomes, after development,

$$\frac{\partial F}{\partial p}\left(\frac{\partial \Phi}{\partial x}+p\frac{\partial \Phi}{\partial z}\right)+\frac{\partial F}{\partial q}\left(\frac{\partial \Phi}{\partial y}+q\frac{\partial \Phi}{\partial z}\right)$$
$$-\frac{\partial \Phi}{\partial p}\left(\frac{\partial F}{\partial x}+p\frac{\partial F}{\partial z}\right)-\frac{\partial \Phi}{\partial q}\left(\frac{\partial F}{\partial y}+q\frac{\partial F}{\partial z}\right)=0.$$

In general, if u and v are any functions of x, y, z, p, q, we shall set

$$\frac{d}{dx}=\frac{\partial}{\partial x}+p\frac{\partial}{\partial z}, \qquad \frac{d}{dy}=\frac{\partial}{\partial y}+q\frac{\partial}{\partial z},$$
$$[u, v]=\frac{\partial u}{\partial p}\frac{dv}{dx}-\frac{\partial v}{\partial p}\frac{du}{dx}+\frac{\partial u}{\partial q}\frac{dv}{dy}-\frac{\partial v}{\partial q}\frac{du}{dy};$$

and we shall call the expression $[u, v]$ *a bracket*. The preceding condition can then be written in an abridged form,

(51) $$[F, \Phi]=0.$$

In order that the two equations (50) shall form a completely integrable system, it must first be possible to solve them for p and q; that is, it must not be possible to derive from them a relation between x, y, z independent of p and of q; and, further, the condition $[F, \Phi]=0$ must be a consequence of the two relations (50). If the bracket $[F, \Phi]$ is identically zero, the two equations $F=a$, $\Phi=b$ form a completely integrable system for any values of the constants a and b. If the relation $[F, \Phi]=0$ is a consequence of the single equation $F=0$, independently of the second equation $\Phi=0$, the two equations $F=0, \Phi=b$ form a completely integrable system for any value of the constant b.

If the two functions F and Φ do not contain z, the expression for the bracket $[F, \Phi]$ is simplified. The following expression, where u and v are any functions of x, y, p, q,

$$(u, v)=\frac{\partial u}{\partial p}\frac{\partial v}{\partial x}-\frac{\partial v}{\partial p}\frac{\partial u}{\partial x}+\frac{\partial u}{\partial q}\frac{\partial v}{\partial y}-\frac{\partial v}{\partial q}\frac{\partial u}{\partial y},$$

is called the *parenthesis* (u, v). The condition that the two equations

$$F(x, y, p, q)=0, \qquad \Phi(x, y, p, q)=0$$

be compatible is, by what precedes, that the equation

$$(F, \Phi)=0$$

shall be satisfied, either identically or as a consequence of the relations $F=0$ and $\Phi=0$ themselves.

III. EQUATIONS OF THE FIRST ORDER IN THREE VARIABLES

82. Complete integrals. We shall now consider the integration of a partial differential equation of the first order, of any form whatever but with only two independent variables, and we shall first present some very important results obtained by Lagrange. Let

$$(52) \qquad F(x, y, z, p, q) = 0$$

be the given equation. The fundamental result obtained by Lagrange is the following: If we know a family of integrals which depend upon *two* arbitrary parameters, we can derive all the other integrals from them by differentiations and eliminations. Let

$$(53) \qquad V(x, y, z, a, b) = 0$$

be a relation which contains two arbitrary constants a and b, and which defines an integral of the equation (52) for any values of those constants. The values of the partial derivatives p and q of that integral are given by the equations

$$(54) \qquad \frac{\partial V}{\partial x} + p\frac{\partial V}{\partial z} = 0, \qquad \frac{\partial V}{\partial y} + q\frac{\partial V}{\partial z} = 0.$$

By hypothesis, the function z always satisfies the equation (52) for any values of a and b; hence the elimination of the two parameters a and b from the three relations (53) and (54) will lead to the equation (52) and to *that one only*.*

We shall now show that this equation (52) expresses the necessary and sufficient condition that the three equations (53) and (54) be satisfied by a system of three functions z, a, b of the two variables x and y, where p and q denote the partial derivatives of z with respect to x and y respectively. When this has been proved, it will be evident that the problem of integrating the single equation (52) is equivalent to the following problem: *To find three functions z, a, b of the two independent variables x and y which satisfy the three equations* (53) *and* (54).

If $z = f_1(x, y)$, $a = f_2(x, y)$, $b = f_2(x, y)$ form a system of solutions of these three equations, the function $f_1(x, y)$ also satisfies the equation (52), which is a consequence of these three relations.

---

* In fact, if the elimination of a and b led to another relation $\Phi(x, y, z, p, q) = 0$ different from $F = 0$, the two simultaneous equations $F = 0$, $\Phi = 0$ would have a common integral $V = 0$ depending upon two arbitrary parameters a and b, which is impossible (§ 78). The given integral would therefore depend in reality upon only a *single* parameter.

Conversely, if $f_1(x, y)$ is an integral of the equation (52), the three equations (53) and (54) are consistent when we replace z by $f_1(x, y)$, and p and q by the partial derivatives of $f_1(x, y)$. Hence we can derive from them as values for a and b two other functions $a = f_2(x, y)$, $b = f_3(x, y)$, which form with $f_1(x, y)$ a system of solutions of the equations (53) and (54).

The new problem, although apparently more complicated than the original, is easily solved. In fact, if we differentiate the relation (53) with respect to x and to y, regarding now z, a, b as unknown functions of x and y, the relations obtained reduce, by (54), to the two equations

$$(55) \qquad \frac{\partial V}{\partial a}\frac{\partial a}{\partial x} + \frac{\partial V}{\partial b}\frac{\partial b}{\partial x} = 0, \qquad \frac{\partial V}{\partial a}\frac{\partial a}{\partial y} + \frac{\partial V}{\partial b}\frac{\partial b}{\partial y} = 0,$$

and the system formed by the equations (53) and (55) is equivalent to the system formed by the equations (53) and (54).

We see at once that this system is satisfied by taking for the unknown functions a and b any two constants. This gives as the value of z the integral already known, which Lagrange called the *complete integral*. In order to treat the problem in a general way, let us observe that the equations (55) are linear and homogeneous in $\partial V/\partial a$, $\partial V/\partial b$. Hence the three equations (53) and (55) are satisfied if we set

$$(56) \qquad V = 0, \qquad \frac{\partial V}{\partial a} = 0, \qquad \frac{\partial V}{\partial b} = 0.$$

If these three equations are consistent, they define three functions z, a, b of the two variables x and y. This gives an integral $z = f_1(x, y)$ of the equation (52) which does not depend upon any arbitrary parameter, and which is commonly called the *singular integral*.

If $\partial V/\partial a$ and $\partial V/\partial b$ are not zero simultaneously, the equations (55) give

$$\frac{D(a, b)}{D(x, y)} = 0,$$

which proves that there exists between the functions a and b at least one relation independent of x and of y. If there exist two relations of that kind, a and b reduce to constants, which gives again the complete integral. If there exists only one relation between a and b, at least one of the functions a and b does not reduce to a constant. Assuming that a is not constant, we can write the relation between a and b in the form

$$(57) \qquad b = \phi(a),$$

and the two equations (55) become

$$\frac{\partial a}{\partial x}\left[\frac{\partial V}{\partial a} + \frac{\partial V}{\partial b}\phi'(a)\right] = 0, \qquad \frac{\partial a}{\partial y}\left[\frac{\partial V}{\partial a} + \frac{\partial V}{\partial b}\phi'(a)\right] = 0.$$

Since a is not a constant by hypothesis, these two relations reduce to a single relation, and the three equations

(58) $\quad V(x, y, z, a, b) = 0, \qquad b = \phi(a), \qquad \dfrac{\partial V}{\partial a} + \dfrac{\partial V}{\partial b}\phi'(a) = 0$

define a new system of solutions of the equations (53) and (54). In particular, the function $z = f_1(x, y)$ defined by (58) is an integral of the given equation (52). It is evident that this integral depends upon the arbitrary function $\phi(a)$. We shall call it the *general integral*.

In order to obtain the relation between x, y, z, the arbitrary parameter a must be eliminated from the two equations

(58') $\quad V[x, y, z, a, \phi(a)] = 0, \qquad \dfrac{\partial V}{\partial a} + \dfrac{\partial V}{\partial \phi(a)}\phi'(a) = 0.$

This elimination can be made only after the function $\phi(a)$ has been chosen, but the equations (58') always enable us to express two of the coördinates of a point of an integral surface as functions of a third coördinate and of a parameter a.

The preceding method is related in a very simple way to the theory of the surface envelopes. Consider, in fact, the family of surfaces S which represent the complete integral (53) and which depend upon two constants a and b. If we choose an arbitrary relation of the form $b = \phi(a)$ between the two parameters a and b, we obtain a family of surfaces which depend upon only one parameter a, and the envelope of this family of surfaces is obtained precisely by eliminating a from the two equations (58'). The process by which we deduce the general integral from the complete integral consists, therefore, in taking the envelope of a one-parameter family of complete integrals obtained by choosing an arbitrary relation between the two parameters a and b. Similarly, the singular integral is obtained by taking the envelope of all the complete integrals, as the two parameters a and b vary independently * (I, § 212, 2d ed.; § 220, 1st ed.).

---

* We have seen above (§ 71) that all considerations founded on the theory of envelopes in the study of differential equations are quite troublesome. All the difficulties pointed out in the study of the singular solutions of an ordinary differential equation of the first order arise again for partial differential equations of the first order. The final conclusion is just as before: a partial differential equation of the first order, given a priori, does not normally have any singular integrals. This

It would seem from what precedes that we ought to distinguish three categories of integrals: the complete integral, the general integral, and the singular integral. But Lagrange's theory itself shows that there exist an infinite number of complete integrals. Indeed, if we establish between the two parameters a and b a relation of a definite form $b = \pi(a, a', b')$, containing two constants a' and b', the corresponding general integral will depend upon these two constants a', b', and may be considered as a new complete integral. The original complete integral will now be included in the general integral, and will correspond to the relation $b = \pi(a, a', b')$ established between the two parameters a' and b'. There is, therefore, no essential distinction between the general integral and the complete integral. On the contrary, the singular integral, as can be seen from its geometric meaning, does not depend upon the choice of the complete integral.

Example 1. Consider the generalized Clairaut's equation

$$z = px + qy + f(p, q).$$

It is easily seen that it has a complete integral of the form

$$z = ax + by + f(a, b).$$

This complete integral is represented by a family of planes which depend upon two arbitrary parameters a and b. These planes envelop a non-developable surface Σ, which is the singular integral surface of the given equation. In order to obtain the general integral, we must choose an arbitrary relation between a and b, say $b = \phi(a)$, and we must find the envelope of the planes thus obtained. This envelope, which is represented by the two equations

$$z = ax + y\phi(a) + f[a, \phi(a)], \quad x + y\phi'(a) + \frac{\partial f}{\partial a} + \frac{\partial f}{\partial \phi(a)} \phi'(a) = 0,$$

is a developable surface tangent to the surface Σ all along a curve Γ. It is evident that we can choose the arbitrary function $\phi(a)$ in such a way that the curve of contact Γ shall be any preassigned curve on Σ.

Example 2. Consider the equation

$$q = f(p),$$

of which a complete integral is

$$z = ax + f(a)y + b.$$

---

conclusion does not contradict the reasoning of the text, for we have assumed that we can apply the theory of implicit functions to the system of three equations (56), and the conclusions are correct only when that condition is satisfied. (See the paper by Darboux, *Sur les solutions singulières des équations aux dérivées partielles du premier ordre* (*Mémoires des Savants étrangers*, Vol. XXVII).)

This equation represents a plane, and the general integral, which is given by the system of two equations

(59) $\qquad z = ax + yf(a) + \phi(a), \qquad 0 = x + yf'(a) + \phi'(a),$

is represented by developable surfaces, which can be defined geometrically in a very simple way. Draw through a fixed point of space (for example, the origin) the planes parallel to the planes which form the complete integral; these planes depend only upon the parameter a, and consequently envelop a cone (T) whose vertex is at the origin. It follows that the edge of regression of the developable surface (59) has its osculating plane constantly parallel to a tangent plane of the cone (T). Hence the generators of this surface are parallel to the generators of the cone just mentioned (I, § 227, 2d ed.; § 224, 1st ed.).

The equations (56), which determine the singular integral, are in this case inconsistent, for the last reduces to $1 = 0$. There is therefore no singular integral.

Example 3. Consider a family of spheres with a given radius R, whose centers remain in a fixed plane. These spheres depend upon two arbitrary parameters, and if we take a system of rectangular axes with the fixed plane for the xy-plane, they are represented by the equation
$$(x - a)^2 + (y - b)^2 + z^2 - R^2 = 0.$$

The corresponding partial differential equation is obtained by eliminating a and b from this equation and the following two,
$$x - a + pz = 0, \qquad y - b + qz = 0,$$
which gives the equation
$$(1 + p^2 + q^2)z^2 - R^2 = 0.$$

Geometrically this equation expresses the fact that the portion of the normal included between any point of the surface and the xy-plane is constant and equal to R. The general integral is a tubular surface, the envelope of a sphere of radius R whose center describes an arbitrary curve in the xy-plane. There is a singular integral surface formed by the two planes $z = \pm R$. It is evident that these two planes are tangent to all the other integral surfaces.

83. Lagrange and Charpit's method. To sum up the preceding, in order to determine all the integrals of an equation of the first order,

(60) $\qquad\qquad\qquad F(x, y, z, p, q) = 0,$

it is sufficient to know a complete integral, that is, an integral depending upon two arbitrary constants. In order to determine a complete integral, let us suppose that, by any means whatever, we have obtained another function $\Phi(x, y, z, p, q)$ such that the two equations

(61) $$F = 0, \qquad \Phi = a$$

can be solved for p and q, and form a completely integrable system, for any value of the constant a. If this is the case, then by solving the two preceding equations for p and q, and substituting these values of p and q in the equation $dz = p\,dx + q\,dy$, we obtain a completely integrable total differential equation

(62) $$dz = f(x, y, z, a)\,dx + \phi(x, y, z, a)\,dy.$$

The integration of this equation introduces a new arbitrary constant b, and in this way we obtain an integral of the given equation which depends upon the two arbitrary constants a and b.

Lagrange and Charpit's method of integration consists precisely in adjoining to the equation $F = 0$ another equation $\Phi = a$ such that the system (61) formed by these two equations is completely integrable. For this it is necessary and sufficient (§ 81) that $[F, \Phi] = 0$, that is, that

(63) $$P\frac{\partial \Phi}{\partial x} + Q\frac{\partial \Phi}{\partial y} + (Pp + Qq)\frac{\partial \Phi}{\partial z} - (X + pZ)\frac{\partial \Phi}{\partial p} - (Y + qZ)\frac{\partial \Phi}{\partial q} = 0,$$

where, for brevity, we have set

$$X = \frac{\partial F}{\partial x}, \quad Y = \frac{\partial F}{\partial y}, \quad Z = \frac{\partial F}{\partial z}, \quad P = \frac{\partial F}{\partial p}, \quad Q = \frac{\partial F}{\partial q}.$$

The auxiliary function $\Phi(x, y, z, p, q)$ must therefore satisfy a *linear* partial differential equation in five independent variables. The integration of this linear equation reduces in turn to that of the system of ordinary differential equations

(64) $$\frac{dx}{P} = \frac{dy}{Q} = \frac{dz}{Pp + Qq} = \frac{-dp}{X + pZ} = \frac{-dq}{Y + qZ}.$$

But, for the purpose which we have in view, it is not necessary to find the general integral of this system (64); it is sufficient to know one first integral $\Phi = a$ of this system, such that we can solve the two equations $F = 0$, $\Phi = a$ for p and q.

We can therefore state the following general rule:

To obtain a complete integral of the equation (60), *we first find one first integral* $\Phi = a$ *of the auxiliary system* (64) *for which the Jacobian* $D(F, \Phi)/D(p, q)$ *is not zero; then we solve the two equations* $F = 0$, $\Phi = a$ *for p and q. Substituting the expressions obtained for p and q in the equation $dz = p\,dx + q\,dy$, we obtain a completely* *integrable total differential equation. The general integral of this equation contains a second arbitrary constant b, and is a complete integral of the equation* (60).

We know in advance one integral of the equation (63); that is, the function F itself. This integral cannot be used directly, but the knowledge of it reduces the integration of the system (64) to the integration of a system of *three* differential equations of the first order. The precise nature of the problem to be solved is thus made clear.

When the function F does not depend upon the unknown function z, we may also suppose that the function Φ does not depend upon z, and the condition that the system (61) be completely integrable is then
$$(F, \Phi) = 0,$$
or

(63') $$P\frac{\partial \Phi}{\partial x} + Q\frac{\partial \Phi}{\partial y} - X\frac{\partial \Phi}{\partial p} - Y\frac{\partial \Phi}{\partial q} = 0.$$

Hence the auxiliary system (64) takes the form

(64') $$\frac{dx}{P} = \frac{dy}{Q} = \frac{-dp}{X} = \frac{-dq}{Y}.$$

If we know a first integral $\Phi = a$ of this system for which
$$\frac{D(F, \Phi)}{D(p, q)}$$
is not zero, we are led to a total differential equation of the form
$$dz = f(x, y, a)dx + \phi(x, y, a)dy,$$
which is integrable by a quadrature. The difficulty of the second part of the problem is therefore diminished in this case. This is also true of the first part, for we know a first integral $F = C$ of the system (64'); we can therefore replace this system by a system of *two* differential equations of the first order.

Example 1. Let us consider an equation containing only one of the three variables x, y, z (for example, the variable y):
$$F(y, p, q) = 0.$$
In this case $X = Z = 0$, and the equations (64) give the integrable combination $dp = 0$. Hence the two equations $F(y, p, q) = 0$, $p = a$ form a completely integrable system, as is easily verified. For if we solve the given equation for q, the total differential equation to be integrated takes the form
$$dz = a\,dx + f(y, a)\,dy.$$
Hence we obtain a complete integral by a quadrature:
$$z = ax + \int f(y, a)\,dy + b.$$

Example 2. An equation of the form $F(z, p, q) = 0$ can be reduced to the preceding form by taking y and z for the independent variables, but we can dispense with this change of variables. For in this case we have $X = Y = 0$, and the equations (64) give

$$\frac{dp}{p} = \frac{dq}{q};$$

whence a first integral is $q = ap$. From the two equations

$$q = ap, \quad F(z, p, q) = 0$$

we then derive

$$p = f(z, a), \quad q = af(z, a),$$

and the total differential equation

$$dz = f(z, a)(dx + a\,dy)$$

can be integrated by a quadrature:

$$\int \frac{dz}{f(z, a)} = x + ay + b.$$

Consider, for example, the equation $pq - z = 0$. Adjoining to it the equation $q = ap$, we derive from them

$$p = \sqrt{\frac{z}{a}}, \quad q = a\sqrt{\frac{z}{a}}, \quad dz = \sqrt{\frac{z}{a}}(dx + a\,dy);$$

hence a complete integral is given by the equation

$$4\,az = (x + ay + b)^2,$$

which represents a family of parabolic cylinders tangent to the xy-plane along the entire length of a generator. The xy-plane represents a singular integral.

The equations (64), in the case where $F = pq - z$, have also the first integral $p - y = a$. Starting with this integral, we are led to the total differential equation

$$dz = (y + a)\,dx + \frac{z\,dy}{y + a},$$

which can also be written in the form

$$dx = d\left(\frac{z}{y + a}\right).$$

This furnishes a new complete integral $z = (y + a)(x + b)$, which represents a family of hyperbolic paraboloids tangent to the xy-plane.

Example 3. Let the equation be of the form $f(x, p) - f_1(y, q) = 0$. The differential equations (64′)

$$\frac{dx}{\dfrac{\partial f}{\partial p}} = \frac{dy}{-\dfrac{\partial f_1}{\partial q}} = \frac{-\,dp}{\dfrac{\partial f}{\partial x}} = \frac{dq}{\dfrac{\partial f_1}{\partial y}}$$

have the first integral $f(x, p) = a$. If we adjoin this equation to the given equation, we derive from the two relations

$$f(x, p) = a, \quad f_1(y, q) = a,$$

the values for p and q, $p = \phi(x, a)$, $q = \phi_1(y, a)$, and the total differential equation

$$dz = \phi(x, a)\,dx + \phi_1(y, a)\,dy$$

can be integrated by two quadratures as follows:

$$z = \int \phi(x, a)\,dx + \int \phi_1(y, a)\,dy + b.$$

When an equation of the first order is of the preceding form, we say that *the variables are separated*. For example, let us consider the equation

$$pq - xy = 0,$$

which can be written in the form

$$\frac{p}{x} = \frac{y}{q}.$$

Equating these two quotients to a constant a, we obtain the total differential equation

$$dz = ax\,dx + \frac{y}{a}\,dy,$$

whence a complete integral is

$$z = \frac{ax^2}{2} + \frac{y^2}{2a} + b.$$

Example 4. Let us propose to find the functions $F(z, y, p, q)$ for which the equations (64) have the first integral $py - qx = a$. For this it is necessary and sufficient that the relation $p\,dy + y\,dp - q\,dx - x\,dq = 0$ shall be a consequence of the relations (64'); that is, that the function F shall itself be an integral of the linear equation

$$p\frac{\partial F}{\partial q} - y\frac{\partial F}{\partial x} + x\frac{\partial F}{\partial y} - q\frac{\partial F}{\partial p} = 0.$$

The corresponding system of differential equations

$$\frac{dx}{-y} = \frac{dy}{x} = \frac{dp}{-q} = \frac{dq}{p}$$

has the three first integrals

$$x^2 + y^2 = C, \qquad p^2 + q^2 = C', \qquad py - qx = C'',$$

and the function F is therefore of the form $F(py - qx, x^2 + y^2, p^2 + q^2)$. The investigation of the equation $F = 0$ for a complete integral is therefore reduced to the integration of two simultaneous equations of the form

$$p^2 + q^2 = f(x^2 + y^2, py - qx), \qquad py - qx = a.$$

Making use of the identity

$$(p^2 + q^2)(x^2 + y^2) = (py - qx)^2 + (px + qy)^2,$$

we derive from the two preceding equations

$$px + qy = \sqrt{(x^2 + y^2)f(x^2 + y^2, a) - a^2} = \phi(x^2 + y^2, a).$$

Solving for p and q, we obtain the values

$$p = \frac{ay + x\phi(x^2 + y^2, a)}{x^2 + y^2}, \qquad q = \frac{-ax + y\phi(x^2 + y^2, a)}{x^2 + y^2},$$

whence we obtain a complete integral by a quadrature,

$$z = -a \arctan\left(\frac{y}{x}\right) + \int \frac{\phi(u, a)}{2u}\,du + b,$$

where $u = x^2 + y^2$.

It is sometimes possible to find a priori, by geometric considerations, certain integrable combinations of the differential equations (64). Suppose, for example, that we wish to find the surfaces S whose tangent plane at any point M meets at a constant angle V the plane passing through M and Oz. It is clear that if a surface S satisfies this condition, all the surfaces obtained from it by a helicoidal movement around the z-axis, for which the pitch of the helix is equal to h, will also satisfy the condition. Hence the surface envelope Σ will also be an integral of the same equation. This envelope Σ is evidently a helicoidal surface of pitch h. Since we may translate it any distance whatever parallel to the z-axis, it follows that the partial differential equation of the problem and the partial differential equation of the helicoidal surfaces $py - qx = a$ (§ 72) have, for any value of a, an infinite number of common integrals which depend upon an arbitrary constant. Consequently the differential equations (64) corresponding to the partial differential equation of the surfaces S have a first integral $py - qx = a$, and the complete integral can be obtained by a quadrature.

Note. It should be noticed that it is not necessary that the relation (63) shall be identically satisfied in order that the system (61) be completely integrable; it is sufficient that it be satisfied by virtue of the relation $F = 0$ itself. We can sometimes make use of this fact in the search for the function Φ. In fact, the problem of finding an integrable combination of the equations (64) reduces essentially to that of finding five functions $\lambda_x, \lambda_y, \lambda_z, \lambda_p, \lambda_q$ of the variables x, y, z, p, q such that

$$\lambda_x dx + \lambda_y dy + \lambda_z dz + \lambda_p dp + \lambda_q dq$$

shall be an exact differential $d\Phi$ and such that we have also

$$P\lambda_x + Q\lambda_y + (Pp + Qq)\lambda_z - (X + pZ)\lambda_p - (Y + qZ)\lambda_q = 0.$$

If this last equation is not satisfied except by virtue of the equation $F = 0$, the function Φ is not, properly speaking, a first integral of the system (64). However, since the multipliers $\lambda_x, \lambda_y, \cdots$ are equal to the partial derivatives of Φ, the two equations $F = 0$, $\Phi = a$ still form a completely integrable system, for the equation (63) is then a consequence of $F = 0$.* A similar remark applies to the sytem (64').

---

* When the equation $F = 0$ can be solved for one of the variables x, y, z, p, q, we may suppose that the function Φ does not contain that variable, and it will also not appear in any of the coefficients X, Y, Z, P, Q. For definiteness, let us take an equation of the form
$$p + f(x, y, z, q) = 0.$$
To find a complete integral, we need only adjoin another equation $\phi(x, y, z, q) = a$, which forms with the first a completely integrable system. In this case the condition $[p + f, \phi] = 0$ takes the form

$$\frac{\partial \phi}{\partial x} + \frac{\partial f}{\partial q}\frac{\partial \phi}{\partial y} + \left(q\frac{\partial f}{\partial q} - f\right)\frac{\partial \phi}{\partial z} - \left(\frac{\partial f}{\partial y} + q\frac{\partial f}{\partial z}\right)\frac{\partial \phi}{\partial q} = 0,$$

in which the letter p does not appear.

More generally, let us suppose that we can satisfy the relation $F = 0$ by putting
$$p = f(x, y, z, \lambda), \qquad q = \phi(x, y, z, \lambda),$$

84. Cauchy's problem.
Given an equation

(65) $$p = f(x, y, z, q)$$

in which the right-hand side is analytic in the neighborhood of a system of values (x_0, y_0, z_0, q_0), and a function $\phi(y)$ analytic in the neighborhood of the point y_0, such that we have $\phi(y_0) = z_0$, $\phi'(y_0) = q_0$, we proved in § 25 that this equation has an analytic integral in the neighborhood of the point (x_0, y_0) which reduces to the given function $\phi(y)$ for $x = x_0$. Let C be the plane curve represented by the two equations $x = x_0$, $z = \phi(y)$. Geometrically this result may be stated as follows: *There exists one and only one analytic integral surface of the equation* (65) *passing through the curve C.*

This proposition is capable of generalization. Let us first consider an equation of any form,

(66) $$F(x, y, z, p, q) = 0,$$

and let us propose to determine an integral surface passing through a plane curve, such as C, which lies in a plane $x = x_0$ parallel to the yz-plane. Let $z = \phi(y)$ be the equation of the cylinder which projects C upon the yz-plane. Since the function ϕ is analytic in the neighborhood of the point y_0, the equation

(67) $$F(x_0, y_0, z_0, p, q_0) = 0,$$

where $z_0 = \phi(y_0)$, $q_0 = \phi'(y_0)$ and where we regard p as the unknown, has a certain number of roots. Let p_0 be one of them. If the function F is analytic in the neighborhood of the system of values $(x_0, y_0, z_0, p_0, q_0)$, and if also the partial derivative $(\partial F/\partial p)_0$ is not zero for this system of values, the equation (66) has a root $p = f(x, y, z, q)$ which is analytic in the neighborhood of the system of values (x_0, y_0, z_0, q_0) (I, § 193, 2d ed.; § 187, 1st ed.). Hence we are led back to an equation of the form (65), which shows that the equation (66) possesses an integral surface through C. As a matter of fact, the reasoning proves that this equation has m integral surfaces which satisfy the conditions if the equation (67) is of degree m with respect to p. There is no possible exception unless one of the roots

---

where λ denotes an auxiliary parameter. We need only replace λ by a function of x, y, z such that the equation $dz = f dx + \phi dy$ is completely integrable, which again leads to a linear equation for $\lambda (x, y, z)$:

$$\frac{\partial f}{\partial y} + \frac{\partial f}{\partial z}\phi + \frac{\partial f}{\partial \lambda}\left(\frac{\partial \lambda}{\partial y} + \frac{\partial \lambda}{\partial z}\phi\right) = \frac{\partial \phi}{\partial x} + \frac{\partial \phi}{\partial z}f + \frac{\partial \phi}{\partial \lambda}\left(\frac{\partial \lambda}{\partial x} + \frac{\partial \lambda}{\partial z}f\right).$$

(ANTOMARI, *Bulletin de la Société Mathématique*, Vol. XIX, p. 154.)

of the equation (67) satisfies also the relation $\partial F/\partial p = 0$ at all the points of C, since x_0, y_0, z_0 are the coördinates of any point of this curve.

Let us consider finally any curve Γ, represented by a system of two equations

(68) $\qquad x = \lambda(y), \qquad z = \mu(y),$

and let it be required to determine an integral surface of the equation (66) which passes through Γ. This problem, in turn, can be reduced to the preceding by means of a change of variables; for if we put
$$x = X + \lambda(Y), \qquad y = Y,$$
the relation $dz = p\,dx + q\,dy$ becomes
$$dz = p\,dX + p\lambda'(Y)\,dY + q\,dY,$$
and from this we derive
$$p = \frac{\partial z}{\partial X}, \qquad p\lambda'(Y) + q = \frac{\partial z}{\partial Y}.$$

The equation (66) is then replaced by the equation

(66') $\qquad F\left[X + \lambda(Y),\, Y,\, z,\, \dfrac{\partial z}{\partial X},\, \dfrac{\partial z}{\partial Y} - \lambda'(Y)\dfrac{\partial z}{\partial X}\right] = 0,$

and it remains to find an integral of this new equation which reduces to $\mu(Y)$ for $X = 0$. Hence we see that in general *an integral surface of an equation of the first order is determined if we assign a curve lying on that surface.* There may be several integral surfaces satisfying this condition if the equation similar to (67) has several distinct roots, just as an ordinary differential equation of the first order and of degree m in y' has in general m integral curves passing through a given point. We shall return later to the exceptional case in which this reasoning fails.

The problem of determining an integral surface of a partial differential equation of the first order through a given curve has been called *Cauchy's problem.* This name is used to remind us of the close relation just explained existing between this problem and Cauchy's general theory. We shall now show how Cauchy's problem can be solved by an elimination if we know a complete integral, and this will furnish also a verification of the preceding results.

Let
$$V(x, y, z, a, b) = 0$$
be a complete integral, and let Γ be a given curve not situated upon the singular integral surface nor upon one of the integral surfaces

obtained by giving to a and to b constant values. Cauchy's problem reduces to determining the function $\phi(a)$ in such a way that the given curve Γ shall lie upon the surface S defined by the two equations

(69) $\quad V[x, y, z, a, \phi(a)] = 0, \quad \dfrac{\partial V}{\partial a} + \dfrac{\partial V}{\partial \phi(a)} \phi'(a) = 0.$

Let us suppose that the coördinates x, y, z of a point of Γ are expressed as functions of an auxiliary parameter λ,

(70) $\quad x = f_1(\lambda), \quad y = f_2(\lambda), \quad z = f_3(\lambda),$

and let $U(\lambda, a, b)$ be the result obtained by replacing x, y, z in $V(x, y, z, a, b)$ by the preceding expressions. The two simultaneous equations

(71) $\quad U[\lambda, a, \phi(a)] = 0, \quad \dfrac{\partial U}{\partial a} + \dfrac{\partial U}{\partial \phi(a)} \phi'(a) = 0$

determine the values of λ and a which correspond to the points of intersection of the curve Γ with the surface S. If the surface S passes through the curve Γ, these two equations form an indeterminate system. Hence, eliminating λ from these two equations, we obtain an identity. This elimination leads to a relation between a, $\phi(a)$, $\phi'(a)$,

(72) $\quad \Pi[a, \phi(a), \phi'(a)] = 0,$

that is, to a differential equation of the first order for the determination of $\phi(a)$. It would seem, therefore, that the problem has an infinite number of solutions, contrary to Cauchy's result. But it is easy to deduce from the equations (71) another relation not containing $\phi'(a)$. In fact, let us suppose that the curve Γ lies entirely on the surface S. When a point moves on Γ, a is a function of λ which satisfies the two equations (71) simultaneously. Hence, if we differentiate the first of these two equations with respect to λ, it follows from this result and the second that

(73) $\quad \dfrac{\partial U}{\partial \lambda} = 0.$

This equation contains only λ, a, $\phi(a)$. Eliminating λ from the two equations $U = 0$, $\partial U / \partial \lambda = 0$, we obtain an equation which determines the function $\phi(a)$. The method to which we are led has an evident geometric meaning. In fact, the equation $U(\lambda, a, b) = 0$ determines the values of λ which correspond to the points of intersection of the curve Γ with the complete integral. If we also have $\partial U / \partial \lambda = 0$, this equation has a double root, and the complete integral is tangent to Γ.

Eliminating λ from the two equations $U(\lambda, a, b) = 0$, $\partial U/\partial \lambda = 0$, the condition obtained, $\Phi(a, b) = 0$, therefore expresses the fact that the complete integral is tangent to Γ, and *the desired integral surface through Γ may be defined as the envelope of the complete integral surfaces tangent to the curve Γ*. This result is geometrically almost intuitive.*

85. Characteristic curves. Cauchy's method. Cauchy's method is independent of the theory of the complete integral. We shall now present it in a geometric form. For this purpose, let us first consider the meaning of a non-linear partial differential equation

(74) $$F(x, y, z, p, q) = 0.$$

This equation may be regarded as a relation between the direction cosines of the tangent plane to an integral surface S through a given point (x, y, z) of space. Hence this tangent plane cannot be any plane passing through the point (x, y, z). Since the possible tangent planes form only a one-parameter family, they envelop in general a cone (T) whose vertex is the point (x, y, z). It follows that *the tangent plane at any point M of space to each integral surface S passing through this point is also tangent to a certain cone (T) whose vertex is at M.*

The cone (T) depends, of course, upon the function F, and also upon the position of its vertex. In order to obtain the equation of the cone (T) whose vertex is (x, y, z), we must, by its definition, find the envelope of the planes

(75) $$Z - z = p(X - x) + q(Y - y),$$

where the parameters p and q are connected by the relation (74). We must therefore eliminate p and q from these two equations and the new relation (I, note, § 208, 2d ed.; § 202, 1st ed.)

(76) $$(Y - y)\frac{\partial F}{\partial p} - (X - x)\frac{\partial F}{\partial q} = 0.$$

---

* It is easy to obtain the general integral of the differential equation (72). In fact, if we replace λ by an arbitrary constant λ_0, the function $\phi(a)$ defined by the equation

(e) $$U[\lambda_0, a, \phi(a)] = 0,$$

also satisfies the equation

(e') $$\frac{\partial U}{\partial a} + \frac{\partial U}{\partial \phi(a)}\phi'(a) = 0.$$

Hence $\phi(a)$ satisfies also the equation obtained by eliminating λ_0 from (e) and (e'), but the resulting equation is exactly the equation (72). The relation (e) therefore represents the general integral of the equation (72). There is also a singular integral, which is indeed precisely the desired solution of the given problem.

The two equations (75) and (76) represent the characteristic direction, that is, the generator of the cone (T) which is the line of contact of the tangent plane. If we suppose that the axes of coördinates are rectangular, we can obtain immediately the equation of the normal cone (N), which is generated by the normals to the different integral surfaces passing through the point M. For, since the equations of the normals are

$$X - x + p(Z - z) = 0, \qquad Y - y + q(Z - z) = 0,$$

the elimination of p and q gives the equation of the cone (N) in the form

(77) $$F\left(x, y, z, -\frac{X-x}{Z-z}, -\frac{Y-y}{Z-z}\right) = 0.$$

If the given equation (74) is linear in p and q, the cone (N) is a plane and the cone (T) reduces to a straight line Δ. We have seen (§ 76) that the integration reduces in this case to the search for the curves which are tangent in each of their points to the corresponding straight line Δ. We are led to Cauchy's method by extending this process to non-linear equations.

Let S be an integral surface represented by the equation

$$z = f(x, y).$$

At each point M of this surface the tangent plane is also tangent to the cone (T) along a generator (G). We shall give the name *characteristic curve* to every curve C of the surface S which is tangent in each of its points to the corresponding generator G. Through each point of S (excepting the singular points, if there are any) there passes one and only one curve of this kind. The name *characteristic curves* will be justified later (§ 86).

The key to Cauchy's method is that we can determine these curves by a system of ordinary differential equations without knowing the function $f(x, y)$. In the first place, the tangent to the curve C coincides with the straight line G represented by the two equations (75) and (76), which may be written in the form

$$\frac{X-x}{P} = \frac{Y-y}{Q} = \frac{Z-z}{Pp + Qq}$$

in the notation of § 83. Along a characteristic curve x, y, z, p, q are functions of a single independent variable, and we may write the relations between the differentials dx, dy, dz in the form

(78) $$\frac{dx}{P} = \frac{dy}{Q} = \frac{dz}{Pp + Qq} = du,$$

where u is a conventional auxiliary variable which is introduced merely for symmetry. Along this curve C we have also

$$dp = r\,dx + s\,dy, \qquad dq = s\,dx + t\,dy,$$

where r, s, t are the usual second derivatives of the function $f(x, y)$. On the other hand, since $z = f(x, y)$ is an integral of the given equation (74), the partial derivatives r, s, t also satisfy the two relations

$$X + pZ + Pr + Qs = 0, \qquad Y + qZ + Ps + Qt = 0,$$

which are obtained by differentiating (74) with respect to x and with respect to y. Replacing the differentials dx and dy by $P\,du$ and $Q\,du$ respectively, the expressions for dp and dq become

$$dp = (Pr + Qs)\,du, \qquad dq = (Ps + Qt)\,du,$$

or, using the preceding relations,

$$dp = -(X + pZ)\,du, \qquad dq = -(Y + qZ)\,du.$$

Adjoining these equations to the equations (78), we arrive at a system of ordinary differential equations

$$(79) \qquad \frac{dx}{P} = \frac{dy}{Q} = \frac{dz}{Pp + Qq} = \frac{-dp}{X + pZ} = \frac{-dq}{Y + qZ} = du,$$

which is identical with the system (64) to which we are led by Lagrange's method.

This system of differential equations is absolutely independent of the integral considered. We derive from it the following conclusions: Let (x_0, y_0, z_0) be the coördinates of a point M_0 of S, and let p_0 and q_0 be the values of p and q for the tangent plane at this point. If the function F is analytic in the neighborhood of this system of values, and if not all the denominators of the quotients (79) vanish simultaneously for x_0, y_0, z_0, p_0, q_0, the equations (79) have one and only one system of integrals which take on the values x_0, y_0, z_0, p_0, q_0 for $u = 0$. It follows that *if two integral surfaces are tangent at a point* (x_0, y_0, z_0), *they are tangent along the entire length of a common characteristic curve through that point.*

For convenience we shall call every system of values assigned to the five variables x, y, z, p, q an *element*. Thus, an element may be thought of as consisting of the set of a point whose coördinates are (x, y, z) and a plane through that point whose position is defined by the values of p, q. Along an entire characteristic curve, x, y, z, p, and q are functions of an independent variable u. To each point of a characteristic curve there corresponds, therefore, an element composed of this point together with the plane through this point defined by the values of p and q. But from the equations (79) we have

$$\frac{dz}{du} = p\frac{dx}{du} + q\frac{dy}{du},$$

so that this plane contains the tangent to the curve at the point (x, y, z). When the point (x, y, z) describes the characteristic curve, the corresponding plane envelops a developable surface passing through this curve, which is called the *characteristic developable surface*. Thus, to each characteristic curve there corresponds a characteristic developable surface through that curve. We shall hereafter use the words *characteristic strip* to denote the combination of the curve and the developable surface, and we shall refer to the curve as the characteristic curve, to avoid any possibility of ambiguity. With this understanding, a characteristic strip is composed of an infinite number of elements which depend upon an independent variable, and the infinitesimal variations of x, y, z, p, q are connected by the relations (79). A characteristic strip is therefore completely defined if we are given one of its elements, and the theorem stated a moment ago can again be expressed in the following exactly equivalent form:

If two integral surfaces have a common element, they have in common all the elements of the characteristic strip to which the given common element belongs.

The totality of all characteristic strips depends upon *three* arbitrary parameters. In fact, a characteristic strip is determined if one of its elements $(x_0, y_0, z_0, p_0, q_0)$ is given. One of the coördinates, x_0 for example, may be assigned a given numerical value, and, moreover, by definition the relation $F(x_0, y_0, z_0, p_0, q_0) = 0$ is satisfied. Hence only three parameters remain arbitrary.

In order to determine the characteristic strips, let us observe first that $F = $ const. is a first integral of the equations (79). Hence, if $F(x, y, z, p, q)$ vanishes for the initial element $(x_0, y_0, z_0, p_0, q_0)$, F vanishes throughout the entire length of the characteristic strip through that element, as we see also from the derivation of the equations (79). In order to find the characteristic strips of the given equation, we can therefore adjoin to the system (79) the relation $F = 0$ itself, which reduces that system to one of three differential equations of the first order.

Let us suppose that we have obtained the equations of the characteristic strip in finite terms; and, for definiteness, let

(80) $\begin{cases} y = f_1(x, x_0, y_0, z_0, p_0, q_0), & z = f_2(x, x_0, y_0, z_0, p_0, q_0), \\ p = f_3(x, x_0, y_0, z_0, p_0, q_0), & q = f_4(x, x_0, y_0, z_0, p_0, q_0) \end{cases}$

be the equations of the characteristic strip through the element

$$(x_0, y_0, z_0, p_0, q_0).$$

The two first equations of (80) represent the characteristic curve itself, and every integral surface, being a locus of the characteristic curves, will be obtained by supposing that x_0, y_0, z_0, p_0, q_0 are functions of an auxiliary parameter v. We are therefore led to investigate how these five functions of v may be chosen in order that the surface generated by these characteristic curves shall be an integral surface. We shall introduce with Darboux an auxiliary variable u, and write the equations in a symmetric form. Let

(81) $\begin{cases} x = \phi_1(u, x_0, y_0, z_0, p_0, q_0), \\ y = \phi_2(u, x_0, y_0, z_0, p_0, q_0), \\ z = \phi_3(u, x_0, y_0, z_0, p_0, q_0), \end{cases}$

(82) $\begin{cases} p = \phi_4(u, x_0, y_0, z_0, p_0, q_0), \\ q = \phi_5(u, x_0, y_0, z_0, p_0, q_0) \end{cases}$

be the equations which represent the integral of the system (79) which takes on the values x_0, y_0, z_0, p_0, q_0 respectively for $u = 0$. If we replace x_0, y_0, z_0, p_0, q_0 in these expressions by functions of a second auxiliary variable v, the equations (81) represent in general a surface S, u and v being regarded as two independent variables. In order that the surface S be an integral surface, and that the curves $v = $ const. be the characteristic curves, the equations (82) must give precisely the values of p and q which determine the tangent plane to that surface; and, moreover, the relation $F = 0$ must be satisfied at every point of S. Hence the five functions x, y, z, p, q of the two variables u and v must satisfy the three conditions

(83) $$F(x, y, z, p, q) = 0,$$

(84) $$\frac{\partial z}{\partial u} - p \frac{\partial x}{\partial u} - q \frac{\partial y}{\partial u} = 0,$$

(85) $$\frac{\partial z}{\partial v} - p \frac{\partial x}{\partial v} - q \frac{\partial y}{\partial v} = 0.$$

Since the five functions ϕ_i are integrals of the system (79), we have, as remarked above, $F(x, y, z, p, q) = F(x_0, y_0, z_0, p_0, q_0)$. Hence the relation (83) will be satisfied if

(86) $$F(x_0, y_0, z_0, p_0, q_0) = 0.$$

The relation (84) is identically satisfied, for it is a consequence of the differential equations (79). Cauchy transforms the condition (85) as follows: Indicating by H the left-hand side of (85), and differentiating with respect to u, we find

$$\frac{\partial H}{\partial u} = \frac{\partial^2 z}{\partial u \partial v} - p \frac{\partial^2 x}{\partial u \partial v} - q \frac{\partial^2 y}{\partial u \partial v} - \frac{\partial x}{\partial v} \frac{\partial p}{\partial u} - \frac{\partial y}{\partial v} \frac{\partial q}{\partial u}.$$

On the other hand, differentiating the relation (84) with respect to v, we have also

$$0 = \frac{\partial^2 z}{\partial u \partial v} - p \frac{\partial^2 x}{\partial u \partial v} - q \frac{\partial^2 y}{\partial u \partial v} - \frac{\partial p}{\partial v} \frac{\partial x}{\partial u} - \frac{\partial q}{\partial v} \frac{\partial y}{\partial u};$$

whence, subtracting,

$$\frac{\partial H}{\partial u} = \frac{\partial p}{\partial v} \frac{\partial x}{\partial u} + \frac{\partial q}{\partial v} \frac{\partial y}{\partial u} - \frac{\partial x}{\partial v} \frac{\partial p}{\partial u} - \frac{\partial y}{\partial v} \frac{\partial q}{\partial u},$$

or, replacing the derivatives with respect to u by their values obtained from the relations (79),

$$\frac{\partial H}{\partial u} = X \frac{\partial x}{\partial v} + Y \frac{\partial y}{\partial v} + P \frac{\partial p}{\partial v} + Q \frac{\partial q}{\partial v} + Z \left(p \frac{\partial x}{\partial v} + q \frac{\partial y}{\partial v} \right).$$

Finally, observing that the five functions x, y, z, p, q of v satisfy the relation

$$F(x, y, z, p, q) = 0,$$

and that we therefore have also

$$X \frac{\partial x}{\partial v} + Y \frac{\partial y}{\partial v} + Z \frac{\partial z}{\partial v} + P \frac{\partial p}{\partial v} + Q \frac{\partial q}{\partial v} = 0,$$

we may write the preceding value of $\partial H/\partial u$ in the form

(87) $$\frac{\partial H}{\partial u} = Z \left(p \frac{\partial x}{\partial v} + q \frac{\partial y}{\partial v} - \frac{\partial z}{\partial v} \right) = -ZH.$$

We derive from this relation the following value for H,

(88) $$H = H_0 e^{-\int_0^u Z\, du},$$

where H_0 denotes the value of H for $u = 0$, that is, when x, y, z, p, q reduce respectively to x_0, y_0, z_0, p_0, q_0. Since the function F, and consequently also the partial derivative Z, is supposed analytic in the neighborhood of the system of values x_0, y_0, z_0, p_0, q_0, the necessary and sufficient condition that H be zero is that H_0 be zero, that is, that $(\partial z/\partial v)_0 = p_0(\partial x/\partial v)_0 + q_0(\partial y/\partial v)_0$.

Summing up, *to obtain an integral surface* * *it is sufficient to replace* x_0, y_0, z_0, p_0, q_0 *in the equations* (80) *or* (81) *by functions of an auxiliary variable* v *which satisfy the two conditions*

$$(89) \quad F(x_0, y_0, z_0, p_0, q_0) = 0, \quad \frac{\partial z_0}{\partial v} = p_0 \frac{\partial x_0}{\partial v} + q_0 \frac{\partial y_0}{\partial v}.$$

This method leads very easily to the solution of Cauchy's problem. In fact, if we wish to determine an integral surface through a given curve Γ, we may take for x_0, y_0, z_0 the coördinates of any point of that curve expressed as functions of a variable parameter v, and the equations (89) then determine p_0 and q_0. The solution may also be stated in geometric language as follows: The first of the equations (89) expresses the fact that the plane through the point (x_0, y_0, z_0)

---

* The argument presumes, however, that the denominators P, Q, $X+pZ$, $Y+qZ$ are not all zero for the initial values x_0, y_0, z_0, p_0, q_0. In case they are, the equations (81) and (82) reduce to $x = x_0$, $y = y_0$, $z = z_0$, $p = p_0$, $q = q_0$, whereas if we suppress the auxiliary variable u, the equations (79) may have integrals which take on the given initial values (§ 31, *Note*). Hence the integrals of the given equation which satisfy also the four equations

$$P = 0, \quad Q = 0, \quad X + pZ = 0, \quad Y + qZ = 0$$

are not given by the general method. Such integrals, if there are any, are *singular* integrals. There exist normally no such integrals for an equation given a priori and not formed by eliminating constants.

The reasoning can be arranged so as to put in evidence the hypotheses necessary for the validity of the conclusions. Let us suppose first of all that the function $F(x, y, z, p, q)$ is an *analytic* function of x, y, z, p, q. In order to show that every integral $z = f(x, y)$ represents a locus of characteristic curves, it is not necessary to suppose that that integral is analytic; it is sufficient to assume that it has continuous partial derivatives of the second order r, s, t, since only these derivatives appear in the proof. The characteristic curves, being defined by a system of analytic differential equations, are necessarily analytic curves, and, consequently, on every integral surface, whether it is analytic or not, there exists a family of analytic curves, namely, the characteristic curves. The functions ϕ_1, ϕ_2, $\cdots$, ϕ_5, which represent the general integral of the equations (79), are analytic functions of u and of the initial values x_0, y_0, z_0, p_0, q_0 (§ 26). In order that the calculations which follow, and their conclusion, be rigorous, it is sufficient that these initial values be continuous functions of a parameter v, and that they have continuous derivatives, but it is not necessary that they shall be analytic functions of v.

This is quite in accord with the method of the variation of constants. If the complete integral $V(x, y, z, a, b)$ is an analytic function of its arguments, the same will be true of $F(x, y, z, p, q)$, but nothing in the argument requires that the arbitrary function $b = \phi(a)$ shall be an analytic function of a. A similar remark applies to the general integral of a linear equation. For more details on this subject see E. R. HEDRICK, *Ueber den analytischen Character der Lösungen von Differentialgleichungen* (*Inaugural-Dissertation, Göttingen,* 1901).

determined by the values p_0 and q_0 is tangent to the cone (T) whose vertex is that point; and the second of the equations (89) expresses the fact that this plane passes through the tangent to the curve Γ. Hence the whole process may be formulated as follows: *Through the tangent at the point M to the curve Γ pass a plane tangent to the cone (T) whose vertex is M; let C be the characteristic curve through the element thus determined; the surface generated by this characteristic curve, as the point M describes the curve Γ, is an integral surface through the curve Γ.*

There will be as many surfaces fulfilling these conditions as there are tangent planes to the cone (T) through a tangent to the curve Γ. It is also clear that we should associate tangent planes which form a continuous sequence.

Let us consider first the general case where the tangent to the curve Γ is not a generator of the cone (T). Since p_0 and q_0 fix the position of the tangent plane to the cone (T), the direction cosines of the element of contact of (T) with that plane are proportional to P_0, Q_0, $P_0 p_0 + Q_0 q_0$, by the formulæ (75) and (76). Since the difference $P_0(\partial y_0/\partial v) - Q_0(\partial x_0/\partial v)$ is not zero, the values of p_0 and of q_0 derived from the equations (89) are analytic functions of v in the neighborhood of the given point of Γ. On the other hand, we can solve the first two equations of (81) for u and v, for the functional determinant $\partial x/\partial u \, \partial y/\partial v - \partial y/\partial u \, \partial x/\partial v$ reduces for $u = 0$ to

$$\left(\frac{\partial x}{\partial u}\right)_0 \left(\frac{\partial y_0}{\partial v}\right) - \left(\frac{\partial y}{\partial u}\right)_0 \left(\frac{\partial x_0}{\partial v}\right),$$

that is, to $P_0(\partial y_0/\partial v) - Q_0(\partial x_0/\partial v)$. Substituting these values of u and v in the third of the equations (81), we see that z is an analytic function of x and y in the neighborhood of the given point (see § 84).

If the tangent at a *particular* point of the curve Γ coincides with the element of contact of (T) with the plane determined by the values p_0, q_0 at that point, this point is in general a singular point for the corresponding integral. If the same thing happens *at every point of* Γ, we must distinguish two cases, according as the curve Γ is a characteristic curve or not.

If the curve Γ is a characteristic curve, it is tangent at each of its points to an element G of the cone (T) whose vertex is at that point, and the characteristic developable surface is the envelope of the tangent plane to the cone (T) along the generator G when the vertex M describes the curve Γ. The characteristic curve through each of the elements thus determined coincides with the curve Γ itself, and the equations (81) do not define a surface. But it is clear that in this case the problem is indeterminate. For let M be a point of Γ, let P be the plane tangent to the cone (T) whose vertex is M along the tangent G to Γ, and let Γ' be another curve through M whose tangent at M is a straight line of the plane P different from G. From what we have just proved, the integral surface through Γ' contains the curve Γ.

If the given equation (74) is not linear in p and q, as we shall suppose, the curve Γ can be tangent at each of its points to a generator G of the corresponding cone (T) without being a characteristic curve. The family of curves having this property depends, in fact, upon an *arbitrary function*. Let

$$\Phi\left(x, y, z;\ \frac{Y-y}{X-x},\ \frac{Z-z}{X-x}\right)=0$$

be the equation of the cone (T) whose vertex is (x, y, z). In order that a curve Γ be tangent at each of its points to an element of (T), the coördinates x, y, z of a point of that curve must be functions of a variable v satisfying the condition

(90) $$\Phi\left(x, y, z;\ \frac{dy}{dx},\ \frac{dz}{dx}\right)=0.$$

If we take x, for example, as the independent variable, we may choose arbitrarily $y = f(x)$, and then, substituting $f(x)$ for y in the preceding relation, we have a differential equation of the first order for the determination of z as a function of x. Every curve not a characteristic satisfying the condition (90) will be called an *integral curve*.

Now let us suppose that the curve Γ, for which we wish to solve Cauchy's problem, is an integral curve. From each point M of Γ there issues a characteristic curve tangent to Γ, and it follows from the preceding arguments that the surface S generated by these characteristic curves is an integral surface. Indeed, it is sufficient to take for x_0, y_0, z_0 the coördinates of a point of Γ, and for p_0, q_0 the coefficients p and q of the plane tangent to (T) along the tangent to Γ. But this curve Γ is a singular line on the surface S; for if it were not, the derivatives r, s, t would have finite values in a point of Γ, and, since we have $Q_0 dx_0 = P_0 dy_0$, the arguments of page 251 to establish the equations (79) would apply without modification, and we should conclude that the curve Γ is a characteristic curve, which is contrary to the hypothesis. This curve Γ, which is the envelope of the characteristic curves of the surface S, is the analogue of the edge of regression of a developable surface.

Note. Cauchy's method also leads readily to a complete integral; for we can satisfy the conditions (89) by putting $x_0 = a$, $y_0 = b$, $z_0 = c$, where a, b, c are any three constants and where p_0 and q_0 satisfy the relation $F(a, b, c, p_0, q_0) = 0$. The integral surface thus obtained is the locus of the characteristic curves starting from the point (a, b, c), which is evidently a conical point for that surface. If we regard one of the coördinates a, b, c as a numerical constant, we have a complete integral.

Example 1. Let us consider the equation treated by Cauchy, $pq - xy = 0$. Making use of the equation itself, we see that the differential equations of the characteristic curves can be written in the form

$$p\,dx = q\,dy = \frac{dz}{2} = x\,dp = y\,dq.$$

We derive from them successively the integrable combinations

$$\frac{dp}{p} = \frac{dx}{x}, \qquad \frac{dq}{q} = \frac{dy}{y}, \qquad dz = \frac{p}{x} 2x\,dx = \frac{q}{y} 2y\,dy,$$

and the characteristic strip through the element $(x_0, y_0, z_0, p_0, q_0)$ is represented by the equations

$$\frac{p}{p_0} = \frac{x}{x_0}, \qquad \frac{q}{q_0} = \frac{y}{y_0}, \qquad z - z_0 = \frac{p_0}{x_0}(x^2 - x_0^2) = \frac{q_0}{y_0}(y^2 - y_0^2),$$

where x_0, y_0, p_0, q_0 are connected by the relation $p_0 q_0 = x_0 y_0$. In order to obtain the integral which, for $x = x_0$, reduces to $\phi(y)$, we shall put, as in the general method, $y_0 = v$, $z_0 = \phi(v)$. In this case the equations (89) give

$$q_0 = \phi'(v), \qquad p_0 = \frac{x_0 v}{\phi'(v)}.$$

The required integral is therefore represented by the simultaneous system of two equations

$$z - \phi(v) = \frac{v}{\phi'(v)}(x^2 - x_0^2) = \frac{\phi'(v)}{v}(y^2 - v^2),$$

which define v and z as functions of x and y. These two equations may be replaced by the equations

$$[z - \phi(v)]^2 = (x^2 - x_0^2)(y^2 - v^2), \qquad [z - \phi(v)]\phi'(v) = v(x^2 - x_0^2),$$

of which the second may be obtained from the first by differentiating with respect to the parameter v. The desired integral can be obtained by eliminating v, and it follows that this result is quite in accord with Lagrange's theory.

Example 2. Let us consider again the equation of page 240,

$$(1 + p^2 + q^2)z^2 - R^2 = 0,$$

which states that the length of the segment of the normal cut off by the xy-plane is equal to R. Hence, in order to obtain the normal cone (N) at the point M of space, we need only describe about the point M as center a sphere of radius R, and then take the cone of revolution whose vertex is M through the circle in which the xy-plane cuts this sphere. The corresponding tangent cone (T) is the cone of revolution whose vertex is M. We know here a complete integral, the spheres of radius R having their centers in the xy-plane. The characteristic curves, which are the limiting positions of the intersections of two spheres that are an infinitesimal distance apart (see § 86), are therefore circles of radius R, whose planes are parallel to the z-axis and whose centers are in the xy-plane. Every integral curve, as we have seen, may be regarded as the envelope of the characteristic curves on an integral surface. These curves are therefore represented by the system of three equations,

$$(x - a)^2 + [y - \phi(a)]^2 + z^2 - R^2 = 0,$$
$$x - a + [y - \phi(a)]\phi'(a) = 0,$$
$$1 + \phi'^2(a) + \phi(a)\phi''(a) - y\phi''(a) = 0,$$

where $\phi(a)$ is an arbitrary function.

86. The characteristic curves derived from a complete integral.

The concept of characteristic curves can be derived in a very natural manner from Lagrange's theory. We have seen, in fact, that if $V = 0$ is a complete integral of a given equation of the first order, we obtain an integral surface by eliminating a from the two equations

$$(91) \qquad V[x, y, z, a, \phi(a)] = 0, \qquad \frac{\partial V}{\partial a} + \frac{\partial V}{\partial \phi(a)} \phi'(a) = 0,$$

where $\phi(a)$ is an arbitrary function. If we give to the parameter a a constant value, these two equations represent a curve whose locus is the integral surface. The equations of this curve are of the form

$$(92) \qquad V(x, y, z, a, b) = 0, \qquad \frac{\partial V}{\partial a} + \frac{\partial V}{\partial b} c = 0,$$

where a, b, c are arbitrary parameters. These curves form a *complex*, and we see that the integral surfaces are generated by the curves of this complex associated according to a suitable law. The name *characteristic curves* is self-explanatory, since they are the curves of contact of the complete integral with its envelope.

The characteristic developable surfaces also appear in a natural manner. Let us consider a characteristic curve corresponding to the values a_0, b_0, c_0 of the parameters a, b, c. All the integral surfaces obtained by means of functions ϕ, such that we have $b_0 = \phi(a_0)$, $c_0 = \phi'(a_0)$, pass through this curve and are tangent to each other along this entire curve, for the values of p and q, which for any point of an integral surface are given by the relations

$$(93) \qquad \frac{\partial V}{\partial x} + p \frac{\partial V}{\partial z} = 0, \qquad \frac{\partial V}{\partial y} + q \frac{\partial V}{\partial z} = 0,$$

are the same for all these surfaces. It is therefore natural to associate with each characteristic curve a characteristic developable surface passing through this curve. The four equations (92) and (93) enable us to express four of the variables x, y, z, p, q in terms of one of them and of the three arbitrary constants a, b, c. In order to prove the identity of the forms thus defined with those of the characteristic strips deduced from Cauchy's method, let us suppose that the complete integral is represented by an equation of the form

$$z = \Phi(x, y, a, b).$$

The equations (92) and (93) then become

$$(94) \qquad z = \Phi(x, y, a, b), \qquad \frac{\partial \Phi}{\partial a} + \frac{\partial \Phi}{\partial b} c = 0,$$

$$(95) \qquad p = \frac{\partial \Phi}{\partial x}, \qquad q = \frac{\partial \Phi}{\partial y}.$$

The relations (94) and (95) enable us to express the five variables (x, y, z, p, q) in terms of one of them (x, for example) and of the three arbitrary constants a, b, c. The proof reduces to showing that these functions satisfy the differential equations (64). Since the function $\Phi(x, y, a, b)$ is a complete integral of the equation $F = 0$, we have already between these functions the two relations

$$(96) \qquad F(x, y, z, p, q) = 0, \qquad dz = p\,dx + q\,dy.$$

On the other hand, we deduce from the second equation of (94)

(97) $$\left(\frac{\partial^2 \Phi}{\partial a\, \partial x} + c\, \frac{\partial^2 \Phi}{\partial b\, \partial x}\right) dx + \left(\frac{\partial^2 \Phi}{\partial a\, \partial y} + c\, \frac{\partial^2 \Phi}{\partial b\, \partial y}\right) dy = 0.$$

Now if we differentiate with respect to the constants a and b the identity

$$F\left(x, y, \Phi, \frac{\partial \Phi}{\partial x}, \frac{\partial \Phi}{\partial y}\right) = 0,$$

we find

$$Z \frac{\partial \Phi}{\partial a} + P \frac{\partial^2 \Phi}{\partial a\, \partial x} + Q \frac{\partial^2 \Phi}{\partial a\, \partial y} = 0,$$

$$Z \frac{\partial \Phi}{\partial b} + P \frac{\partial^2 \Phi}{\partial b\, \partial x} + Q \frac{\partial^2 \Phi}{\partial b\, \partial y} = 0;$$

and consequently, by eliminating Z, we have

(98) $$P\left(\frac{\partial^2 \Phi}{\partial a\, \partial x} + c\, \frac{\partial^2 \Phi}{\partial b\, \partial x}\right) + Q\left(\frac{\partial^2 \Phi}{\partial a\, \partial y} + c\, \frac{\partial^2 \Phi}{\partial b\, \partial y}\right) = 0.$$

A comparison of the two relations (97) and (98) shows that we have $dx/P = dy/Q$. The remaining equations of (64) are established as in § 85, by comparing the relations

$$dp = \frac{\partial^2 \Phi}{\partial x^2} dx + \frac{\partial^2 \Phi}{\partial x\, \partial y} dy, \qquad dq = \frac{\partial^2 \Phi}{\partial x\, \partial y} dx + \frac{\partial^2 \Phi}{\partial y^2} dy,$$

which are deduced from the equations (95), with the relations

$$X + Z \frac{\partial \Phi}{\partial x} + P \frac{\partial^2 \Phi}{\partial x^2} + Q \frac{\partial^2 \Phi}{\partial x\, \partial y} = 0,$$

$$Y + Z \frac{\partial \Phi}{\partial y} + P \frac{\partial^2 \Phi}{\partial x\, \partial y} + Q \frac{\partial^2 \Phi}{\partial y^2} = 0,$$

which in turn are obtained by differentiating the identity

$$F\left(x, y, \Phi, \frac{\partial \Phi}{\partial x}, \frac{\partial \Phi}{\partial y}\right) = 0$$

with respect to the variables x and y.

Note. The theory of the complete integral applies to linear equations as well as to the non-linear equations. It seems at first sight, on the contrary, that Cauchy's method is altogether different for linear equations and for non-linear equations. In fact, the characteristic curves of a linear equation, or of an equation which separates into several linear equations, form a congruence and not a complex. But if we associate with each characteristic curve a characteristic developable surface, the paradox disappears. Each characteristic curve belongs, in fact, to an infinite number of characteristic developable surfaces which depend upon an arbitrary constant, so that this family of characteristic strips does depend upon three arbitrary constants. Let us consider, for example, the equation of the cones $px + qy - z = 0$. The equation $z = ax + by$ represents a complete integral formed by all planes P through the origin. The characteristic curves are the straight lines passing through the origin, and the characteristic developable surfaces are the planes P themselves. We shall therefore obtain a characteristic strip by associating with a straight line through the origin a plane through that straight line; this set actually depends upon three arbitrary constants.

87. Extension of Cauchy's method. Cauchy's method can be extended without difficulty to an equation in any number of independent variables,

(99) $\quad F(x_1, x_2, \cdots, x_n; z, p_1, p_2, \cdots, p_n) = 0.\quad \left(p_i = \dfrac{\partial z}{\partial x_i}\right)$

Let $z = \Phi(x_1, x_2, \cdots, x_n)$ be any integral of the equation (99); we shall designate as an *element* of this integral the set which consists of a system of particular values $x_1^0, x_2^0, \cdots, x_n^0$ of the independent variables, together with the corresponding values $z^0, p_1^0, \cdots, p_n^0$ of the function Φ and its partial derivatives. Let us suppose that an element of the integral, starting with certain initial values x_i^0, z^0, p_i^0, varies so as always to satisfy the differential equations

(100) $\quad \dfrac{dx_1}{P_1} = \dfrac{dx_2}{P_2} = \cdots = \dfrac{dx_n}{P_n},$

where, as in § 83,

$$X_i = \dfrac{\partial F}{\partial x_i}, \qquad P_k = \dfrac{\partial F}{\partial p_k}, \qquad Z = \dfrac{\partial F}{\partial z}.$$

It is clear that these equations determine completely a family of curves (or one-dimensional manifolds) on each integral. For if z is known as a function of $x_1, x_2, \cdots, x_n$, the same thing is true of the partial derivatives p_i, and consequently of the functions P_i. These relations (100) form, therefore, a system of $(n-1)$ differential equations of the first order between the n variables $(x_1, x_2, \cdots, x_n)$. By the theory of differential equations, through each point of the integral surface there passes in general one and only one of these manifolds. If to each point $(x_1, x_2, \cdots, x_n, z)$ of one of these manifolds we associate the corresponding values of $p_1, p_2, \cdots, p_n$, we have a simply infinite sequence of elements, which we may again call a characteristic strip. We shall show that, without knowing the expression for the function z, we can adjoin to the relations (100) other differential equations enabling us to define completely the variation of the variables x_i, z, p_k along a characteristic curve.

Let us start from an element of the integral (x_i^0, z^0, p_k^0), and let us consider the characteristic strip through this element. Along this characteristic strip the variables x_i, z, p_k are functions of a single independent variable satisfying the relation $F = 0$, whose differentials satisfy the equations (100) and also the relations

$$dz = p_1 dx_1 + \cdots + p_n dx_n, \qquad dp_i = p_{i1} dx_1 + \cdots + p_{in} dx_n,$$

$$p_{ik} = \dfrac{\partial^2 z}{\partial x_i \partial x_k}, \qquad\qquad (i = 1, 2, \cdots, n)$$

which result from the definition. Differentiating the relation $F = 0$ with respect to the variable x_i, we find

$$X_i + p_i Z + P_1 p_{i1} + \cdots + P_n p_{in} = 0.$$

Indicating by du the common value of the quotients (100), and replacing P_i in the preceding relation by dx_i/du, we find

$$(X_i + p_i Z) du + p_{i1} dx_1 + \cdots + p_{in} dx_n = 0,$$

that is,
$$(X_i + p_i Z) du + dp_i = 0.$$

This shows us that the elements of an integral satisfy, along the entire length of a characteristic strip, the system of differential equations,

$$(101) \quad \frac{dx_i}{P_i} = \frac{dz}{P_1 p_1 + \cdots + P_n p_n} = \frac{-dp_k}{X_k + Z p_k} = du. \quad (i, k = 1, 2, \cdots, n)$$

These equations do not depend upon the function Φ; hence we can determine the successive elements of a characteristic strip, provided that we know a single element (x_i^0, z^0, p_k^0). We conclude from this, just as before (§ 85), that *if two integrals have a common element, they have in common all the elements of the characteristic strip through that element.*

If, as we shall assume, the denominators of the equations (101) remain finite and are not all zero for the initial values, we derive from these equations

$$(102) \quad \begin{cases} x_i = f_i(u, x_i^0, z^0, p_k^0), \\ p_k = \phi_k(u, x_i^0, z^0, p_k^0), \\ z = \psi(u, x_i^0, z^0, p_k^0), \end{cases}$$

where x_i^0, z^0, p_k^0 denote the initial values corresponding to the initial value $u = 0$ of the auxiliary variable u, and where the functions f_i, ϕ_k, ψ are continuous differentiable functions of u and of the initial values, at least within certain limits.

Since each integral is a locus of characteristic curves, it is clear that every integral will be represented by the equations (102), where x_i^0, z^0, p_k^0 must be functions of $n-1$ independent variables, so that these equations represent, in fact, a manifold of n dimensions. But in addition these $2n+1$ functions x_i, z, p_k of n independent variables must satisfy the relations

$$(103) \quad \begin{cases} F(x_i, z, p_k) = F(x_1, \cdots, x_n, z; p_1, \cdots, p_n) = 0, \\ dz - p_1 dx_1 - p_2 dx_2 - \cdots - p_n dx_n = 0. \end{cases}$$

Since the differential equations (101) have the integrable combination $dF = 0$, the first of the relations of (103) will surely be satisfied if $F(x_i^0, z^0, p_k^0) = 0$. On the other hand, we also have from the equations (101)

$$\frac{dz}{du} = p_1 \frac{dx_1}{du} + \cdots + p_n \frac{dx_n}{du}.$$

Since the initial values x_i^0, z^0, p_k^0 are functions of $n-1$ independent variables $v_1, v_2, \cdots, v_{n-1}$, we must also have

$$U = \delta z - p_1 \delta x_1 - \cdots - p_n \delta x_n = 0,$$

where the letter δ denotes the differentials corresponding to arbitrary increments $\delta v_1, \cdots, \delta v_{n-1}$ of these variables. By proceeding as in the case for $n = 2$, we have necessarily

$$dU = d\delta z - p_1 d\delta x_1 - \cdots - p_n d\delta x_n - dp_1 \delta x_1 - \cdots - dp_n \delta x_n,$$
$$dz = p_1 dx_1 + \cdots + p_n dx_n,$$
$$\delta dz = p_1 \delta dx_1 + \cdots + p_n \delta dx_n + \delta p_1 dx_1 + \cdots + \delta p_n dx_n,$$

and, since we may interchange the order of the operations d and δ,

$$dU = \sum_{i=1}^{n} \{\delta p_i dx_i - dp_i \delta x_i\}$$
$$= \sum_{i=1}^{n} \{P_i \delta p_i + (X_i + p_i Z) \delta x_i\} du.$$

Since z, x_i, p_k satisfy the equation $F = 0$, we have

$$\sum_i (P_i \delta p_i + X_i \delta x_i) = -Z \delta z$$

and, consequently,

$$dU = -ZU\, du.$$

From this we find the following expression for U:

$$U = U_0 e^{-\int_0^u Z du}.$$

In order that U shall be zero, it is necessary and sufficient that U_0 be zero, that is, that we have

$$\delta z^0 - p_1^0 \delta x_1^0 - \cdots - p_n^0 \delta x_n^0 = 0.$$

To sum up, *in order that the equations* (102) *represent an integral, it is necessary and sufficient that the initial values* (x_i^0, z^0, p_k^0) *be* *functions of $n-1$ independent variables satisfying identically the conditions*

(104) $$F(x_i^0, z^0, p_k^0) = 0,$$
(105) $$\delta z^0 - p_1^0 \delta x_1^0 - p_2^0 \delta x_2^0 - \cdots - p_n^0 \delta x_n^0 = 0.$$

Every system of $2n+1$ functions (x_i^0, z^0, p_k^0) of $n-1$ variables satisfying these conditions defines an $(n-1)$-dimensional manifold of elements. Again, we may say that every integral of the equation $F = 0$ is generated by the characteristic curves through the different elements of a manifold of this kind.

In particular, to obtain Cauchy's integral, which for $x_1 = x_1^0$ reduces to a given function $\Phi(x_2, \cdots, x_n)$, if we take $x_2^0, x_3^0, \cdots, x_n^0$ for independent variables (x_1^0 being supposed constant), the relation (105) gives the values of $z^0, p_2^0, \cdots, p_n^0$,

$$z^0 = \Phi(x_2^0, \cdots, x_n^0), \qquad p_2^0 = \frac{\partial \Phi}{\partial x_2^0}, \quad \cdots, \quad p_n^0 = \frac{\partial \Phi}{\partial x_n^0}.$$

The value of p_1^0 can be obtained from the relation (104). If P_1^0 is not zero (as we must assume in order to apply the general existence theorem of I, § 194, 2d ed.; § 188, 1st ed.), p_1^0 will be an analytic function of $x_2^0, \cdots, x_n^0$ in a certain region, and the equations (102) will give, for z, x_i, p_k, analytic functions of $u, x_2^0, \cdots, x_n^0$. Moreover, the Jacobian

$$\frac{D(x_1, x_2, \cdots, x_n)}{D(u, x_2^0, \cdots, x_n^0)}$$

is not zero, for it reduces to P_1^0 for $u = 0$. Hence we can solve the first n equations (102) for $u, x_2^0, \cdots, x_n^0$, and, putting these expressions in the last of the equations (102), we obtain for z an analytic function of the variables $x_1, x_2, \cdots, x_n$.

Note. It may happen that the application of the preceding general rule does not lead to an integral. For example, it might turn out that the manifold of elements defined by the equations (102) does not really depend upon n arbitrary parameters. This is what would happen if the manifold formed by the elements (x_i^0, z^0, p_k^0) were composed of characteristic strips; in this case, in fact, the manifold defined by the equations (102) would coincide with the manifold of the elements (x_i^0, z^0, p_k^0).

Disregarding this case, it may also happen that the elimination of the parameters $u, v_1, \cdots, v_{n-1}$ from the equations (102) leads to *several* distinct relations between the variables $x_1, \cdots, x_n, z$. In order not to reject such solutions, we agree with Sophus Lie to enlarge the definition of the integral and to designate as an *integral* of the equation $F = 0$ every system of ∞^n elements (x_i, z, p_k) satisfying the relations

(106) $$F(x_i, z, p_k) = 0, \qquad dz = p_1 dx_1 + \cdots + p_n dx_n.$$

IV. SIMULTANEOUS EQUATIONS*

88. Linear homogeneous systems. Let us consider a system of q linear homogeneous equations in one unknown, f,

$$(107) \quad \begin{cases} X_1(f) = a_{11}\dfrac{\partial f}{\partial x_1} + a_{21}\dfrac{\partial f}{\partial x_2} + \cdots + a_{n1}\dfrac{\partial f}{\partial x_n} = 0, \\ X_2(f) = a_{12}\dfrac{\partial f}{\partial x_1} + a_{22}\dfrac{\partial f}{\partial x_2} + \cdots + a_{n2}\dfrac{\partial f}{\partial x_n} = 0, \\ \cdots\cdots\cdots\cdots\cdots\cdots\cdots\cdots\cdots\cdots\cdots, \\ X_q(f) = a_{1q}\dfrac{\partial f}{\partial x_1} + a_{2q}\dfrac{\partial f}{\partial x_2} + \cdots + a_{nq}\dfrac{\partial f}{\partial x_n} = 0, \end{cases}$$

where the coefficients a_{ik} are functions of the n independent variables $x_1, x_2, \cdots, x_n$ and do not contain the unknown function f. The q equations (107) are said to be *independent* if there does not exist any identical relation of the form

$$\lambda_1 X_1(f) + \cdots + \lambda_q X_q(f) = 0,$$

where $\lambda_1, \lambda_2, \cdots, \lambda_q$ are functions of $x_1, x_2, \cdots, x_n$ *not all zero*. It is clear that every system of q equations that are not independent can be replaced by a system of q' independent equations ($q' < q$) equivalent to the first, and that no system can contain more than n independent equations.

We can therefore always suppose the q equations (107) independent and $q \leqq n$.

If $q = n$, and if the equations (107) are independent, the determinant of the coefficients a_{ik} is not zero, and these equations have no other common integral than the trivial solution $f = C$, which we shall hereafter discard. If q is less than n, we can always find the integrals common to the equations (107) by successive integrations. In fact, let us suppose that we have integrated one of these equations (the first, for example), and let $y_1, y_2, \cdots, y_{n-1}$ be a system of $n-1$ independent integrals. Again, let y_n be another function such that the Jacobian $D(y_1, y_2, \cdots, y_n)/D(x_1, x_2, \cdots, x_n)$ is not zero. Then we may take $y_1, y_2, \cdots, y_n$ for new independent variables, and the equation $X_1(f) = 0$ becomes $\partial f / \partial y_n = 0$ by this change of variables, while the equation $X_i(f) = 0$ ($i > 1$) is replaced by an equation

---

* We shall limit ourselves to an indication of the principal methods in their essential features. For further details the reader is referred to E. GOURSAT, *Sur l'intégration des équations aux dérivées partielles du premier ordre* (Paris, Hermann et fils, 1892).

of the same form, in which the term in $\partial f/\partial y_n$ may be suppressed. This new equation may be written in the form

$$Y_i(f) = b_{1i}\frac{\partial f}{\partial y_1} + \cdots + b_{n-1,i}\frac{\partial f}{\partial y_{n-1}} = 0,$$

where the coefficients b_{ki} are functions of $y_1, y_2, \cdots, y_n$. If we suppose the coefficients b_{ki} arranged according to powers of y_n, this equation can be written in the form

$$Y_i(f) = \left(c_{1i}\frac{\partial f}{\partial y_1} + \cdots + c_{n-1,i}\frac{\partial f}{\partial y_{n-1}}\right)$$
$$+ y_n\left(c'_{1i}\frac{\partial f}{\partial y_1} + \cdots + c'_{n-1,i}\frac{\partial f}{\partial y_{n-1}}\right) + y_n^2(\cdots),$$

where the coefficients $c_{1i}, c'_{1i}, \cdots$ are independent of y_n. Since the unknown function f must be independent of y_n, this function must satisfy all the linear equations which are obtained by equating to zero all the coefficients of the different powers of y_n. Suppose that we proceed in this way with all the equations $X_i(f) = 0$ $(i > 1)$. If the system formed by all the independent equations which we thus obtain contains $n - 1$ equations, the only solution is $f = C$. If not, the system will be composed of r linear independent equations ($r < n - 1$). We may operate in the same way on an equation of the new system, and so on in the same manner. Since at each operation the number of independent variables is diminished by unity, it is easy to see that the given system has no other integral than $f = C$, or else it reduces to a system composed of a single linear equation.

This method, which may be easily applied in certain cases, is evidently very imperfect from a theoretical point of view, since it does not enable us to determine a priori whether the equations (107) have common integrals other than $f = C$. We shall now show that this question can be settled without any integrations.

Let f be an integral common to the equations (107). This function satisfies the two relations $X_i(f) = 0$, $X_k(f) = 0$, where i and k are any two of the indices $1, 2, \cdots, q$. We also have

$$X_i[X_k(f)] = X_i(0) = 0, \qquad X_k[X_i(f)] = X_k(0) = 0,$$

and, consequently,

$$X_i[X_k(f)] - X_k[X_i(f)] = 0.$$

We have already observed that this new equation contains only derivatives of the first order (§ 36), and that it may be written in the form

$$X_i[X_k(f)] - X_k[X_i(f)] = \sum_{h=1}^n \{X_i(a_{hk}) - X_k(a_{hi})\}\frac{\partial f}{\partial x_h} = 0.$$

Suppose that we form all the equations, similar to the preceding, obtained by combining any two of the given equations. These equations have all the integrals of the system (107). Let us indicate by

$$X_{q+1}(f) = 0, \qquad X_{q+2}(f) = 0, \qquad \cdots, \qquad X_{q+s}(f) = 0$$

all those of these new equations which are independent of each other and which form with the equations (107) a system

(108) $\qquad \begin{cases} X_1(f) = 0, & \cdots, & X_q(f) = 0, \\ X_{q+1}(f) = 0, & \cdots, & X_{q+s}(f) = 0 \end{cases}$

of independent equations. If $q + s = n$, the system (108), and consequently the system (107), has only the solution $f = C$. If $q + s < n$, we repeat on the system (108) the operations performed on the first system, and so on in the same manner. Continuing in this fashion, we finally obtain either a system of n independent equations, in which case the system (107) will have only the solution $f = C$, or else a system of r independent equations ($r < n$) such that all the combinations $X_i[X_k(f)] - X_k[X_i(f)]$ are linear combinations of $X_1(f), \cdots, X_r(f)$. Such a system has been called by Clebsch a *complete system*.

It follows, then, that *the search for the integrals of a system of the form* (107) *leads to the integration of a complete system*.

Since it is clear that every system of n linear independent equations is a complete system, we may say that every linear system reduces to a complete system.

89. Complete systems. The theory of complete systems rests upon the following properties:

1) *Every complete system is transformed into a complete system by any change of variables.*

Let
$$x_i = \phi_i(y_1, y_2, \cdots, y_n) \qquad (i = 1, 2, \cdots, n)$$

be the formulæ that define a change of variables such that we can express also the variables $y_1, y_2, \cdots, y_n$ in terms of the variables $x_1, x_2, \cdots, x_n$. By means of such a transformation every symbol of the type

$$X(f) = a_1 \frac{\partial f}{\partial x_1} + \cdots + a_n \frac{\partial f}{\partial x_n},$$

where $a_1, a_2, \cdots, a_n$ are functions of $x_1, x_2, \cdots, x_n$, changes into an expression of the same form, $Y(f) = b_1 \partial f/\partial y_1 + \cdots + b_n \partial f/\partial y_n$, where $b_1, b_2, \cdots, b_n$ are functions of $y_1, \cdots, y_n$. We have identically $X(f) = Y(f)$, where the letter f on the left-hand side denotes any function of $x_1, x_2, \cdots, x_n$, and on the right-hand side the same function expressed in terms of the variables $y_1, y_2, \cdots, y_n$.

Now let

(109) $\qquad X_1(f) = 0, \quad \cdots, \quad X_r(f) = 0$

be a complete system. By means of such a transformation this system goes over into the system

(110) $\qquad Y_1(f) = 0, \quad \cdots, \quad Y_r(f) = 0,$

where $X_i(f) = Y_i(f)$ identically, with the understanding just mentioned concerning the interpretation of f on the two sides. *This new system is also a complete system.* For, since we have identically

$$X_i(f) = Y_i(f), \qquad X_k(f) = Y_k(f)$$

for any function f, we also have

$$X_i[X_k(f)] = Y_i[X_k(f)] = Y_i[Y_k(f)],$$
$$X_k[X_i(f)] = Y_k[X_i(f)] = Y_k[Y_i(f)],$$

and, consequently,

$$X_i[X_k(f)] - X_k[X_i(f)] = Y_i[Y_k(f)] - Y_k[Y_i(f)].$$

Since by hypothesis the system (109) is complete, we have for any two indices i and k

$$X_i[X_k(f)] - X_k[X_i(f)] = \lambda_1 X_1(f) + \cdots + \lambda_r X_r(f).$$

Hence, after the transformation, we have

$$Y_i[Y_k(f)] - Y_k[Y_i(f)] = \lambda'_1 Y_1(f) + \cdots + \lambda'_r Y_r(f),$$

where $\lambda'_1, \cdots, \lambda'_r$ indicate the results obtained by replacing $x_1, x_2, \cdots, x_n$ in $\lambda_1, \cdots, \lambda_r$ by their expressions in terms of $y_1, \cdots, y_n$. The new system is therefore a complete system.

2) *Every system equivalent to a complete system is also a complete system.*

A system of r linear homogeneous equations in $\partial f/\partial x_i$,

(109′) $\qquad Z_1(f) = 0, \quad \cdots, \quad Z_r(f) = 0,$

is said to be *equivalent* to the system (109) if we have r identities of the form

$$Z_k(f) = A_{1k} X_1(f) + A_{2k} X_2(f) + \cdots + A_{rk} X_r(f), \qquad (k = 1, 2, \cdots, r)$$

where the coefficients A_{ik} are functions of $x_1, x_2, \cdots x_n$ whose determinant is not zero. In that case we can express $X_1(f), \cdots, X_r(f)$ linearly in terms of $Z_1(f), \cdots, Z_r(f)$, and the name *equivalent systems* is self-explanatory. The difference $Z_i[Z_k(f)] - Z_k[Z_i(f)]$ can now be written in the form

$$\sum_{\lambda=1}^{r} A_{\lambda i} X_\lambda \left[\sum_{l=1}^{r} A_{lk} X_l(f) \right] - \sum_{l=1}^{r} A_{lk} X_l \left[\sum_{\lambda=1}^{r} A_{\lambda i} X_\lambda(f) \right];$$

hence it is equal to a sum of terms of the form

$$A_{\lambda i} A_{lk} \{ X_\lambda[X_l(f)] - X_l[X_\lambda(f)] \} + A_{\lambda i} X_\lambda(A_{lk}) X_l(f) - A_{lk} X_l(A_{\lambda i}) X_\lambda(f).$$

If the system (109) is complete, this difference will therefore be a linear function of $X_1(f), \cdots, X_r(f)$, since all the differences

$$X_\lambda[X_l(f)] - X_l[X_\lambda(f)]$$

are, by hypothesis, linear functions of $X_1(f), \cdots, X_r(f)$. Since the two systems (109) and (109') are equivalent, all the differences

$$Z_i[Z_k(f)] - Z_k[Z_i(f)]$$

can be expressed linearly in terms of $Z_1(f), Z_2(f), \cdots, Z_r(f)$.

It is clear that every complete system can be replaced by an equivalent system in an infinite number of ways. We say that the complete system (109) is a *Jacobian system* if all the expressions $X_i[X_k(f)] - X_k[X_i(f)]$ are identically zero. We shall now show that *every complete system is equivalent to a Jacobian system.*

Since the r equations (109) are independent by hypothesis, we can solve them for r of the derivatives of f, for example, for the derivatives $\partial f/\partial x_1, \cdots, \partial f/\partial x_r$. Since the system thus obtained,

$$(111) \begin{cases} Z_1(f) = \dfrac{\partial f}{\partial x_1} + b_{11} \dfrac{\partial f}{\partial x_{r+1}} + \cdots + b_{1, n-r} \dfrac{\partial f}{\partial x_n} = 0, \\ Z_2(f) = \dfrac{\partial f}{\partial x_2} + b_{21} \dfrac{\partial f}{\partial x_{r+1}} + \cdots + b_{2, n-r} \dfrac{\partial f}{\partial x_n} = 0, \\ \cdots \cdots \cdots \cdots \cdots \cdots \cdots \cdots \cdots \cdots \cdots \cdots \\ Z_r(f) = \dfrac{\partial f}{\partial x_r} + b_{r1} \dfrac{\partial f}{\partial x_{r+1}} + \cdots + b_{r, n-r} \dfrac{\partial f}{\partial x_n} = 0, \end{cases}$$

is equivalent to the system (109), it also is a complete system. Now if we form the expressions $Z_i[Z_k(f)] - Z_k[Z_i(f)]$, it is clear that only the derivatives $\partial f/\partial x_{r+1}, \cdots, \partial f/\partial x_n$ will appear, and consequently the new equations $Z_i[Z_k(f)] - Z_k[Z_i(f)] = 0$ can be linear combinations of the equations (111) only if the left-hand sides of these new equations are identically zero. The system (111) is therefore a Jacobian system.

The reasoning proves that every complete system of the special form (111) is a Jacobian system, but it is clear that a Jacobian system is not necessarily of that form.

3) *Every complete system of r equations in n independent variables can be reduced by the integration of one of the equations of the system to a complete system of $r-1$ equations in $n-1$ independent variables.*

Suppose that we have integrated one of the equations of the system, for example, the equation $X_1(f) = 0$, and that we choose a new system of independent variables $(y_1, y_2, \cdots, y_n)$, as in the preceding paragraph, in such a way that $y_2, y_3, \cdots, y_n$ are $n-1$ integrals of $X_1(f) = 0$. The system (109) is replaced by a new complete system in which the first equation reduces to $\partial f/\partial y_1 = 0$. Solving the $r-1$ remaining equations for the $r-1$ derivatives $\partial f/\partial y_2, \cdots, \partial f/\partial y_r$, for example, we obtain a complete system,

$$(112) \quad \begin{cases} Y_1(f) = \dfrac{\partial f}{\partial y_1} = 0, \\ Y_2(f) = \dfrac{\partial f}{\partial y_2} + c_{21}\dfrac{\partial f}{\partial y_{r+1}} + \cdots + c_{2,n-r}\dfrac{\partial f}{\partial y_n} = 0, \\ \quad\cdots\cdots\cdots\cdots\cdots\cdots\cdots\cdots\cdots\cdots, \\ Y_r(f) = \dfrac{\partial f}{\partial y_r} + c_{r1}\dfrac{\partial f}{\partial y_{r+1}} + \cdots + c_{r,n-r}\dfrac{\partial f}{\partial y_n} = 0, \end{cases}$$

which is of the special form (111) and which is therefore a Jacobian system. Now we have

$$Y_1[Y_i(f)] - Y_i[Y_1(f)] = \frac{\partial c_{i1}}{\partial y_1}\frac{\partial f}{\partial y_{r+1}} + \cdots + \frac{\partial c_{i,n-r}}{\partial y_1}\frac{\partial f}{\partial y_n},$$

and since this expression must be identically zero, we see that the coefficients c_{ik} of the new system are independent of the variable y_1. Moreover, for $i > 1$, $k > 1$ we have identically

$$Y_i[Y_k(f)] - Y_k[Y_i(f)] = 0;$$

consequently the $r-1$ equations

$$(113) \quad Y_2(f) = 0, \quad Y_3(f) = 0, \quad \cdots, \quad Y_r(f) = 0$$

form a Jacobian system of $r-1$ equations in $n-1$ independent variables $y_2, y_3, \cdots, y_n$, which establishes the proposition.

The system (113) can in turn be reduced to a complete system of $r-2$ equations in $n-2$ independent variables, and so on in this way. Continuing in this manner, we finally reduce the given complete system to *one* linear equation in $n-r+1$ independent variables. We conclude from this that *every complete system of r equations in n independent variables has $n-r$ independent integrals, and the general integral of the system is an arbitrary function of these $n-r$ particular integrals.*

The preceding reasoning shows also what are the integrations to be carried out in order to obtain these integrals. Moreover, it is clear that this method can be applied in a variety of ways. We may, in fact, replace the given complete system by any other equivalent system, and begin by integrating any one of the equations of this new system. For example, if we replace the complete system by a Jacobian system of the form (111), we know at once $r-1$ particular integrals $x_2, \ldots, x_r$ of the equation $Z_1(f) = 0$, and it is sufficient to integrate a system of $n-r$ ordinary differential equations in order to have the general integral. For complete details of other methods of integration of complete systems, the reader is referred to special treatises.

Example. Let it be required to integrate the system

$$(114) \quad \begin{cases} X_1(f) = \dfrac{\partial f}{\partial x_1} + (x_2 + x_4 - 3x_1)\dfrac{\partial f}{\partial x_3} + (x_3 + x_1 x_2 + x_1 x_4)\dfrac{\partial f}{\partial x_4} = 0, \\ X_2(f) = \dfrac{\partial f}{\partial x_2} + (x_3 x_4 - x_2)\dfrac{\partial f}{\partial x_3} + (x_1 x_3 x_4 + x_2 - x_1 x_2)\dfrac{\partial f}{\partial x_4} = 0. \end{cases}$$

Forming the combination $X_1[X_2(f)] - X_2[X_1(f)]$, we are led to add to the given equations a new equation $\partial f/\partial x_3 + x_1 \partial f/\partial x_4 = 0$, and the system of three equations thus obtained is equivalent to the system

$$(115) \quad \frac{\partial f}{\partial x_1} + (x_3 + 3x_1^2)\frac{\partial f}{\partial x_4} = 0, \quad \frac{\partial f}{\partial x_2} + x_2 \frac{\partial f}{\partial x_4} = 0, \quad \frac{\partial f}{\partial x_3} + x_1 \frac{\partial f}{\partial x_4} = 0,$$

which is a Jacobian system. The system (114) has therefore only one independent integral. The general integral of the last equation of this system is an arbitrary function of x_1, x_2, and $x_4 - x_1 x_3$. If we take for independent variables x_1, x_2, x_3, and $u = x_4 - x_1 x_3$, every function $f(x_1, x_2, x_3, x_4)$ changes into a corresponding function $\phi(x_1, x_2, x_3, u)$, and the system (115) is replaced by the system

$$(116) \quad \frac{\partial \phi}{\partial x_1} + 3x_1^2 \frac{\partial \phi}{\partial u} = 0, \quad \frac{\partial \phi}{\partial x_2} + x_2 \frac{\partial \phi}{\partial u} = 0, \quad \frac{\partial \phi}{\partial x_3} = 0.$$

The first two equations of (116) form a new Jacobian system of two equations in three independent variables x_1, x_2, u. The general integral of the second is an arbitrary function of x_1 and of $u - x_2^2/2$.

Let us now take for independent variables x_1, x_2, and $u - x_2^2/2 = v$. Every function $\phi(x_1, x_2, u)$ changes into a corresponding function $\psi(x_1, x_2, v)$, and the first two equations of (116) become

$$\frac{\partial \psi}{\partial x_1} + 3x_1^2 \frac{\partial \psi}{\partial v} = 0, \quad \frac{\partial \psi}{\partial x_2} = 0.$$

The general integral of the first is an arbitrary function of $v - x_1^3$, and, consequently, returning to the original variables, we see that the general integral of the system (114) is an arbitrary function of

$$x_4 - x_1 x_3 - \frac{x_2^2}{2} - x_1^3.$$

90. Generalization of the theory of the complete integrals.

Let us consider an equation

(117) $\quad V(x_1, x_2, \cdots, x_n, z; a_1, a_2, \cdots, a_{n-r+1}) = 0$

defining a function z of the n independent variables $x_1, x_2, \cdots, x_n$, which depends also upon $(n - r + 1)$ arbitrary parameters $a_1, a_2, \cdots, a_{n-r+1}$. If we suppose that definite values have been assigned to these parameters, and if we eliminate them from the relation (117) and the relations obtained by successive differentiations,

(118) $\quad \dfrac{\partial V}{\partial x_i} + p_i \dfrac{\partial V}{\partial z} = 0, \quad p_i = \dfrac{\partial z}{\partial x_i}, \quad (i = 1, 2, \cdots, n)$

we obtain *in general* only r independent relations between $z, x_1, \cdots, x_n, p_1, \cdots, p_n$,

(119) $\quad F_1(x_1, \cdots, x_n, z, p_1, \cdots, p_n) = 0, \quad F_2 = 0, \quad \cdots, \quad F_r = 0.$

Limiting ourselves to this case, which is the general case, we shall say, as above (§ 82), that the function z defined by the relation (117) is a *complete integral* of the system of partial differential equations (119). We shall show that, in this case also, the knowledge of a complete integral of the system (119) enables us to find all other integrals. In fact, since the equations (119) arise from the elimination of $a_1, a_2, \cdots, a_{n-r+1}$ between the equations (117) and (118), finding an integral common to these r equations (119) reduces to finding a system of functions $z, a_1, \cdots, a_{n-r+1}$ of the variables $x_1, x_2, \cdots, x_n$ satisfying the equations (117) and (118). It is obvious that we can replace the system of equations (117) and (118) by the system consisting of the equation (117) and the equation

(120) $\quad \dfrac{\partial V}{\partial a_1} da_1 + \dfrac{\partial V}{\partial a_2} da_2 + \cdots + \dfrac{\partial V}{\partial a_{n-r+1}} da_{n-r+1} = 0,$

which is obtained by differentiating the equation (117) and making use of the equations (118). We can satisfy the equations (117) and (120) in a variety of ways:

1) By supposing that $a_1, a_2, \cdots, a_{n-r+1}$ are constants, which gives precisely the complete integral.

2) By putting

$$V = 0, \quad \dfrac{\partial V}{\partial a_1} = 0, \quad \cdots, \quad \dfrac{\partial V}{\partial a_{n-r+1}} = 0.$$

The elimination of $a_1, a_2, \cdots, a_{n-r+1}$ from these equations, if it is possible, furnishes an integral which does not contain any arbitrary constant, and which we shall call, as before, a *singular integral*.

3) If all the coefficients $\partial V/\partial a_i$ are not zero simultaneously, there exists at least one relation between the unknown functions $a_1, a_2, \cdots, a_{n-r+1}$ of the variables x_i (I, § 55, 2d ed.; § 28, 1st ed.). Suppose that there exist k and only k independent relations between these functions,

(121) $\quad f_1(a_1, \cdots, a_{n-r+1}) = 0, \quad \cdots, \quad f_k(a_1, a_2, \cdots, a_{n-r+1}) = 0.$

Since the relation (120) must be a consequence of the relations $df_i = 0 \, (i = 1, 2, \cdots, k)$, there exist k coefficients $\lambda_1, \lambda_2, \cdots, \lambda_k$ such that we have identically

$$\frac{\partial V}{\partial a_1} da_1 + \cdots + \frac{\partial V}{\partial a_{n-r+1}} da_{n-r+1} = \lambda_1 df_1 + \cdots + \lambda_k df_k.$$

This relation is equivalent to $n - r + 1$ distinct relations,

(122) $\quad \begin{cases} \dfrac{\partial V}{\partial a_1} = \lambda_1 \dfrac{\partial f_1}{\partial a_1} + \cdots + \lambda_k \dfrac{\partial f_k}{\partial a_1}, \\ \cdots \cdots \cdots \cdots \cdots \cdots \cdots \cdots \cdots, \\ \dfrac{\partial V}{\partial a_{n-r+1}} = \lambda_1 \dfrac{\partial f_1}{\partial a_{n-r+1}} + \cdots + \lambda_k \dfrac{\partial f_k}{\partial a_{n-r+1}}. \end{cases}$

The elimination of $a_1, a_2, \cdots, a_{n-r+1}, \lambda_1, \lambda_2, \cdots, \lambda_k$ from the equations (117), (121), and (122) will lead, in general, to a single relation between $x_1, x_2, \cdots, x_n$, and z, that is, to an integral common to the equations (119), which depends upon the arbitrary functions chosen. The set of integrals thus obtained, by making the number k vary from 1 to $n - r$, and by taking the functions $f_1, f_2, \cdots, f_k$ arbitrarily, constitutes the *general integral* of the system (119). It will be observed that the complete integral will be obtained by supposing $k = n - r + 1$.

If $r = 1$, the system (119) reduces to a single equation. Conversely, given any equation of the first order $F(x_i, z; p_k) = 0$, it follows from the general existence theorems that it always has an infinite number of integrals which depend upon as many arbitrary parameters as we wish, and consequently an infinite number of complete integrals. The preceding method, which is a direct generalization of that of § 82, enables us to find all the other integrals of the equation $F = 0$ when we know one complete integral.

If $r > 1$, the system (119) is not the most general of its kind, for a system of r equations of the first order with a single dependent variable does not necessarily have any integrals. We shall show in the following paragraphs how to determine whether such a system is consistent, and how to find the integrals when they exist.

91. Involutory systems. Let

(123) $\quad F_1(x_1, x_2, \cdots, x_n; p_1, p_2, \cdots, p_n) = 0, \quad F_2 = 0, \quad \cdots, \quad F_r = 0$

be a system of r independent partial differential equations of the first order, not containing the dependent variable z. The general case can always be reduced to this particular case by the device used in § 75. The problem of finding an integral common to the r equations (123) is equivalent to the following problem: *To find n functions $p_i = \phi_i(x_1, \cdots, x_n)$ satisfying the relations* (123) *and the conditions* $\partial p_i/\partial x_k = \partial p_k/\partial x_i$.

If we know a system of n functions $\phi_i(x_1, \cdots, x_n)$ satisfying these conditions, we can derive from them, by quadratures, an integral of the equations (123) which depends upon an arbitrary constant.

Let F and H be any two functions of the $2n$ variables x_i, p_i. Using the notation (see § 81)

$$(F, H) = \sum_{i=1}^{n} \left(\frac{\partial F}{\partial p_i} \frac{\partial H}{\partial x_i} - \frac{\partial H}{\partial p_i} \frac{\partial F}{\partial x_i} \right),$$

we shall call the expression (F, H) a *Poisson parenthesis*. We now have the following theorem: *If the two equations $F = 0, H = 0$ have a common integral, that integral also satisfies the equation $(F, H) = 0$.*

For let us suppose that $p_1, p_2, \cdots, p_n$ are functions of the n variables $x_1, \cdots, x_n$ satisfying the two equations $F = 0, H = 0$ and the conditions $\partial p_i/\partial x_k = \partial p_k/\partial x_i$. Differentiating the relation $F = 0$ with respect to x_i, we find

$$\frac{\partial F}{\partial x_i} + \sum_{k=1}^{n} \frac{\partial F}{\partial p_k} \frac{\partial p_k}{\partial x_i} = 0.$$

Multiplying this equation by $\partial H/\partial p_i$ and adding all the similar resulting equations, we find

$$\sum_{i=1}^{i=n} \frac{\partial F}{\partial x_i} \frac{\partial H}{\partial p_i} + \sum_{i=1}^{i=n} \sum_{k=1}^{k=n} \frac{\partial H}{\partial p_i} \frac{\partial F}{\partial p_k} \frac{\partial p_k}{\partial x_i} = 0.$$

Permuting the letters F and H and observing that we may permute the indices i and k in the double sum, we have also

$$\sum_{i=1}^{i=n} \frac{\partial H}{\partial x_i} \frac{\partial F}{\partial p_i} + \sum_{i=1}^{i=n} \sum_{k=1}^{k=n} \frac{\partial F}{\partial p_k} \frac{\partial H}{\partial p_i} \frac{\partial p_i}{\partial x_k} = 0.$$

Subtracting the two results term by term, it follows that

(124) $\quad (F, H) + \sum_{i=1}^{n} \sum_{k=1}^{n} \frac{\partial F}{\partial p_k} \frac{\partial H}{\partial p_i} \left(\frac{\partial p_i}{\partial x_k} - \frac{\partial p_k}{\partial x_i} \right) = 0,$

If $p_1, \cdots, p_n$ are the partial derivatives of the same function, we have, for any two indices i and k, $\partial p_i/\partial x_k = \partial p_k/\partial x_i$ and, consequently, $(F, H) = 0$.

This theorem contains as a particular case the one which was proved above (§ 88) for linear homogeneous equations in $p_1, \cdots, p_n$, and its logical consequences are also analogous to those of § 88. For every integral of the equations (123) is also an integral of all the equations $(F_\alpha, F_\beta) = 0$ which can be formed from pairs of the equations (123). Hence we can adjoin to the given system all of these new equations which form with the original equations a system of independent equations. Continuing in this way, we must finally obtain either a system of independent equations whose number exceeds n, in which case the system has no integral in general, or else a system of m equations ($m \leqq n$) such that all the equations $(F_\alpha, F_\beta) = 0$ are satisfied identically or are algebraic consequences of the preceding.

Such systems are similar to complete systems. It is always possible either to show that the given equations are inconsistent or to reduce them to a system for which all the parentheses (F_α, F_β) are identically zero. In fact, let us suppose that we have solved the r equations (123) for r of the variables $p_1, \cdots, p_n$, which must always be possible, for otherwise the elimination of $p_1, \cdots, p_n$ from these r equations would lead to a relation between the variables $x_1, \cdots, x_n$, and the given system would evidently be inconsistent. Let

(125) $\quad p_1 - f_1(p_{r+1}, \cdots, p_n; x_1, \cdots, x_n) = 0, \cdots, p_r - f_r(\cdots) = 0$

be the equivalent system thus obtained. The parenthesis

$$(p_\alpha - f_\alpha, p_\beta - f_\beta)$$

does not contain any of the variables $p_1, \cdots, p_r$; hence the equations obtained by equating these parentheses to zero cannot be consequences of the first, and they furnish new equations if the parentheses are not identically zero. Solving these new equations for certain of the quantities $p_{r+1}, \cdots, p_n$, and continuing in the same way, we finally either demonstrate the impossibility of the problem or else obtain a system of m equations of the first order ($m \leqq n$),

(126) $\qquad F_1 = 0, \quad \cdots, \quad F_m = 0,$

such that all the parentheses (F_α, F_β) are identically zero.

Such systems, which are similar to the linear Jacobian systems, are called *involutory systems*. It follows from what precedes that the search for the common integrals of a system of equations of the first order reduces to the integration of an involutory system.

This integration is immediate if $m = n$, as the following proposition shows: *Let $F_1, F_2, \cdots, F_n$ be functions of the $2n$ variables x_i, p_k, such that all the parentheses (F_α, F_β) are identically zero, and such that the Jacobian $\Delta = D(F_1, \cdots, F_n)/D(p_1, \cdots, p_n)$ is not zero. If we solve the n equations*

(127) $\quad F_1 = a_1, \quad F_2 = a_2, \quad \cdots, \quad F_n = a_n,$

where $a_1, a_2, \cdots, a_n$ are any constants, for $p_1, p_2, \cdots, p_n$, the expression $p_1 dx_1 + \cdots + p_n dx_n$ is an exact differential for the resulting values of the p's.

For we have
$$(F_\alpha - a_\alpha, F_\beta - a_\beta) = (F_\alpha, F_\beta) = 0,$$

and, by what precedes, these n functions $p_1, p_2, \cdots, p_n$ of the n variables $x_1, x_2, \cdots, x_n$ defined by the n equations (127) must satisfy all the relations
$$\sum_{i=1}^{i=n} \sum_{k=1}^{k=n} \frac{\partial F_\alpha}{\partial p_i} \frac{\partial F_\beta}{\partial p_k} \left(\frac{\partial p_k}{\partial x_i} - \frac{\partial p_i}{\partial x_k} \right) = 0. \quad (\alpha, \beta = 1, 2, \cdots, n)$$

Let us take all the n relations of this kind in which the index β retains the same value. These relations can be written in the form
$$\sum_{i=1}^{n} \frac{\partial F_\alpha}{\partial p_i} \sum_{k=1}^{n} \frac{\partial F_\beta}{\partial p_k} \left(\frac{\partial p_k}{\partial x_i} - \frac{\partial p_i}{\partial x_k} \right) = 0.$$

If we take for unknowns the n expressions
$$\sum_{k=1}^{n} \frac{\partial F_\beta}{\partial p_k} \left(\frac{\partial p_k}{\partial x_i} - \frac{\partial p_i}{\partial x_k} \right), \quad (i = 1, 2, \cdots, n)$$

the determinant of the coefficients of these unknowns is precisely the determinant Δ, which, by hypothesis, does not vanish identically. It follows that we have, for any two indices i and β,
$$\sum_{k=1}^{n} \frac{\partial F_\beta}{\partial p_k} \left(\frac{\partial p_k}{\partial x_i} - \frac{\partial p_i}{\partial x_k} \right) = 0.$$

Similarly, taking the n equations of this kind in which the index i has a definite value, we evidently have $\partial p_i/\partial x_k = \partial p_k/\partial x_i$, which proves the proposition.

The function

(128) $\quad \begin{cases} z = \Phi(x_1, \cdots, x_n; a_1, \cdots, a_{n+1}) \\ = \int (p_1 dx_1 + \cdots + p_n dx_n) + a_{n+1}, \end{cases}$

where a_{n+1} is a new arbitrary constant, represents the complete integral of the involutory system (127). If we regard the r constants $a_1, a_2, \cdots, a_r$ as having definite values, while the constants

$a_{r+1}, \cdots, a_{n+1}$ remain arbitrary, the formula (128) represents a complete integral of the involutory system formed from the first r equations of (127). This is a true complete integral, for, from the way in which we have obtained the function Φ, the equations

$$p_1 = \frac{\partial \Phi}{\partial x_1}, \quad \cdots, \quad p_n = \frac{\partial \Phi}{\partial x_n}$$

form a system equivalent to the system (127), and the only independent relations not containing $a_{r+1}, \cdots, a_n$ which can be deduced from them are evidently the first r equations of this system.

92. Jacobi's method. Let us consider an involutory system of r equations ($r < n$),

(129) $\quad F_1(x_1, \cdots, x_n; p_1, \cdots, p_n) = a_1, \quad \cdots, \quad F_r(x_1, \cdots, p_n) = a_r,$

where the constants $a_1, \cdots, a_r$ have definite values. To obtain a complete integral of this system, it is sufficient to adjoin to it $n - r$ new functions $F_{r+1}, \cdots, F_n$, such that the Jacobian

$$\frac{D(F_1, \cdots, F_n)}{D(p_1, \cdots, p_n)}$$

is not zero, and such that the new system

(130) $\quad F_1 = a_1, \cdots, F_r = a_r, \quad F_{r+1} = a_{r+1}, \quad \cdots, \quad F_n = a_n$

is itself involutory. Indeed, the complete integral of this system (130) will furnish, as we have just seen, a complete integral of the system (129). If $r = 1$, this method is merely the extension of the method of Lagrange and Charpit to an equation in n variables.

Jacobi's method for solving this problem depends upon a noted identity due to Poisson. Let f, ϕ, ψ be any three functions of the $2n$ variables x_i, p_k; then we have *identically the relation*

(131) $\quad ((f, \phi), \psi) + ((\phi, \psi), f) + ((\psi, f), \phi) = 0.$

In fact, each term on the left-hand side is the product of a partial derivative of the second order and two partial derivatives of the first order. Hence, to show that it vanishes, it is sufficient to show that it does not contain any derivative of the second order of the function f, for example, since the three functions f, ϕ, ψ appear in it symmetrically. The terms containing the second derivatives of f can arise only from $((f, \phi), \psi) + ((\psi, f), \phi) = (\psi, (\phi, f)) - (\phi, (\psi, f))$. Observing that (ϕ, f) and (ψ, f) are two linear homogeneous expressions in the derivatives of f, and setting

$$(\phi, f) = X(f), \quad (\psi, f) = Y(f),$$

the preceding expression can be written in the form
$$Y[X(f)] - X[Y(f)].$$
Now we saw in § 88 that this expression does not contain any second derivatives of f. It follows that all the terms of the left-hand side of the equation (131) cancel each other in pairs.

Finally, in order to integrate the involutory system (129), let us first try to find a function Φ independent of $F_1, \cdots, F_r$ satisfying r linear homogeneous partial differential equations of the first order,

(132) $(F_1, \Phi) = 0, \quad (F_2, \Phi) = 0, \quad \cdots, \quad (F_r, \Phi) = 0.$

These r equations form a Jacobian system. For if we set
$$X_i(\Phi) = (F_i, \Phi),$$
Poisson's identity,
$$((F_\alpha, F_\beta), \Phi) + ((F_\beta, \Phi), F_\alpha) + ((\Phi, F_\alpha), F_\beta) = 0,$$
becomes
$$X_\alpha[X_\beta(\Phi)] - X_\beta[X_\alpha(\Phi)] = 0,$$
since $(F_\alpha, F_\beta) = 0$.

Let F_{r+1} be an integral of this Jacobian system which forms with $F_1, \cdots, F_r$ a system of independent functions of $p_1, \cdots, p_n$. We next proceed to form the new Jacobian system of $r+1$ equations,

$$(F_1, \Phi) = 0, \quad \cdots, \quad (F_{r+1}, \Phi) = 0,$$

and to find an integral of this system which is independent of $F_1, \cdots, F_{r+1}$ as functions of the p_i; and we continue in the same way. Finally, when we have found an integral of the last Jacobian system,

$$(F_1, \Phi) = 0, \quad \cdots, \quad (F_{n-1}, \Phi) = 0,$$

we can obtain a complete integral of the given system by a quadrature, as we have seen above.

V. GENERALITIES ON THE EQUATIONS OF HIGHER ORDER

93. Elimination of arbitrary functions. The study of partial differential equations of the first order in a single dependent variable has led us to the following conclusions: 1) The integration of an equation of this form reduces to the integration of a system of ordinary differential equations. 2) All the integrals of this equation are represented by one or more systems of equations in which appear explicitly one or more arbitrary functions and their derivatives.

These properties are not extensible to the most general partial differential equations of order higher than the first. The problem of the integration of such an equation cannot, in general, be reduced to the integration of a system of ordinary differential equations.

We can easily generalize, however, the method of the elimination of arbitrary functions which leads to a partial differential equation of the first order (§§ 77 and 82), but the equations of higher order which we obtain in this way form only a very special class.

Thus we have seen that the general integral of a linear equation in two independent variables $Pp + Qq = R$ is obtained by associating the curves of a congruence according to an arbitrary law. Let us now consider a family of curves Γ that depends upon $n+1$ arbitrary parameters $a_1, a_2, \cdots, a_{n+1}, (n > 1)$,

(133) $\quad F(x, y, z, a_1, \cdots, a_{n+1}) = 0, \qquad \Phi(x, y, z, a_1, \cdots, a_{n+1}) = 0.$

If we establish n *relations of arbitrary form* between these $n+1$ parameters, we obtain a family of curves Γ that depends upon only one parameter. These curves generate a surface S, and all these surfaces S satisfy, whatever may be the n relations established between the $n+1$ parameters, a partial differential equation of the nth order, which is called the *partial differential equation* of the family of surfaces S. To prove this, let us observe that instead of establishing n relations between the $n+1$ parameters a_i, it amounts to the same thing to take for these parameters arbitrary functions $a_i(\lambda)$ of an auxiliary variable λ. The two equations (133) then define two implicit functions $z = f(x, y)$, $\lambda = \phi(x, y)$, and we have to prove that the function $z = f(x, y)$ satisfies a partial differential equation of the nth order, independent of the form of the arbitrary functions $a_i(\lambda)$. Differentiating the first of the equations (133) with respect to x and then with respect to y, we obtain the two relations

$$\frac{\partial F}{\partial x} + \frac{\partial F}{\partial z} p + \left[\frac{\partial F}{\partial a_1} a_1'(\lambda) + \cdots + \frac{\partial F}{\partial a_{n+1}} a_{n+1}'(\lambda)\right] \lambda_x' = 0,$$

$$\frac{\partial F}{\partial y} + \frac{\partial F}{\partial z} q + \left[\frac{\partial F}{\partial a_1} a_1'(\lambda) + \cdots + \frac{\partial F}{\partial a_{n+1}} a_{n+1}'(\lambda)\right] \lambda_y' = 0,$$

from which we derive

$$\frac{\lambda_y'}{\lambda_x'} = \frac{\dfrac{\partial F}{\partial y} + \dfrac{\partial F}{\partial z} q}{\dfrac{\partial F}{\partial x} + \dfrac{\partial F}{\partial z} p}.$$

From the second of the equations (133) we derive, similarly, an expression for the quotient λ_y'/λ_x', which is deduced from the preceding by replacing in it F by Φ. Equating these two expressions, we adjoin to the equations (133) a new equation containing $x, y, z, p, q, a_1, a_2, \cdots, a_{n+1}$,

(134) $\quad \Psi_1(x, y, z, p, q, a_1, a_2, \cdots, a_{n+1}) = 0.$

Operating on this new equation as on the equation $F = 0$, we derive from it an expression for λ'_y/λ'_x which depends upon $x, y, z, p, q, r, s, t, a_1, \cdots, a_{n+1}$. Equating this new expression to one of the expressions already obtained for this same quotient, we obtain a new relation containing the second derivatives of z,

(135) $\quad \Psi_2(x, y, z, p, q, r, s, t, a_1, a_2, \cdots, a_{n+1}) = 0.$

After n similar operations we adjoin to the system (133) a system of n relations containing $a_1, a_2, \cdots, a_{n+1}, x, y, z$, and the derivatives of z up to those of the nth order. The elimination of $a_1, a_2, \cdots, a_{n+1}$ from these $n + 2$ equations will lead, in general, to one and only one equation between x, y, z, and the partial derivatives of z up to those of the nth order. This is the partial differential equation of the surfaces generated by the curves Γ.

Example 1. If the curves Γ are the straight lines parallel to the xy-plane, the equations (133) are
$$z = a_1, \quad y = a_2 x + a_3.$$

Applying the general method, let us suppose that a_1, a_2, a_3 are functions of a parameter λ. From the two preceding equations we derive for the quotient λ'_x/λ'_y the two values p/q and $-a_2$, which leads to the relation $p/q + a_2 = 0$.

Differentiating this last relation with respect to x and then with respect to y, and dividing the corresponding sides of the resulting equations, we find

$$\frac{\lambda'_y}{\lambda'_x} = \frac{pt - qs}{ps - qr}.$$

Equating this value of the quotient to the expression q/p already obtained, we find again the partial differential equation of the ruled surfaces which have the xy-plane for the directing plane (I, Ex. § 39, 2d ed.; § 24, 1st ed.),

$$q^2 r - 2pqs + p^2 t = 0.$$

Example 2. If the curves Γ are all possible straight lines, the equations (133) can be written in the form
$$x = a_1 z + a_2, \quad y = a_3 z + a_4.$$

Applying the general method, we derive from them successively

$$\frac{\lambda'_y}{\lambda'_x} = \frac{a_1 q}{a_1 p - 1} = \frac{a_3 q - 1}{a_3 p} = -\frac{a_1}{a_3},$$

or
$$a_1 p + a_3 q - 1 = 0.$$

From this new equation we then derive

$$\frac{\lambda'_y}{\lambda'_x} = \frac{a_1 s + a_3 t}{a_1 r + a_3 s} = -\frac{a_1}{a_3},$$

or

(A) $\qquad a_1^2 r + 2 a_1 a_3 s + a_3^2 t = 0.$

This last equation gives in turn

$$\frac{\lambda'_y}{\lambda'_x} = \frac{a_1^2 p_{21} + 2 a_1 a_3 p_{12} + a_3^2 p_{03}}{a_1^2 p_{30} + 2 a_1 a_3 p_{21} + a_3^2 p_{12}} = -\frac{a_1}{a_3}, \qquad p_{ih} = \frac{\partial^{i+h} z}{\partial x^i \, \partial y^h}$$

or, clearing of fractions,

(B) $\qquad a_1^3 p_{30} + 3 a_1^2 a_3 p_{21} + 3 a_1 a_3^2 p_{12} + a_3^3 p_{03} = 0.$

Eliminating the quotient a_1/a_3 from the relations (A) and (B), we obtain the partial differential equation of all *ruled* surfaces. We see that this equation contains only derivatives of the second and third orders. By its derivation we see that it states that at each point of the surface one of the asymptotic tangents has contact of the third order with the surface (I, § 223, 2d ed.; § 238, 1st ed.).

Example 3. Let us consider the plane curves Γ represented by the two equations
$$z = f(x, y, a_1, \cdots, a_n), \qquad y = a_{n+1}.$$
Instead of applying the general method, let us suppose that $a_1, a_2, \cdots, a_n$ are functions of the last parameter a_{n+1}. The surface S generated by these curves Γ has for its equation
$$z = f[x, y, \phi_1(y), \cdots, \phi_n(y)],$$
where $\phi_1, \cdots, \phi_n$ are arbitrary functions of y. The elimination of these n functions from the preceding relation and the relations which give $\partial z/\partial x$, $\partial^2 z/\partial x^2$, $\cdots$, $\partial^n z/\partial x^n$ leads to a partial differential equation of the nth order,

(136) $\qquad \dfrac{\partial^n z}{\partial x^n} = F\left(x, y, \dfrac{\partial z}{\partial x}, \cdots, \dfrac{\partial^{n-1} z}{\partial x^{n-1}}\right),$

in which only derivatives with respect to x appear.

Conversely, every partial differential equation of this type can be integrated as an ordinary differential equation containing a parameter. If $z = f(x, y, C_1, C_2, \cdots, C_n)$ is the general integral of such a differential equation, it will suffice to replace $C_1, C_2, \cdots, C_n$ in it by arbitrary functions of y in order to have the general integral of the same equation, considered as a partial differential equation in two independent variables x and y.

The general integral of a partial differential equation of the first order, of any form, in two independent variables, is obtained by taking the envelope of a two-parameter family of surfaces when we establish an arbitrary relation between these two parameters (§ 82). To generalize this result, let us consider a family of surfaces Σ which depends upon $n + 1$ parameters,

(137) $\qquad F(x, y, z, a_1, a_2, \cdots, a_{n+1}) = 0. \qquad (n > 1)$

If we establish n arbitrary relations between these $n+1$ parameters, or, what amounts to the same thing, if we replace $a_1, a_2, \cdots, a_{n+1}$ by arbitrary functions of an auxiliary variable λ, we have a family of surfaces Σ which depends upon a single parameter. The envelope of this family of surfaces is a surface S which satisfies a partial differential equation of the nth order, independent of the form of the arbitrary functions $a_i(\lambda)$. For we should obtain the equation of this surface by eliminating λ from the two equations (137) and (138)

$$(138) \qquad \frac{\partial F}{\partial a_1} a_1'(\lambda) + \cdots + \frac{\partial F}{\partial a_{n+1}} a_{n+1}'(\lambda) = 0.$$

But these two equations may be considered as defining two functions $z = f(x, y)$ and $\lambda = \phi(x, y)$ of the two variables x and y. The partial derivatives p and q are given by the two equations (I, § 41, 2d ed.; § 25, 1st ed.)

$$(139) \qquad \frac{\partial F}{\partial x} + \frac{\partial F}{\partial z} p = 0, \qquad \frac{\partial F}{\partial y} + \frac{\partial F}{\partial z} q = 0.$$

Applying to this system (139) the method applied to the system (133), we can adjoin to it, step by step, $n-1$ new relations between $a_1, a_2, \cdots, a_{n+1}, x, y, z$, and the partial derivatives of z of orders $2, 3, \cdots, n$. The elimination of $a_1, a_2, \cdots, a_{n+1}$ from these $n-1$ equations and the equations (137) and (139) will lead, in general, to a single relation independent of $a_1, a_2, \cdots, a_{n+1}$, in which will appear x, y, z, and the partial derivatives of z up to those of the nth order.

Example. If the surface Σ is a plane, we find again the equation of the developable surfaces $s^2 - rt = 0$. If the surface Σ is a sphere with the constant radius R, the equations (137) and (139) become

$$(140) \qquad \begin{cases} (x-a_1)^2 + (y-a_2)^2 + (z-a_3)^2 - R^2 = 0, \\ x - a_1 + (z - a_3)p = 0, \qquad y - a_2 + (z - a_3)q = 0. \end{cases}$$

Suppose that a_1, a_2, a_3 are functions of a parameter λ. Equating the values of the quotient λ_y'/λ_x' derived from the last two equations (140), we obtain the relation

$$(141) \quad (rt - s^2)(z - a_3)^2 + [(1+p^2)t + (1+q^2)r - 2pqs](z - a_3) + 1 + p^2 + q^2 = 0.$$

We shall obtain the desired equation by eliminating a_1, a_2, a_3 from (140) and (141). From the first we derive $z - a_3 = R/\sqrt{1 + p^2 + q^2}$, and, replacing $z - a_3$ by this value in (141), we obtain the partial differential equation of the *tubular surfaces*,

$$(142) \quad (rt - s^2)R^2 + [(1+p^2)t + (1+q^2)r - 2pqs]R\sqrt{1+p^2+q^2} + (1+p^2+q^2)^2 = 0.$$

The geometric meaning of this equation is easily verified. It states that one of the principal radii of curvature of the surface is equal to R (I, § 242, 2d ed.; § 241, 1st ed.).

Note. Given a function of several variables which depends upon one or more arbitrary functions, it is not always possible, as in the two cases which have just been examined, to deduce from them *one and only one* relation, independent of the form of the arbitrary functions, between the independent variables, the function z and its partial derivatives up to a given order. Let us consider, for example, a function $z = F(x, y, X, Y)$, where F is a given function of the four arguments which appear in it, and where X and Y are arbitrary functions of the variables x and y respectively. The five derivatives p, q, r, s, t of the first and second orders depend upon X, X', X'', Y, Y', Y'', and it is in general impossible to eliminate these six quantities from the six equations. But if we continue up to derivatives of the third order, we have, in all, *ten* relations containing eight arbitrary quantities, $X, X', X'', X''', Y, Y', Y'', Y'''$, and the elimination will lead to a system of *two* equations of the third order.*

94. General existence theorem. The proof given for a system of partial differential equations of the first order (§ 25) can be extended readily to the most general systems of the normal form, studied by Madame Kovalevsky,†

$$(143) \begin{cases} \dfrac{\partial^{r_1} z_1}{\partial x_1^{r_1}} = F_1(x_1, x_2, \cdots, x_n;\ z_1, z_2, \cdots, z_p, \cdots), \\ \dfrac{\partial^{r_2} z_2}{\partial x_1^{r_2}} = F_2(x_1, x_2, \cdots, x_n;\ z_1, z_2, \cdots, z_p, \cdots), \\ \cdots\cdots\cdots\cdots\cdots\cdots\cdots\cdots\cdots\cdots\cdots\cdots\cdots, \\ \dfrac{\partial^{r_p} z_p}{\partial x_1^{r_p}} = F_p(x_1, x_2, \cdots, x_n;\ z_1, z_2, \cdots, z_p, \cdots), \end{cases}$$

in which the right-hand sides contain the independent variables x_1, $x_2, \cdots, x_n$, the dependent functions $z_1, \cdots, z_p$, the partial derivatives of z_1 up to and including those of order r_1, the partial derivatives of z_2 of orders up to and including those of order $r_2, \cdots$, and so on,

---

* See HERMITE, *Cours d'Analyse*, pp. 215–229.

† *Journal de Crelle*, Vol. LXXX. In her proof, Madame Kovalevsky reduces the general case to the case of a *linear* system of the first order, but for us it will be sufficient to reduce the general case to the case of a system of the first order of any form whatever.

but none of the derivatives $\partial^{r_1}z_1/\partial x_1^{r_1}$, $\partial^{r_2}z_2/\partial x_1^{r_2}$, $\cdots$, $\partial^{r_p}z_p/\partial x_1^{r_p}$. We may then state the general theorem as follows:

Regarding the quantities $x_1, x_2, \cdots, x_n, z_1, z_2, \cdots, z_p$,

$$\frac{\partial^{a_1+a_2+\cdots+a_n} z_i}{\partial x_1^{a_1} \partial x_2^{a_2} \cdots \partial x_n^{a_n}},$$

which appear in the functions F_i *as independent variables, let*

$$a_1, a_2, \cdots, a_n, b_1, b_2, \cdots, b_p, b^i_{a_1, a_2, \cdots, a_n}$$

be any system of values of these variables in whose neighborhood the functions F_i *are analytic. On the other hand, let*

(144) $$\begin{cases} \phi_1, \phi_1^1, \phi_1^2, \cdots, \phi_1^{r_1-1}, \\ \phi_2, \phi_2^1, \phi_2^2, \cdots, \phi_2^{r_2-1}, \\ \cdots \cdots \cdots \cdots \cdots \cdots, \\ \phi_p, \phi_p^1, \phi_p^2, \cdots, \phi_p^{r_p-1} \end{cases}$$

be functions of the $n-1$ *variables* $x_2, x_3, \cdots, x_n$, *regular in the neighborhood of the point* $a_2, \cdots, a_n$, *and such that we have*

$$\phi_i = b_i, \qquad \frac{\partial^{a_2+\cdots+a_n} \phi_i^{a_1}}{\partial x_2^{a_2} \cdots \partial x_n^{a_n}} = b^i_{a_1, a_2, \cdots a_n}$$

for $x_2 = a_2, \cdots, x_n = a_n$. *Then the equations* (143) *have one and only one system of integrals, analytic in the neighborhood of the point* $(a_1, a_2, \cdots, a_n)$, *and such that we have, for* $x_1 = a_1$,

$$z_i = \phi_i, \quad \frac{\partial z_i}{\partial x_1} = \phi_i^1, \quad \cdots, \quad \frac{\partial^{r_i-1} z_i}{\partial x_1^{r_i-1}} = \phi_i^{r_i-1}. \qquad (i = 1, 2, \cdots, p)$$

To prove this we observe first that the equations (143), and those which we obtain from them by successive differentiations, enable us to express all the partial derivatives of the dependent variables in terms of the independent variables, the dependent variables, and the partial derivatives $\partial^{a_1+\cdots+a_n} z_i/\partial x_1^{a_1} \cdots \partial x_n^{a_n}$, where $a_1 < r_1$ for $i = 1$, $a_1 < r_2$ for $i = 2, \cdots, a_1 < r_p$ for $i = p$. This follows, step by step, by a process of reasoning exactly like that of § 25. Now the initial conditions determine immediately for $x_1 = a_1, \cdots, x_n = a_n$ the numerical values of the derivatives in terms of which all the others are expressible. Hence the coefficients of the developments in power series of the integrals whose existence we wish to prove can be calculated by the operations of addition and multiplication alone, in terms of the coefficients of the developments of the functions F_i and of the functions of the array (144).

To finish the proof, it remains to establish the convergence of the power series thus obtained when the absolute values of the differences $x_i - a_i$ are sufficiently small. We have already proved this convergence when all the numbers $r_1, r_2, \cdots, r_p$ are equal to unity. We shall now show how to reduce the general case to this particular case by considering as dependent variables the functions $z_1, \cdots, z_p$, and their partial derivatives up to those of order $r_i - 1$, inclusive, for z_i ($i = 1, 2, \cdots, p$).

Let us put

$$\frac{\partial^{\alpha_1 + \alpha_2 + \cdots + \alpha_n} z_i}{\partial x_1^{\alpha_1} \partial x_2^{\alpha_2} \cdots \partial x_n^{\alpha_n}} = z^i_{\alpha_1, \alpha_2, \cdots, \alpha_n}. \qquad (z^i_{0,0,\cdots,0} = z_i)$$

The right-hand sides of the equations (143) contain the variables $x_1, \cdots, x_n$, the dependent variables $z_1, \cdots, z_p$, the new dependent variables, and certain derivatives of the first order of these new dependent variables. But, by hypothesis, the derivatives of the variable z_i of order r_i which can appear are different from the derivative $z^i_{r_i, 0, 0, \cdots, 0}$. Hence at least one of the numbers $\alpha_2, \alpha_3, \cdots, \alpha_n$ is different from zero. If, for example, $\alpha_2 > 0$, we can replace $z^i_{\alpha_1, \alpha_2, \cdots, \alpha_n}$ by

$$\frac{\partial z^i_{\alpha_1, \alpha_2-1, \alpha_3, \cdots, \alpha_n}}{\partial x_2}$$

when $\alpha_1 + \alpha_2 + \cdots + \alpha_n = r_i$, and similarly for the others. We can therefore write the given equations (143) in the equivalent form

$$(145) \quad \frac{\partial z^i_{r_i-1, 0, 0, \cdots, 0}}{\partial x_1} = \Phi_i(x_1, \cdots, x_n; z_1, \cdots, z_p \cdots), \quad (i = 1, 2, \cdots, p)$$

the right-hand sides containing only the independent variables and the dependent variables with some of the partial derivatives of the first order taken with respect to one of the variables $x_2, \cdots, x_n$. To these equations must be adjoined those which give the derivatives with respect to x_1 of the new dependent variables, other than those which we have already written. If we have $\alpha_1 + \alpha_2 + \cdots + \alpha_n \leqq r_i - 2$, we can write immediately

$$(146) \quad \frac{\partial z^i_{\alpha_1, \alpha_2, \cdots, \alpha_n}}{\partial x_1} = z^i_{\alpha_1+1, \alpha_2, \cdots, \alpha_n},$$

and we have $\alpha_1 + 1 + \alpha_2 + \cdots + \alpha_n \leqq r_i - 1$, so that the right-hand side is one of the dependent variables. If we have

$$\alpha_1 + \alpha_2 + \cdots + \alpha_n = r_i - 1,$$

we must suppose $a_1 < r_i - 1$, and, consequently, one at least of the numbers $a_2, \cdots, a_n$ is different from zero. If, for example, we have $a_2 > 0$, we shall write

(147) $$\frac{\partial z^i_{a_1, a_2, \cdots, a_n}}{\partial x_1} = \frac{\partial z^i_{a_1+1, a_2-1, \cdots, a_n}}{\partial x_2},$$

and the right-hand side is the derivative with respect to x_2 of one of the auxiliary dependent variables. The equations (145), (146), and (147) form a normal system of equations of the first order. The initial conditions which must be satisfied by the integrals of this new system result immediately from the initial conditions imposed upon the integrals of the original system, and it is clear that the power series obtained for the integrals $z_1, z_2, \cdots, z_p$ of the new system will be identical with the power series obtained for the integrals of the given system. These series are therefore convergent (see § 26) in the neighborhood of the point $(a_1, a_2, \cdots, a_n)$.

For example, the equation of the second order $r = f(x, y, z, p, q, s, t)$ can be replaced by a system of three equations of the first order in the normal form,

$$\frac{\partial z}{\partial x} = p, \qquad \frac{\partial p}{\partial x} = f\left(x, y, z, p, q, \frac{\partial p}{\partial y}, \frac{\partial q}{\partial y}\right), \qquad \frac{\partial q}{\partial x} = \frac{\partial p}{\partial y}.$$

If it is required that $z = \phi(y)$, $\partial z/\partial x = \psi(y)$, for $x = x_0$, the integrals of the auxiliary system must reduce respectively, for $x = x_0$, to the functions $\phi(y)$, $\psi(y)$, $\phi'(y)$.

This general theorem does not furnish a reply to all the questions which can be proposed on the existence of integrals of any system whatever of partial differential equations, for it applies only to systems in the normal form considered. The most general systems have been the subject of a great number of studies, the most recent of which, due to Tresse, Riquier, and Delassus, have led to the general solution of the following problem: Given a system of m partial differential equations of any order in any number of independent and any number of dependent variables, to determine whether this system has any integrals and, if it has, to define the arbitrary quantities (constants or functions) upon which the integrals depend.*

To sum up, every partial differential equation of any order in which both sides are analytic functions of their arguments has an infinite number of analytic integrals, but we cannot say, in general, as in the case of ordinary differential equations (§ 26), that all the

---

* The investigations of Riquier have been collected by him in his work *Sur les systèmes d'équations aux dérivées partielles* (1910).

integrals are analytic functions of the independent variables. We have seen above (p. 255, ftn.) that it is not true for an equation of the first order. It is, moreover, easy to see this by elementary examples such as the equation $p = 0$, whose general integral is *any* function of y.

The methods of the calculus of limits do not apply to the non-analytic equations. Let us consider, for example, the equation

(148) $$p + qf(x, y) = 0,$$

where $f(x, y)$ is a continuous non-analytic function satisfying the Lipschitz condition with respect to y. We have proved in §§ 27–30 that the differential equation

$$\frac{dy}{dx} = f(x, y)$$

has an infinite number of integrals which depend upon an arbitrary constant C. In order to conclude from this, as in § 31, the existence of an integral of the equation (148), it would be necessary to prove that all these integrals are defined by an equation of the form $\phi(x, y) = C$, where the function ϕ possesses continuous derivatives of the first order. We shall return to this question in the next volume.

EXERCISES

1. Integrate the partial differential equations

$$ax^4 p + (x^4 z + ax^3 y - ax^2 y^2) q = 2 ax^2 yz - za^2 y^3,$$

$$\left(\frac{y}{x} + \frac{z^2}{y^2}\right) p - \left(\frac{x}{y} + \frac{z^2}{x^2} + 1\right) q = \left(\frac{y}{x^2} - \frac{x}{y^2}\right) z,$$

$$(x - 6y) p + (10x - y) q = 6y^2 - 4x^2 - 36xy.$$

2. Find the general equation of the surfaces which cut at right angles the spheres represented by the equation

$$x^2 + y^2 + z^2 + 2az = 0,$$

where a is a variable parameter.

Deduce from the result obtained some systems of three families of orthogonal surfaces.

3. It is required to find the partial differential equation of the surfaces described by a straight line which moves so that it always meets a fixed straight line at a given angle. Integrate this partial differential equation.

[*Licence*, Paris, July, 1878.]

4. Given a plane P and a point O in the plane, find the general equation of all the surfaces such that, if we draw the normal mn at any point m of one of them meeting the plane P at n, and then the perpendicular mp to this plane, the area of the triangle Onp will be equal to a given constant.

[*Licence*, Paris, November, 1871.]

5. The same question as in Ex. 4, supposing, however, that the angle nOp is constant.
[*Licence*, Rennes, 1883.]

6. Determine all the surfaces which satisfy the condition

$$Op \times \overline{mn} = \lambda \overline{Om}^2,$$

where λ denotes a given constant, O the origin of coördinates, m any point of one of the surfaces, p the foot of the perpendicular dropped from O upon the tangent plane at m, and n the trace of the normal on the plane xOy.
[*Licence*, Paris, 1875.]

7. Find the general equation of the surfaces such that if we draw the normal mn from any point m of one of them terminating in the xy-plane, the length mn will be equal to the distance On.
[*Licence*, Poitiers, 1883.]

8. Find the integral surfaces of the equation

$$xy^2 p + x^2 y q = z(x^2 + y^2).$$

Determine the arbitrary function in such a way that the characteristic curves form a family of asymptotic lines of the integral surfaces, and find the orthogonal trajectories of the surfaces thus obtained.
[*Licence*, Paris, July, 1904.]

9. Consider a family of skew curves Γ represented by the two equations

$$x^2 + 2y^2 = az^2, \qquad x^2 + y^2 + z^2 = bz,$$

where a and b are two variable parameters.

1) Prove that these curves are the orthogonal trajectories of a one-parameter family of surfaces S;

2) Find the lines of curvature of these surfaces S;

3) Show that these surfaces form part of a triply orthogonal system, and find the other two families of this system.
[*Licence*, Paris, July, 1901.]

10. Form the partial differential equation which has the complete integral $y^2(x^2 - a) = (z + b)^2$, and integrate this equation.

11. Determine the surfaces such that the segment mn of the normal included between the surface and the point of intersection n with a fixed plane P projects upon this plane into a segment of constant length.

12. Let n be the point where the normal at m to a surface meets the xy-plane. Find the surfaces such that the straight line On will be parallel to the tangent plane at m.
[*Licence*, Poitiers, July, 1884.]

13. It is required to determine the surfaces which cut at a given angle V all the planes passing through a fixed straight line. Show that the characteristic curves are the lines of curvature of the integral surfaces.

14. The integral curves of the partial differential equation for which a complete integral is
$$(1 - a^2)x + k(1 + a^2)z + 2ay + b = 0,$$
where a and b are two arbitrary constants, satisfy the relation

$$dx^2 + dy^2 = k^2 dz^2.$$

15*. Every integral curve of a partial differential equation $F(x, y, z, p, q) = 0$, tangent at a point M to a generator G of the cone T with its vertex at M, has contact of the second order with every integral surface tangent at M to the plane tangent along the generator G to the cone T.

[SOPHUS LIE.]

16. From a point M of a surface S a perpendicular MP is dropped upon the fixed axis OO', then from the point P a perpendicular PN upon the normal to the surface at M. It is required to determine the surfaces S such that the length MN will be a given constant a.

Study in particular the surfaces S which are helicoids having OO' for axis.

[*Licence*, Paris, October, 1908.]

17. It is required to find the general form of the functions $F(x, y, z, p, q)$ such that the differential equations of the characteristic curves of the equation $F = 0$ will have the integrable combination $d(q/p) = 0$.

Application. Determine the surfaces S such that the distance of any point M of one of them to the xy-plane is equal to the distance from the point O to the tangent plane to the surface at the point M.

18. Given the partial differential equation

(I) $$Pp + Qq = Rz^2 + Sz + T,$$

where P, Q, R, S, T depend only upon the variables x and y, show that the anharmonic ratio u of any four particular integrals of the equation (I) satisfies the equation

$$P\frac{\partial u}{\partial x} + Q\frac{\partial u}{\partial y} = 1.$$

Knowing four particular integrals z_1, z_2, z_3, z_4 of the equation (I), can we derive from them the general integral?

19. *Parallel surfaces.* Let $\theta(x, y, z)$ be an integral of the equation

(E) $$\left(\frac{\partial \theta}{\partial x}\right)^2 + \left(\frac{\partial \theta}{\partial y}\right)^2 + \left(\frac{\partial \theta}{\partial z}\right)^2 = 1.$$

Prove that the equation $\theta(x, y, z) = C$ represents, in rectangular coördinates, a family of parallel surfaces.

Note. We observe that the equation (E) has the complete integral

$$\theta = \sqrt{(x-a)^2 + (y-b)^2 + (z-c)^2},$$

and the general integral is obtained by finding the envelope of the sphere of radius θ whose center describes a surface or a curve. It is clear that by making the radius θ vary we obtain a family of parallel surfaces.

Conversely, in order that the equation $u(x, y, z) = C$ shall represent a family of parallel surfaces, it is necessary and sufficient (Ex. 9, p. 42) that $u(x, y, z)$ satisfy an equation of the form

$$\left(\frac{\partial u}{\partial x}\right)^2 + \left(\frac{\partial u}{\partial y}\right)^2 + \left(\frac{\partial u}{\partial z}\right)^2 = \phi(u),$$

which we may reduce to the form (E) by putting $\theta = \psi(u)$.

20. In order that the expression $dz + A\,dx + B\,dy$ shall have an integrating factor independent of z, it is necessary and sufficient that it be of the form

$$dz + z\,d\phi + e^{-\phi}\,d\psi,$$

where ϕ and ψ are functions of x and y.

21. Apply the method of J. Bertrand (p. 232) to the equation

$$P\,dx + Q\,dy + R\,dz = 0,$$

where P, Q, R are linear functions of x, y, z satisfying the condition of integrability.

22*. Given a *completely integrable* system of the form

$$\begin{aligned} dz &= p\,dx + q\,dy, \\ dp &= (a_1 p + a_2 q + a_3 z)\,dx + (c_1 p + c_2 q + c_3 z)\,dy, \\ dq &= (c_1 p + c_2 q + c_3 z)\,dx + (b_1 p + b_2 q + b_3 z)\,dy, \end{aligned}$$

where a_i, b_i, c_i are functions of x and y, the general integral is of the form $z = C_1 z_1 + C_2 z_2 + C_3 z_3$, where z_1, z_2, z_3 are three linearly independent integrals, and where C_1, C_2, C_3 are arbitrary constants.*

23. Find the necessary and sufficient conditions in order that the equations

$$r = f_1(x, y), \qquad s = f_2(x, y), \qquad t = f_3(x, y)$$

be consistent.

Application. Find what condition the functions $A(x, y)$, $B(x, y)$, $C(x, y)$ must satisfy in order that the integral curves of the differential equation $A\,dx^2 + 2B\,dx\,dy + C\,dy^2 = 0$ be the projections on the xy-plane of the two families of asymptotic lines of a surface.

---

* APPELL, *Journal de Liouville*, 3d series, Vol. VIII, p. 192.

INDEX

[Titles in italic are proper names; numbers in italic are page numbers; and numbers in roman type are paragraph numbers.]

Abel: *28*, 15
Abelian integral: *18*, 11
Abel's theorem, applied to Euler's equation: *28*, 15
Adjoint equation, polynomial: *115*, 42; *116*, 42
Adjoint system of equations: *156*, 57; *166*, 62
Algebraic critical points: *173*, 63; *183*, 67; *199*, 71; *201*, 71
Algebraic differential equations: *180*, 66; *182*, 67
Analytic extension of integrals: *101*, 37; *152*, 56; *180*, 66; *182*, 67; *188*, 68
Analytic integrals: *45*, 22; *49*, 22; *50*, ftn.; *51*, 23; *53*, 24; *54*, 25; *57*, 25; *59*, 26; *60*, 26; *61*, 26; *67*, 29; *100*, 37; *175*, 64; *246*, 84; *284*, 94
Anharmonic ratio of integrals: *13*, 7
Antomari: *246*, ftn.
Appell: *41*, ftn.; *115*, 42; *290*, ftn.
Approximate integration of differential equations: *64*, ftn.
Arbitrary constants: *74*, 31; *see also* Elimination of constants
Asymptotic lines: *43*, ex. 18; *91*, 35; *206*, ex. 6
Auxiliary equation, polynomial: *12*, 6; *117*, 43; *124*, 45; *163*, 60; roots of: *119*, 43; for a system of equations: *158*, 58; *163*, 60

Bernoulli: *11*, 5
Bernoulli's equation: *11*, 5
Bertrand: *41*, ftn.; *232*, 80; *290*, ex. 21
Bertrand's method: *232*, 80; *290*, ex. 21
Bessel: *126*, 46; *142*, 52; *169*, ex. 8
Bessel's equation: *126*, 46; *142*, 52; *169*, ex. 8
Boole: *212*, ex. 1
Bounitzky: *44*, ex. 21
Bracket [u, v]: *234*, 81; *241*, 83

Briot and Bouquet: *45*, 21; *50*, ftn.; *59*, 26; *173*, 64; *175*, 64; *176*, 64; *177*, 64; *178*, 65; *193*, ftn.
Briot and Bouquet's equation: *173*, 64
Briot and Bouquet's method, analytic integrals: *50*, ftn.; *59*, 26
Briot and Bouquet's theorem: *175*, 64; *176*, 64; *177*, 64; *178*, 65
Calculus of limits: *45*, 21 and 22; *65*, ftn.; *137*, 50; (system of equations): *48*, 22; equations of the nth order: *49*, 22; *100*, 37; non-linear equations: *174*, 64; partial differential equations: *53*, 25; (system of): *56*, 25; *283*, 94; *287*, 94; system of linear equations: *50*, 23; total differential equation: *51*, 24; (system of): *53*, 24
Canonical form, of substitutions: *131*, 48; *132*, 48; of a system of linear equations: *161*, 59; *165*, 61; *179*, 65
Cauchy: *35*, 18; *45*, 21; *46*, 22; *61*, 27; *68*, 30; *73*, 30; *74*, 30; *108*, 39; *109*, ftn.; *128*, 46; *154*, ftn.; *172*, 63; *183*, 67; *198*, 71; *202*, 71; *214*, 75; *217*, 75; *246*, 84; *249*, 85; *254*, 85; *257*, 85; *257*, Note; *259*, 86; *260*, Note; *261*, 87; *264*, 87
Cauchy-Lipschitz method: *61*, 27; *68*, 30; *74*, 30
Cauchy's equation: *257*, ex. 1.
Cauchy's first proof: *68*, 30; *73*, 30
Cauchy's method: non-homogeneous linear equations: *108*, 39; *109*, ftn.; (system of): *154*, ftn.; partial differential equations: *249*, 85; *257*, Note; *259*, 86; *260*, Note; (extended): *261*, 87
Cauchy's problem: *246*, 84; *264*, 87
Cauchy's theorem: *45*, 22; *172*, 63; *183*, 67; *198*, 71; *202*, 71; (system of equations): *48*, 22; *217*, 75; partial differential equations: *54*, 25

292 INDEX

Center, of integral curves: *180*, 65; of similitude, *8*, 3

Characteristic curves: *219*, 76; *224*, 77; *249*, 85; *250*, 85; *259*, 86; *261*, 87; Cauchy's method: *249*, 85; *257*, Note; *259*, 86; *260*, Note; *261*, 87; congruence of: *219*, 76; *220*, 76; *222*, 77; derivation from complete integral: *259*, 86; differential equations of: *219*, 76; *222*, 77; *224*, 77; *251*, 85; see also Characteristic strip

Characteristic developable surface: *252*, 85; *259*, 86; *260*, Note

Characteristic direction: *250*, 85

Characteristic equation: *130*, 47; *139*, 50; *140*, 51; *143*, 53; *147*, 54; *165*, 61; *166*, 62; *179*, 65; elementary divisors: *132*, ftn.; roots of: *130*, 47; *131*, 48; *139*, 50; *149*, Note 2

Characteristic exponents: *147*, 54; *150*, 55

Characteristic numbers: *147*, 54

Characteristic strip: *252*, 85; *259*, 86; *260*, Note; *261*, 87; differential equation of: *262*, 87

Circles, differential equation of: *5*, 1; of double contact with a conic: *206*, ex. 4

Cissoid: *206*, ex. 5

Clairaut: *17*, 10; *41*, ftn.; *44*, ex. 20; *205*, 72; *212*, 74; *239*, ex. 1

Clairaut's equation: *17*, 10; *41*, ftn.; *44*, ex. 20; *205*, 72; generalized: *212*, 74; *239*, ex. 1

Clebsch: *267*, 88

Complete integral: *236*, 82; *239*, 82; *241*, 83; *247*, 84; *260*, Note; *277*, 91; *278*, 92; generalization of theory: *272*, 90; geometric interpretation: *238*, 82; of involutory systems: *277*, 91; see also Cauchy's method and Lagrange's theory

Complete systems: *267*, 88 and 89; equivalent: *268*, 89; Jacobian systems: *269*, 89; *270*, 89; *271*, ex.; *275*, 91; *278*, 92; method of integration: *270*, 89; change of variables: *267*, 89

Completely integrable total differential equations: *52*, 24; *225*, 73; system of equations: *53*, 24; see also Condition for integrability

Complex of curves: *259*, 86

Condition for incompressibility of a fluid: *86*, 33

Condition for integrability of total differential equations: *52*, 24; *225*, 78; *230*, 80; the bracket $[u, v]$: *234*, 81; *241*, 83; invariance of: *231*, 80; involutory systems, Poisson's parenthesis: *274*, 91; the parenthesis (u, v): *234*, 81

Conformal representation: *22*, 13

Congruence of curves: *209*, 74; *219*, 76; *222*, 77; focal points of, focal surface: *209*, 74; *224*, 77; see also Characteristic curves and Edge of regression

Conical point: *257*, 85

Conics, differential equation of: *5*, 1; having circles of double contact: *206*, ex. 4

Conoids: *220*, ex. 1

Constant coefficients in differential equations: *117*, 43; (system of equations): *157*, 58; *160*, 58; D'Alembert's method: *122*, 44; *161*, 58

Constants of integration: *74*, 31; see also Elimination of constants

Continuous one-parameter groups: *87*, 34; see also Groups

Corresponding homogeneous linear equation: *107*, 39

Cotton: *64*, ftn.

Covariant: *80*, Note 2

Cremona transformation: *198*, ftn.

Critical points, algebraic: *173*, 63; *183*, 67; *199*, 71; *201*, 71; infinite number of: *185*, ftn.; linear equations: *129*, 47; non-linear equations: *173*, 63; permutation of integrals about: *129*, 47; *133*, 49; transcendental: *197*, 70

Curves, asymptotic lines: *43*, ex. 18; *91*, 35; *206*, ex. 6; circles: see Circles; cissoid: *206*, ex. 5; complex of: *259*, 86; congruence of: see

INDEX 293

Congruence of curves; conics: *see* Conics; cycloid: *41*, 20; edge of regression: *209*, 74; *212*, 74; *240*, ex. 2; *257*, 85; elastic space curve: *99*, ex. 7; ellipse: *18*, 10; envelope: *see* Envelope; family of: *3*, 1; helices: *220*, ex. 2; isothermal: *43*, ex. 12; orthogonal: *14*, 7; *33*, 17; *220*, ex. 3; *223*, 78; parabola: *6*, 1; parallel: *42*, ex. 9; similar: *8*, 3; straight lines: *4*, 1; trajectories: *14*, 7; *34*, 17; *93*, 36; unicursal quartic: *19*, ex. 2; *205*, 72; *see also* Cusps Integral curves, Lines of Curvature, Locus

Cusps of integral curves: *41*, 20; *201*, 71; *202*, 71; *208*, 73; *212*, 74; *213*, ex. 2; *see also* Locus of cusps

Darboux: *29*, 16; *41*, ftn.; *45*, 21; *79*, ftn.; *116* ftn.; *205*, ftn.; *213*, ex. 2; *239*, ftn.; *253*, 85

Darboux's theorems: *29*, 16

D'Alembert: *122*, 44; *161*, 58

D'Alembert's method: *122*, 44; *161*, 58

Definite integrals as solutions, of Bessel's equation: *126*, 46; *169*, ex. 8; of Laplace's equation: *124*, 46

Delassus: *286*, 94

Depression of order: *36*, 19; *109*, 40

Derivative in non-linear equations, infinite: *172*, 63; indeterminate: *173*, 64; *177*, 65; *see also* Briot and Bouquet's equation *and* Briot and Bouquet's theorem

Developable surfaces: *240*, 82; *257*, 85; *282*, ex.; *see also* Characteristic developable surfaces

Differential equations: *3*, 1; admitting a group of transformations: *89*, 35; *91*, 35; *95*, 36; *96*, 36; *97*, 36; *98*, ex. 4; algebraic: *180*, 66; *182*, 67; algebraic, of deficiency zero or one: *18*, 11; Bernoulli's: *11*, 5; Bessel's: *126*, 46; *142*, 52; *169*, ex. 8; Briot and Bouquet's: *173*, 64; Cauchy's: *257*, ex. 1; of characteristic curves: *219*, 76; *222*, 77; *224*, 77; *251*, 85; of characteristic strip: *262*, 87; of circles: *5*, 1; Clairaut's: *see* Clairaut's equation; of conics (Halphen's method): *5*, 1; Darboux's theorems: *29*, 16; depression of order of: *36*, 19; *109*, 40; differential notation: *7*, 2; elastic space curve: *99*, ex. 7; equations $F(x, y')= 0$, $F(y, y') = 0$: *18*, 11; Euler's: *see* Euler's equation; Euler's linear: *123*, 45; existence theorems: *see* Existence theorems; of first order: *6*, 2; *180*, 66; Gauss's: *140*, 51; geometric representation of: *14*, 8; of higher order: *35*, 18; *196*, 70; homogeneous: *8*, 3; *16*, ftn.; *38*, 19; *90*, 35; of incompressible fluid: *84*, 33; integrals of: *see* Integral curves, Integral surfaces, *and* Integrals; of isothermal curves: *43*, ex. 12; Jacobi's: *see* Jacobi's equation; Lagrange's: *16*, 9; *204*, 72; *205*, 72; Lamé's: *146*, 53; Laplace's linear: *124*, 46; Legendre's: *112*, ex.; linear: *9*, 4; *90*, 35; Liouville's: *79*, ex. 3; of the nth order: *4*, 1; *6*, 2; *49*, 22; *100*, 37; order of: *4*, 1; of orthogonal trajectories: *14*, 7; *33*, 17; *223*, 78; Painlevé's: *196*, 70; *197*, 70; of parabolas: *6*, 1; with periodic coefficients: *see* Periodic coefficients; Picard's: *143*, 53; raising order of: *41*, Note; regular: *134*, 50; Riccati's: *see* Riccati's equation; of similar curves: *8*, 3; singular points of: *see* Singular points; of straight lines: *4*, 1; of trajectories: *see* Trajectories; *see also* special classes of differential equations *and* systems of equations

Differential notation: *7*, 2

Differential operators: *97*, 36; *102*, 38; *113*, 41; bracket $[u, v]$: *234*, 81; *241*, 83; the parenthesis (u, v): *234*, 81; Poisson's parenthesis: *274*, 91; $X[Y(f)] - Y[X(f)]$: *97*, 36; *266*, 88; *278*, 92

Dixon: *44*, ex. 21

Dominant functions: *45*, 21; *47*, 22; *51*, 23; *52*, 24; *55*, 25; *138*, 50; *174*, 64

INDEX

Doubly periodic functions of the second kind: *145*, 53

Edge of regression: *209*, 74; *212*, 74; *240*, ex. 2; *257*, 85
Elastic space curve: *99*, ex. 7
Element: *251*, 85; *261*, 87
Elementary divisors: *132*, ftn.
Elimination, of arbitrary functions: *222*, 77; *258*, 82; *259*, 86; *273*, 90; *278*, 93; of constants: *3*, 1; *208*, 74; *236*, 82; *255*, ftn.; *272*, 90
Ellipsoid, lines of curvature of: *41*, Note
Elliptic functions: *23*, 14; as coefficients of a linear equation: *144*, 53; *146*, 54; existence proof from Euler's equation: *23*, 14; *194*, 69; as integrals: *19*, ex. 3; *39*, 20; *144*, 53; *192*, 68; Picard's equation: *144*, 53
Envelope, of asymptotic lines: *206*, ex. 6; of integral curves: *17*, 10; *203*, 71; *204*, ftn.; *205*, 72; *209*, 74; *213*, ex. 8; of integral surfaces: *238*, 82; *281*, 93; of straight lines: *13*, 10
Equations of first order, higher order: see Differential equations *and* special classes of equations
Equivalent complete systems: *268*, 89
Essentially singular points: *131*, 47; *134*, 49; movable: *196*, 70
Euler: *19*, 12; *23*, 14; *27*, 14; *28*, 15; *29*, 16; *41*, ftn.; *43*, ex. 17; *117*, 43; *123*, 45; *194*, 69; *205*, 72; *221*, ex. 3
Euler's equation: *23*, 14; *28*, 15; *41*, ftn.; *194*, 69; *205*, 72; Abel's theorem: *28*, 15; existence of elliptic functions: *23*, 14; *194*, 69; Lagrange's integral of: *43*, ex. 17; Stieltjes's general integral: *27*, 14
Euler's linear equation: *123*, 45
Euler's relation for homogeneity: *221*, ex. 3
Exceptional initial values: *172*, 63; *173*, 64; *177*, 65
Existence theorems: *45*, 22; *98*, ex. 1; analytic integrals: see Analytic

integrals *and* Briot and Bouquet's method; calculus of limits: see Calculus of limits; for elliptic functions: *23*, 14; *194*, 69; for integrating factors: *57*, 26; successive approximations: see Successive approximations; for systems of partial differential equations in normal form: *283*, 94; see also Exceptional initial values
Extended group: *94*, 36

First integrals: *74*, 31; *76*, 31; *81*, 32; *83*, 32; *157*, 57; *216*, 75
Fixed singular points: *131*, 66; *182*, 67
Floquet: *151*, ftn.
Focal point: *209*, 74
Focal surface: *209*, 74; *224*, 77
Focus: *180*, 65
Fuchs: *134*, 50; *139*, ftn.; *150*, ftn.; *194*, ftn.
Fuchs' theorem: *134*, 50
Functions defined by differential equations: *182*, 67
Fundamental characteristic equation: *139*, 50; see also Characteristic equation
Fundamental system of integrals: *103*, 38; *105*, 38; *129*, 47; *130*, 47; *147*, 54; for a system of linear equations: *153*, 56

Gauss: *140*, 51
Gauss's equation: *140*, 51
General integral: *3*, 1; *12*, 7; *59*, 26; *74*, 31; of homogeneous linear equations: *103*, 38; *105*, 38; of partial differential equations: *217*, 75; *238*, 82; *273*, 90; of a system of equations: *57*, 26; *152*, 56
Goursat: *83*, ftn.; *170*, exs. 14, 15; *208*, ftn.; *265*, ftn.
Group, differential equations admitting a: *89*, 35; *91*, 35; *95*, 36; *96*, 36; *97*, 36; *98*, ex. 4; differential equations of a: *83*, 34
Groups, one-parameter continuous: *86*, 34; *91*, 36; application to differential equations: *89*, 35; *96*, 36;

97, 36; functions admitting: *93*, 36; of infinitesimal transformations: *91*, 36; *93*, 36; invariants: *93*, 36; similar: *88*, 34; of translations: *89*, 34; *see also* Transformations

Halphen: *5*, 1; *115*, 42
Hedrick: *255*, ftn.
Helices: *220*, ex. 2; *245*, 83
Helicoid: *220*, ex. 2; *245*, 83; lines of curvature of: *91*, 35
Hermite: *99*, ex. 7; *146*, 53; *169*, ex. 8; *193*, ftn.; *283*, ftn.
Homogeneity of functions, Euler's relation: *221*, ex. 3
Homogeneous equations: *8*, 3; *16*, ftn.; *53*, 19; *90*, 35
Homogeneous linear equations: *102*, 38; *107*, 39; adjoint equation, polynomial: *116*, 42; analogies with algebraic equations: *113*, 41; analogies with the Galois theory, with symmetric functions of roots: *115*, 41; auxiliary equation, polynomial: *117*, 43; Bessel's equation: *126*, 46; *142*, 52; *169*, ex. 8; common integrals of two equations: *114*, 41; constant coefficients: *117*, 43; (D'Alembert's method): *122*, 44; corresponding: *107*, 39; critical points: *129*, 47; depression of order: *109*, 40; elliptic coefficients: *144*, 53; *146*, 54; Euler's linear equation: *123*, 45; Fuchs' theorem: *134*, 50; fundamental system of integrals: *103*, 38; *105*, 38; *129*, 47; Gauss's equation: *140*, 51; general integral: *103*, 38; *105*, 38; greatest common divisor: *113*, 41; group of substitutions: *132*, 48; *134*, 48; invariants: *115*, 41; Lamé's equation: *146*, 53; Laplace's equation: *124*, 46; Legendre's equation: *112*, ex.; linearly independent integrals: *103*, 38; *105*, 38; periodic coefficients: *128*, 47; *146*, 54; *150*, ex.; *151*, ftn.; permutations of integrals around a critical point: *129*, 47; Picard's equation: *143*, 53; ratio of two integrals: *169*, ex. 10; regular: *134*, 50; regular integrals: *128*, 47; *131*, 47; *134*, 49; relation to Riccati's equation: *111*, 40; *112*, ftn.; roots of integrals, Sturm's theorem: *111*, ftn.; solution as a definite integral: *see* Definite integrals; system of: *see* System of homogeneous linear equations; Wronskian: *129*, 47; *see also* Characteristic equation, Characteristic numbers, *and* Characteristic exponents
Houël: *212*, ex. 1
Hyperelliptic functions: *193*, 68
Hypergeometric series: *140*, 51; degenerate cases: *142*, 52

Identical transformation: *88*, 34; *91*, 36
Incompressible fluid, condition for: *86*, 33; invariant integrals: *84*, 33
Independent equations: *265*, 88
Independent integrals: *81*, 31; linearly: *103*, 38; *105*, 38
Infinitesimal transformations: *86*, 34; *91*, 36; *93*, 36; *98*, 36
Initial conditions: *45*, 22; *48*, 22; *49*, 22; *50*, 23; *52*, 24; *53*, 24; *61*, 26; defining an integral: *100*, 37; partial differential equations: *54*, 25; *57*, 25; *214*, 74; *221*, 76; *246*, 84; *284*, 94; *see also* Cauchy's problem, Derivatives in non-linear equations, *and* Exceptional initial values
Integrable combination: 77, 31; *78*, exs. 1, 2; *220*, 76; *245*, 83
Integral curves: *4*, 1; *60*, 26 and ftn.; *61*, 26; *79*, Note 1; *173*, ftn.; *179*, 65; *199*, 71; center: *180*, 65; cusps: *see* Cusps; envelope of: *17*, 10; *203*, 71; *204*, ftn.; *205*, 72; *209*, 74; *213*, ex. 8; focus: *180*, 65; in parametric form: *16*, 9; of a partial differential equation: *257*, 85; *258*, ex. 2; *289*, ex. 15; saddleback: *179*, 65; *see also* Integrals
Integral equation: *61*, 27
Integral surfaces: *218*, 76; *219*, 76; *227*, 78; *246*, 84; *250*, 85; *255*, 85;

INDEX

envelope of: *238*, 82; *281*, 93; *see also* Cauchy's problem and Integrals
Integrals: Abelian: *18*, 11; analytic: *see* Analytic extension *and* Analytic integrals; anharmonic ratio of: *13*, 7; Cauchy's problem: *246*, 84; *264*, 87; complete: *see* Complete integral; common to two linear equations: *114*, 41; defined by initial conditions: *100*, 37; in form of definite integrals: *see* Definite Integrals; elements of: *251*, 85; *261*, 87; elliptic functions: *19*, ex. 3; *39*, 20; *144*, 53; *192*, 68; of equations of higher order: *196*, 70; existence of: *see* Existence theorems; first: *see* First integrals; fundamental system of: *see* Fundamental system of integrals; general: *see* General integral; general properties of: *100*, 37; hypergeometric series: *140*, 51; *142*, 52; independent: *81*, 31; initial conditions: *see* Initial conditions; invariant: *see* Invariant integrals; Legendre's polynomials: *112*, ex.; Lie's enlarged definition: *264*, Note; linearly independent: *103*, 38; *105*, 38; non-analytic: *see* Non-analytic integrals; particular: *3*, 1; *12*, 7; *14*, 7; *20*, 12; *107*, 39; *109*, 40; periodic: *192*, 68; permutation of integrals around a critical point: *129*, 47; *133*, 49; rational functions: *144*, 53; *192*, 68; rational functions of constants: *10*, 4; *12*, 7; *186*, 67; regular: *128*, 47; *131*, 47; *134*, 49; roots of, Sturm's theorem: *111*, ftn.; singular: *see* Singular integrals; singular points: *see* Singular points; Wronskian: *129*, 47; *see also* Integrable combination, Integral curves, Integral surfaces, *and* special types of equations
Integrating factors: *19*, 12; *81*, 32; *83*, 32; *96*, 36; *98*, exs. 3, 4; *115*, 42; *231*, 80; *290*, ex. 20; existence of: *57*, 26
Integration by raising order: *41*, Note
Invariance of condition of integrability: *231*, 80

Invariant functions: *93*, 36
Invariant integral: *83*, 33; of homogeneous linear equations: *115*, 41; of incompressible fluid: *84*, 33; line and surface: *84*, 33; multiple: *85*, 33; volume: *86*, 33
Involutory systems: *274*, 91; complete integral: *277*, 91; Jacobi's method: *277*, 92; Poisson's parenthesis: *274*, 91
Isothermal curves: *43*, ex. 12

Jacobi: *11*, 6; *25*, 14; *32*, 16; *74*, 31; *81*, 32; *163*, 60; *269*, 89; *270*, 89; *271*, ex.; *275*, 91; *277*, 92; *278*, 92
Jacobi's equation: *11*, 6; *32*, 16; relation to a system of homogeneous linear equations: *163*, 60
Jacobi's method, involutory systems: *277*, 92
Jacobi's multipliers: *74*, 31; *81*, 32
Jacobian system: *269*, 89; *270*, 89; *271*, ex.; *275*, 91; *278*, 92

Kovalevsky, Madame: *45*, 21; *283*, 94 and ftn.

Lagrange: *16*, 9; *41*, ftn.; *43*, ex. 17; *107*, 39; *109*, ftn.; *115*, 42; *203*, 71; *204*, 72; *205*, 72; *213*, ex. 4; *236*, 82; *239*, 82; *240*, 83; *241*, 83; *251*, 85; *255*, ftn.; *258*, ex. 1; *259*, 86; *277*, 92
Lagrange and Charpit's method: *240*, 83; *277*, 92
Lagrange's equation: *16*, 9; *204*, 72; *205*, 72
Lagrange's integral of Euler's equation: *43*, ex. 17
Lagrange's method: *241*, 83; *251*, 85
Lagrange's method of the variation of constants: *107*, 39; *109*, ftn.; *255*, ftn.
Lagrange's theory of the complete integral: *236*, 82; *239*, 82; *258*, ex. 1; *259*, 86
Laguerre: *115*, 41
Lamé: *146*, 53
Lamé's equation: *146*, 53

Laplace: *124*, 46; *127*, Note
Laplace's linear equation: *124*, 46; *127*, Note
Legendre: *16*, ftn.; *28*, 15; *112*, ex.
Legendre's equation: *112*, ex.
Legendre's polynomials: *112*, ex.
Legendre's transformation: *16*, ftn.
Leibnitz: *118*, 43
Leibnitz's formula: *118*, 43
Liapunof: *151*, 55 and ftn.; *166*, 62
Lie: *43*, ex. 12; *86*, ftn.; *98*, 36; *264*, Note; *289*, ex. 15
Lie's enlarged definition of the integral: *264*, Note
Lie's theory of differential equations: *86*, 34; *see also* Groups
Lindelöf: *61*, 27; *98*, ex. 1
Linear equations: *9*, 4; *90*, 35; *100*, 37; *106*, 39; *186*, 67; coefficients depending upon a parameter: *65*, Note; depression of order: *109*, 40; general properties of integrals: *100*, 37; *see also* Homogeneous linear equations, Integrals, Non-homogeneous linear equations, Partial differential equations, and Singular points
Linearly independent functions: *103*, 38; integrals: *103*, 38; *105*, 38
Lines of curvature: *206*, 72; of an ellipsoid: *41*, Note; of a helicoid: *91*, 35
Liouville: *79*, ex. 3
Liouville's equation: *79*, ex. 3
Lipschitz: *68*, 30; *287*, 94
Lipschitz condition: *68*, 30; *287*, 94
Locus, of characteristic curves: *219*, 76; of cusps of integral curves: *201*, 71; *202*, 71; *206*, 72; *208*, 73; *212*, 74; *213*, ex. 2; of points of inflection of integral curves: *213*, ex. 2.

Mayer: *229*, 79
Mayer's method: *229*, 79
Méray: *45*, 21
Moigno: *68*, 30; *212*, ex. 1
Monge: *41*, Note
Monge's method of finding the lines of curvature of an ellipsoid: *41*, Note

Movable singular points: *181*, 66; *185*, 67; *197*, 70 and ftn.; for equations of higher order: *196*, 70; essentially singular: *196*, 70; lines of: *197*, 70; poles: *197*, ftn.; transcendental critical points: *197*, 70
Multipliers: *74*, 31; *81*, 32; *85*, 33

Non-analytic integrals: *50*, 22; *255*, ftn.; Briot and Bouquet's theorem: *175*, 64; *176*, 64; *177*, 64; *178*, 65; *see also* Analytic integrals *and* Briot and Bouquet's method
Non-homogeneous linear equations: *100*, 37; *106*, 39; analytic extension of integrals: *101*, 37; Cauchy's method: *108*, 39; *109*, ftn.; constant coefficients: *120*, 43; corresponding homogeneous equation: *107*, 39; depression of order: *110*, 40; general integral: *107*, 39; Lagrange's method of the variation of constants: *107*,39; *109*,ftn.; singular points: *100*,37; system of equations: *see* Systems of non-homogeneous linear equations
Non-linear differential equations, *172*, 63; *179*, 65; algebraic equations of the first order: *180*, 66; *182*, 67; Briot and Bouquet's problem: *193*, ftn.; having single-valued integrals: *187*, 68; *192*, 68; *193*, ftn.; exceptional initial values: *172*, 63; (derivative infinite): *172*, 63; (derivative indeterminate): *175*, 64; *177*, 65; integrals: *see* Envelope of integrals, Integral curves, Integrals, Locus of cusps, and Singular integrals; functions defined by $y' = R(x, y)$: *182*, 67; non-analytic integrals: *see* Non-analytic integrals; singular points: *see* Critical points, Fixed singular points, Movable singular points, Singular points; systems of: *208*, 74; *see also* Equations of Briot and Bouquet, Clairaut, Euler, Lagrange, and Riccati
Normal form of a system of partial differential equations: *283*, 94

Order of a differential equation: *4*, 1;
depression of: *36*, 19; *109*, 40; first
order: *6*, 2; *180*, 66; higher order,
nth order: *4*, 1; *6*, 2; *35*, 18; *196*,
70; integration by raising order:
41, Note

Orthogonal trajectories: *14*, 7; *33*,
17; *220*, ex. 3; *223*, 78

Orthogonal surfaces: *223*, 77

Painlevé: *59*, ftn.; *74*, 30; *196*, 70;
197, 70; *213*, ex. 7

Painlevé's equation: *196*, 70; *197*,
70

Parallel curves: *42*, ex. 9

Parallel surfaces: *289*, ex. 19

Parenthesis (u, v): *234*, 81; Poisson's:
274, 91

Partial differential equations: *76*, 31;
of first order: *see* Partial differential
equations of the first order; of
higher order: *278*, 93; (system of
equations): *283*, 94; of ruled surfaces: *280*, ex. 1; *281*, ex. 2; of
tubular surfaces: *240*, ex. 3; *282*,
ex.; *see also* Systems of differential
equations *and* Existence theorems

Partial differential equations of the
first order, linear: *75*, 31; *214*, 75;
characteristic curves: *see* Characteristic curves; of conoids: *220*, ex. 1;
general integral: *217*, 75; geometric interpretation: *218*, 76; general
method of integration: *214*, 75; of
helicoids: *220*, ex. 2; *245*, 83; initial
conditions, *221*, 76; integral surface:
218, 76; *219*, 76; singular integral,
surface: *224*, 77; *see also* Systems
of differential equations

Partial differential equations of the
first order, non-linear: any number of variables: *261*, 87; Cauchy's
equation: *257*, ex. 1; Cauchy's
method: *249*, 85; *259*, 86; *260*,
Note; (extended): *261*, 87; Cauchy's
problem: *246*, 84; characteristic
curves, characteristic developable
surface, characteristic direction,
characteristic strip: *see* these titles;
Clairaut's equation, generalized:
239, ex. 1; complete integral: *236*,
82, and *see also* Lagrange's theory;
element: *251*, 85; envelope of surfaces: *238*, 82; general integral:
238, 82; integral, Lie's enlarged
definition: *264*, Note; integral
curves: *257*, 85; *289*, ex. 15; Lagrange and Charpit's method: *240*,
83; *277*, 92; separation of variables:
244, ex. 3; singular integrals: *224*,
77; *237*, 82; *238*, ftn.; *272*, 90;
three variables: *236*, 82; *see also*
Involutory systems

Particular integral, solution: *3*, 1;
12, 7; *14*, 7; *20*, 12; *107*, 39;
109, 40

Periodic coefficients: *128*, 47; *146*,
54; *150*, ex.; *151*, ftn.; elliptic:
144, 53; *146*, 54; Picard's equation: *144*, 53; system of linear
equations: *164*, 61; *166*, 62

Picard: *59*, ftn.; *61*, 27; *74*, 30; *115*,
42; *144*, 53; *177*, 64

Picard's equation: *144*, 53

Picard's method of successive approximations: *see* Successive approximations

Poincaré: *83*, 33; *126*, ftn.; *151*, ftn.;
177, 64; *180*, 65; *194*, ftn.

Poisson: *274*, 91; *277*, 92

Poisson's identity: *277*, 92

Poisson's parenthesis: *274*, 91

Poles of integrals: *143*, 53; *183*, 67;
185, 67; movable: *186*, 67; *197*, ftn.

Properties, of differential equations of
higher order: *196*, 70; of e^x, $\tan x$:
213, ex. 6

Quadratures: *7*, 2; *10*, 4; *12*, 7; *13*,
7; *14*, 7; *16*, 9; *19*, 12; *78*, 31; *79*,
Note 1; *83*, 32; *90*, 35; *108*, 39;
110, 40; *111*, 40; *154*, 56; *278*, 92

Raffy: *44*, ex. 21

Ratio, of similitude: *8*, 3; of two
integrals: *169*, ex. 10

Rational functions, of constants as integrals: *10*, 4; *12*, 7; *186*, 67; of

INDEX   299

variables as integrals: *144*, 53; *192*, 68
Reducible systems: *165*, 62
Regular differential equations: *134*, 50
Regular integrals: *128*, 47; *131*, 47; *134*, 49; Fuchs' theorem: *134*, 50; substitutions: *132*, 48; *134*, 48
Riccati: *12*, 7; *13*, ftn.; *43*, ex. 13; *79*, ex. 2; *111*, 40; *112*, 40 and ftn.; *143*, Note; *157*, 57; *169*, ex. 9; *170*, ex. 15; *186*, 67; *187*, 67 and ftn.; *194*, ftn.; *197*, ftn.; *213*, ex. 7
Riccati's equation: *12*, 7; *43*, ex. 13; *79*, ex. 2; *111*, 40; *112*, 40 and ftn.; *143*, Note; *157*, 57; *169*, ex. 9; *170*, ex. 15; *181*, 67 and ftn.; *186*, 67; *194*, ftn.; *213*, ex. 7; generalization of: *197*, ftn.; linear transformation of: *13*, ftn.; relation to linear equations: *111*, 40; *112*, ftn.
Riquier: *45*, 21; *286*, 94 and ftn.
Roots of characteristic equation: *130*, 47; *131*, 48; *139*, 50; *149*, Note 2; elementary divisors: *132*, ftn.
Roots of integrals: Sturm's theorem: 111, ftn.
Ruled surfaces: *280*, ex. 1; *281*, ex. 2

Saddleback: *179*, 65
Sauvage: *132*, ftn.
Schlömilch: *212*, ex. 1
Separation of variables: *6*, 2; *8*, 3; *19*, 12; *244*, ex. 3
Serret: *212*, ex. 1; *213*, ex. 3
Similar curves: *8*, 3
Similar groups: *88*, 34
Single-valued integrals: *144*, 53; of $(y')^m = R(y)$, classification of equations: *187*, 68; *192*, 68; *193*, ftn.
Singular integral, curve, surface: *17*, 10; *27*, 14; *76*, ftn.; *198*, 71; *202*, 71; *205*, 72; *206*, 72; *208*, 74; *210*, 74; *224*, 77; *237*, 82; *255*, ftn.; as an envelope: *203*, 71; *238*, 82; geometric interpretation: *207*, 73
Singular integral, curves and surfaces: determination of: *205*, 72; of first-order equations: *198*, 71; *202*, 71; *206*, 72; geometric interpretation: *207*, 73; of partial differential equations: *224*, 77; *237*, 82; *238*, ftn.; *272*, 90; of a system of equations: *208*, 74; *210*, 74
Singular lines, movable: *197*, 70
Singular points: algebraic critical points: *173*, 63; *183*, 67; *184*, 67; *201*, 71; Briot and Bouquet's theorem: *176*, 64; center: *180*, 65; of equations of the first order: *180*, 66; essentially: *131*, 47; *134*, 49; essentially singular movable: *196*, 70; fixed: *181*, 66; *182*, 67; focus: *180*, 65; of linear equations: *65*, 28; *100*, 37; *129*, 47; *140*, 51; *142*, 52; *143*, 53; indeterminate derivative: *173*, 64; infinite derivative: *172*, 63; infinite number of critical points: *185*, ftn.; movable: see Movable singular points; poles: *131*, 47; *143*, 53; *144*, 53; *183*, 67; *184*, 67; *185*, 67; *197*, ftn.; saddleback: *179*, 65
Solution: see Integral
Star: *67*, 29
Stationary flow: *86*, 33
Stieltjes: *27*, 14
Stieltjes's general integral of Euler's equation: *27*, 14
Straight lines, differential equation of: *4*, 1
Sturm: *111*, ftn.
Sturm's theorem: *111*, ftn.
Substitutions: linear equations: *129*, 47; *132*, 48; *134*, 48; canonical form: *131*, 48; *132*, 48; system of linear equations, canonical form: *165*, 61; Wronskian: *129*, 47
Successive approximations: *61*, 27; analytic functions: *66*, 29; *102*, 37; *175*, 64; Cauchy-Lipschitz method: *61*, 27; *68*, 30; *74*, 30; Cauchy's first proof: *68*, 30; *73*, 30; coefficients functions of a parameter: *65*, Note; Lindelöf's addition: *61*, 27; linear equations: *64*, 28; Lipschitz condition: *68*, 30; *287*, 94; non-analytic integrals: *175*, 64; real variables:

C1, 27; *C2*, 27; *68*, 30; *73*, 30; *150*, 55; star: *67*, 29

Surfaces, conoids: *220*, ex. 1; developable: *240*, 82; *257*, 85; *282*, ex.; ellipsoid: *41*, Note; focal: *209*, 74; *224*, 77; helicoids: *220*, ex. 2; *245*, 83; orthogonal: *223*, 77; parallel: *289*, ex. 10; ruled: *280*, ex. 1; *281*, ex. 2; tubular: *240*, ex. 3; *282*, ex.; *see also* Characteristic developable surfaces, Envelopes, and Integral surfaces

Symbolic polynomial: *113*, 41; *116*, 42; *118*, 43; divisor: *114*, 41; greatest common divisor: *113*, 41

Systems of differential equations: *60*, 26; *74*, 31; *79*, Note 1; covariant: *80*, Note 2; existence theorem: *see* Existence theorem; first integrals: *see* First integrals; general integral: *57*, 26; integral curve: *60*, 26; invariant integrals: *see* Invariant integrals; multipliers: *74*, 31; *81*, 32; *85*, 33; singular integrals: *208*, 74; *see also* Integrable combination, Systems of homogeneous linear equations, *and* Systems of non-homogeneous linear equations

Systems of homogeneous linear equations: *152*, 56; adjoint system: *156*, 57; *166*, 62; auxiliary equation: *158*, 58; canonical form: *161*, 59; *165*, 61; *179*, 65; constant coefficients: *157*, 58; *160*, 58; (D'Alembert's method): *161*, 58; fundamental system of integrals: *153*, 56; periodic coefficients: *164*, 61; *166*, 62; reducible systems: *165*, 62; relation to Jacobi's equation: *163*, 60; substitutions: *165*, 61

Systems of non-homogeneous linear equations: *154*, 56; Cauchy's method: *154*, ftn.; existence theorem: *50*, 23

Systems of partial differential equations: of first order: *272*, 90; normal form, general existence theorem: *283*, 94; *see also* Existence theorems, Involutory systems, *and* Systems of homogeneous linear partial differential equations of the first order

Systems of partial differential equations, homogeneous linear equations of the first order: *265*, 88; independent equations: *265*, 88; $X[Y(f)] - Y[X(f)]$: *266*, 88; *see also* Complete systems

Tannery: *139*, ftn.
Taylor: *35*, 18
Total differential equations: *51*, 24; *225*, 78; *241*, 83; *276*, 91; Bertrand's method: *232*, 80; *290*, ex. 21; completely integrable: *52*, 24; *225*, 78; existence theorem: *51*, 24; geometric interpretation: *227*, 78; integral surface: *227*, 78; Mayer's method: *229*, 79; method of integration: *225*, 78; *232*, 80; $Pdx + Qdy + Rdz = 0$: *230*, 80; *see also* Condition of integrability

Trajectories: *13*, 7; *14*, 7; *34*, 17; *93*, 36

Transcendental critical points: *197*, 70

Transformations: *82*, 32; *83*, 32; *84*, 33; admitting a group of: *89*, 35; *96*, 36; of complete systems: *267*, 89; covariants: *80*, Note 2; Cremona: *198*, ftn.; extended group of: *94*, 36; identical: *88*, 34; *91*, 36; infinitesimal: *86*, 34; *91*, 36; *93*, 36; *98*, 36; inverse: *89*, 34; Legendre's: *16*, ftn.; of linear equations: *115*, 41; *162*, 59; of Riccati's equation: *13*, ftn.; *see also* Groups and Invariants

Tresse: *286*, 94
Tubular surfaces: *240*, ex. 3; *282*, ex.

Unicursal quartic: *19*, ex. 2; *205*, 72

Variation of constants: *107*, 39; *109*, ftn.; *255*, ftn.

Weierstrass: *45*, 21; *132*, 48 and ftn.
Weierstrass's elementary divisors: *132*, ftn.
Wronskian: *129*, 47

www.ingramcontent.com/pod-product-compliance
Lightning Source LLC
Chambersburg PA
CBHW022106230426
43672CB00008B/1301